Human-Machine Interface for Intelligent Vehicles
Design Methodology and Cognitive Evaluation

Human-Machine Interface for Intelligent Vehicles
Design Methodology and Cognitive Evaluation

Editors-in-Chief

Fang You
*Car Interaction Design Lab, College of Arts and Media,
Tongji University, Shanghai, P.R. China*

Jianmin Wang
*Car Interaction Design Lab, College of Arts and Media,
Tongji University, Shanghai, P.R. China*

Edited by

Fusheng Jia
College of Design and Innovation, Tongji University, Shanghai, P.R. China

Huiyan Chen
Shenzhen Research Institute, Sun Yat-Sen University, Shenzhen, GD, P.R. China

Qianwen Fu
College of Design and Innovation, Tongji University, Shanghai, P.R. China

Elsevier
Radarweg 29, PO Box 211, 1000 AE Amsterdam, Netherlands
125 London Wall, London EC2Y 5AS, United Kingdom
50 Hampshire Street, 5th Floor, Cambridge, MA 02139, United States

Copyright © 2024 Tongji University Press Co., Ltd. Published by Elsevier Inc. All rights are reserved, including those for text and data mining, AI training, and similar technologies.

Publisher's note: Elsevier takes a neutral position with respect to territorial disputes or jurisdictional claims in its published content, including in maps and institutional affiliations.

No part of this publication may be reproduced or transmitted in any form or by any means, electronic or mechanical, including photocopying, recording, or any information storage and retrieval system, without permission in writing from the publisher. Details on how to seek permission, further information about the Publisher's permissions policies and our arrangements with organizations such as the Copyright Clearance Center and the Copyright Licensing Agency, can be found at our website: www.elsevier.com/permissions.

This book and the individual contributions contained in it are protected under copyright by the Publisher (other than as may be noted herein).

Notices

Knowledge and best practice in this field are constantly changing. As new research and experience broaden our understanding, changes in research methods, professional practices, or medical treatment may become necessary.

Practitioners and researchers must always rely on their own experience and knowledge in evaluating and using any information, methods, compounds, or experiments described herein. In using such information or methods they should be mindful of their own safety and the safety of others, including parties for whom they have a professional responsibility.

To the fullest extent of the law, neither the Publisher nor the authors, contributors, or editors, assume any liability for any injury and/or damage to persons or property as a matter of products liability, negligence or otherwise, or from any use or operation of any methods, products, instructions, or ideas contained in the material herein.

ISBN: 978-0-443-23606-8

For Information on all Elsevier publications
visit our website at https://www.elsevier.com/books-and-journals

Publisher: Matthew Deans
Acquisitions Editor: Glyn Jones
Editorial Project Manager: Naomi Robertson
Production Project Manager: Sujithkumar Chandran
Cover Designer: Christian Bilbow

Typeset by MPS Limited, Chennai, India

Contents

List of contributors xxi
Acknowledgments xxv

PART 1 Intelligent cockpit HMI information perception and understanding

CHAPTER 1 Interface color design of intelligent vehicle central consoles 3
Fang You, Yaru Li, Preben Hansen, Liping Li, Mengting Fu,
Yifan Yang, Xin Jin and Jianmin Wang
- 1.1 Introduction 3
- 1.2 Experimental design 4
  - 1.2.1 Preliminary preparations 4
  - 1.2.2 Simulator settings 5
  - 1.2.3 Driving scene settings 5
- 1.3 Experiment execution 6
- 1.4 Experimental results 6
  - 1.4.1 Data collection 6
  - 1.4.2 Subjective data 6
  - 1.4.3 Objective data 8
- 1.5 Discussion 9
- Further reading 9

CHAPTER 2 Icon design recommendations for central consoles of intelligent vehicles 11
Fang You, Yifan Yang, Mengting Fu, Jifang Wang, Xiaojun Luo,
Liping Li, Preben Hansen and Jianmin Wang
- 2.1 Introduction 11
- 2.2 Method 12
  - 2.2.1 Participants 12
  - 2.2.2 Experimental environment 12
  - 2.2.3 Task 13
  - 2.2.4 Procedure 13
  - 2.2.5 Questionnaires and data analysis 14
- 2.3 Results 15
  - 2.3.1 Influence of icon size on usability and workload 15
  - 2.3.2 Impact of icon thickness on usability and workload 16
- 2.4 Conclusion and discussion 16
- Further reading 17

CHAPTER 3 Design guidelines for the size and length of Chinese characters displayed in the intelligent vehicle's central console interface 19

Fang You, Yifan Yang, Mengting Fu, Jun Zhang and Jianmin Wang

- 3.1 Introduction 19
- 3.2 Related works 20
  - 3.2.1 Classification of visual signals and human-computer interaction model 20
  - 3.2.2 Standards and guidelines about size and length of text 21
  - 3.2.3 Driving cognitive model 21
- 3.3 Method 22
  - 3.3.1 Participants 22
  - 3.3.2 Experimental environment 22
  - 3.3.3 Task and stimuli 23
  - 3.3.4 Procedure 24
  - 3.3.5 Questionnaires and data analysis 24
- 3.4 Results 25
  - 3.4.1 Reaction time 25
  - 3.4.2 Visual subjective rating 26
  - 3.4.3 Usability scores and workload 27
- 3.5 Guidelines and discussion 29
- 3.6 Limitations 31
- 3.7 Conclusions 31
- Further reading 31

CHAPTER 4 The research on basic visual design of the head-up display of an automobile based on driving cognition 35

Fang You, Jinghui Zhang, Jianmin Wang, Mengting Fu and Zhenghe Lin

- 4.1 Introduction 35
- 4.2 Method 36
  - 4.2.1 Experimental environment 36
  - 4.2.2 Subjects 38
  - 4.2.3 Test contents and steps 38
- 4.3 Results 38
  - 4.3.1 Influence of weather conditions on reaction time 38
  - 4.3.2 Influence of age on response time 40
  - 4.3.3 Trends of workload 41
- 4.4 Discussion 42
- Further reading 43

CHAPTER 5 A novel cooperation-guided warning system utilizing augmented reality head-up display to enhance driver's perception of invisible dangers 45
Fang You, Jun Zhang, Jie Zhang, Lian Shen, Weixuan Fang, Wei Cui and Jianmin Wang
   5.1 Introduction 46
   5.2 Context and theoretical framework 47
        5.2.1 Human-vehicle teaming and cooperative interface 47
        5.2.2 Situational awareness in cooperative driving 48
        5.2.3 Cognitive information design 50
        5.2.4 Automation surprises and trust 51
   5.3 Designing HMI 52
        5.3.1 HMI for Case 1 52
        5.3.2 HMI for Case 2 54
   5.4 Method and materials 55
        5.4.1 Case Study 1 55
        5.4.2 Case study 2 59
   5.5 Results and analysis 61
        5.5.1 Results of Case 1 61
        5.5.2 Results of Case 2 64
   5.6 Discussion 66
        5.6.1 Case 1 66
        5.6.2 Case 2 67
        5.6.3 Limitations 69
   5.7 Conclusion 70
        References 70

PART 2 Intelligent cockpit HMI information decision and control

CHAPTER 6 Acting like a human: teaching an autonomous vehicle to deal with traffic encounters 75
Jianmin Wang, Jiawei Lu, Fang You and Yujia Wang
   6.1 Introduction 75
   6.2 Socially incapable autonomous vehicles 75
   6.3 Method 76
   6.4 Results and analysis 76
   6.5 Discussion 78
   6.6 Conclusion 79
        Further reading 79

CHAPTER 7 Research on transparency design based on shared situation awareness in semiautomatic driving 81
Fang You, Huijun Deng, Preben Hansen and Jun Zhang

- **7.1** Introduction 82
- **7.2** Related works 82
  - 7.2.1 Shared situation representation 82
  - 7.2.2 Situation awareness 83
  - 7.2.3 Lyons's task model of transparency 83
  - 7.2.4 Relationship between transparency, performance, and trust 83
  - 7.2.5 Human-machine interface requirements of system transparency 83
  - 7.2.6 Handovers in automated driving 84
- **7.3** Study 84
  - 7.3.1 Scenario 84
  - 7.3.2 Framework 85
  - 7.3.3 Transparency design 86
- **7.4** Experiment 87
  - 7.4.1 Participants 88
  - 7.4.2 Prototype 89
  - 7.4.3 Procedure 90
  - 7.4.4 Apparatus and simulated environment 90
  - 7.4.5 Materials 90
- **7.5** Results 91
  - 7.5.1 Transparency 91
  - 7.5.2 Situation awareness 92
  - 7.5.3 Usability 92
  - 7.5.4 Workload 93
- **7.6** Discussion 94
- **7.7** Conclusions 95
- Reference 96
- Further reading 96

CHAPTER 8 Human-machine interface design based on transparency in autonomous driving scenes 99
Jianmin Wang, Qiaofeng Wang and Jun Zhang

- **8.1** Introduction 99
- **8.2** Theoretical basis 100
  - 8.2.1 Situational awareness and transparency 100
  - 8.2.2 Trust 102
  - 8.2.3 Human-machine interface design case study 102

8.3 Human-machine interface design 104
    8.3.1 Scenario analysis 104
    8.3.2 Interface element design 104
8.4 Experimental evaluation 108
    8.4.1 Participants 108
    8.4.2 Experimental environment 108
    8.4.3 Experimental process 109
8.5 Analysis of experimental results 110
    8.5.1 Scale data 110
    8.5.2 Correlation analysis 111
    8.5.3 Interview data 112
    8.5.4 Experimental conclusions 112
8.6 Conclusion 114
    References 114
    Further reading 114

CHAPTER 9 Drivers' trust of the human-machine interface of an adaptive cruise control system 115
    Jianmin Wang, Wenjuan Wang, Xiaomeng Li and Fang You
9.1 Introduction 115
9.2 Research question 116
9.3 Method 116
    9.3.1 Participants 116
    9.3.2 Experimental design 116
    9.3.3 Experimental procedure 117
9.4 Results 119
9.5 Limitations 120
9.6 Conclusion and discussion 120
    Further reading 121

CHAPTER 10 Take-over requests analysis in conditional automated driving 123
    Fang You, Yujia Wang, Jianmin Wang, Xichan Zhu and Preben Hansen
10.1 Introduction 123
10.2 Background 124
10.3 Highway hazard scenario analysis 125
    10.3.1 Scenario analysis 125
    10.3.2 Lane changing visual scanning analysis 126
    10.3.3 Decision-making phase vision scanning analysis 127
    10.3.4 Execution phase vision scanning analysis 128

	10.3.5 Adjustment phase vision scanning analysis	128
10.4	Application and evaluation	129
	10.4.1 Evaluation procedure	129
	10.4.2 Results	130
10.5	Discussion and conclusion	131
	References	131
	Further reading	132

PART 3 Research on intelligent cockpit HMI design under automated driving scenarios

CHAPTER 11 Interactive framework of a cooperative interface for collaborative driving 135

Jun Zhang, Yujia Liu, Preben Hansen, Jianmin Wang and Fang You

11.1	Introduction	135
11.2	Framework of cooperative interface	137
	11.2.1 Human autonomy team	138
	11.2.2 Level of interaction	143
	11.2.3 Dynamic external factors	144
11.3	Conclusion	144
	References	145
	Further reading	146

CHAPTER 12 Design methodologies for human-artificial systems: an automotive augmented reality headup display design case study 149

Cuiqiong Cheng, Fang You, Preben Hansen and Jianmin Wang

12.1	Introduction	149
12.2	Related works	150
12.3	Interaction design method framework	150
	12.3.1 Six-phase model of interaction design method framework	150
	12.3.2 Three perspectives of interaction design method framework	151
12.4	Automotive augmented reality head-up display design case study	152
	12.4.1 Project description	152
	12.4.2 Project implementation process	152
12.5	Conclusion	154
	References	154

CHAPTER 13 Design factors of shared situation awareness interface in human-machine co-driving 155
Fang You, Xu Yan, Jun Zhang and Wei Cui

13.1 Introduction 156
13.2 Related works 156
    13.2.1 Situation awareness 156
    13.2.2 Shared situation awareness 157
    13.2.3 Interface designs of human-machine co-driving 158
13.3 Design methods 158
    13.3.1 Abstraction hierarchy analysis 158
    13.3.2 Abstraction hierarchy analysis of shared situation awareness human-machine interaction interfaces 159
    13.3.3 Four factors of shared situation awareness interface 160
13.4 Simulation experiment 160
    13.4.1 Experimental subjects 160
    13.4.2 Experimental design 161
    13.4.3 Selection of experimental scenarios and requirement analysis of shared situation awareness information 161
    13.4.4 Experimental environment 163
    13.4.5 Experimental process 164
    13.4.6 Experimental evaluation method 164
13.5 Results 165
    13.5.1 Eye movement 166
    13.5.2 Situation awareness 167
    13.5.3 Usability 167
    13.5.4 Task response time and task accuracy 168
    13.5.5 Interview 169
13.6 Discussion 170
13.7 Conclusions 172
References 172
Further reading 172

CHAPTER 14 Automotive head-up display interaction design based on a lane-changing scenario 175
Chenxi Jin, Fang You and Jianmin Wang

14.1 Introduction 175
14.2 Background 176
14.3 Research on lane-changing scenario 176
    14.3.1 Preliminary investigations 176

		14.3.2 Study on environment, behavior, and psychology 177
		14.3.3 Observation 178
	14.4	Automotive head-up display interaction design 178
		14.4.1 Head-up display information filtration and organization 179
		14.4.2 Head-up display interface design 179
	14.5	Conclusion 181
		Further reading 181

CHAPTER 15 Human-computer collaborative interaction design of intelligent vehicles—a case study of HMI of adaptive cruise control 183
Yujia Liu, Jun Zhang, Yang Li, Preben Hansen and Jianmin Wang

	15.1	Introduction 184
	15.2	Related work 184
		15.2.1 Taxonomy of automated driving systems 184
		15.2.2 Human-computer collaborative interaction in automated driving 185
	15.3	Human-engaged automated driving framework 186
		15.3.1 Full human 186
		15.3.2 Full automation 187
		15.3.3 Driver assistance 187
		15.3.4 Human supervision 187
		15.3.5 Collaboration driving 188
	15.4	Case 188
		15.4.1 Application scenario design 188
		15.4.2 Information architecture design 190
		15.4.3 Interface design 193
	15.5	Interface experiment 194
		15.5.1 Experimental design 194
		15.5.2 Participants 195
		15.5.3 Apparatus 195
		15.5.4 Task and procedure 195
		15.5.5 Measures 196
		15.5.6 Results 196
		15.5.7 Discussion 198
	15.6	Conclusions 199
		References 199
		Further reading 199

PART 4 Research on interaction design of on-board robots

CHAPTER 16 Research hotspots and trends of social robot interaction design: a bibliometric analysis 203
Jianmin Wang, Yongkang Chen, Siguang Huo, Liya Mai and Fusheng Jia

- 16.1 Introduction 203
- 16.2 Research design 204
  - 16.2.1 Data sources 204
  - 16.2.2 Research methods 205
- 16.3 Bibliometric results and analysis 205
  - 16.3.1 Trend analysis of annual outputs of social robot interaction design literature 205
  - 16.3.2 Research hotspots of social robot interaction design 207
  - 16.3.3 Evolution and trend of social robot interaction design research hotspots 210
  - 16.3.4 Knowledge base of SR-human-robot interaction research .. 211
  - 16.3.5 Distribution of social robot interaction design literature sources 215
  - 16.3.6 High-impact countries and research institutions 216
  - 16.3.7 High-impact author analysis 217
- 16.4 Discussion 218
  - 16.4.1 More realistic research scenarios 218
  - 16.4.2 Effective measurement of research indicators 218
  - 16.4.3 Longitudinal trial 219
  - 16.4.4 More specific design strategies 219
  - 16.4.5 Uncovering the psychological cognitive mechanisms behind user behavior 219
- 16.5 Conclusions 219
- References 220
- Further reading 220

CHAPTER 17 Experimental study on abstract expression of human-robot emotional communication 225
Jianmin Wang, Yuxi Wang, Yujia Liu, Tianyang Yue, Chengji Wang, Weiguang Yang, Preben Hansen and Fang You

- 17.1 Introduction 225
- 17.2 Literature review 226
  - 17.2.1 Development of emotion expression design in virtual images .. 226
  - 17.2.2 Facial emotion evaluation model 227
- 17.3 Emotion expression design of virtual image 228

17.4	Evaluation of facial expression design of virtual image	229
17.5	Method	233
	17.5.1 Participants	233
	17.5.2 Design of experiment	233
	17.5.3 Procedure	235
	17.5.4 Results	236
17.6	Conclusions	239
	References	239
	Further reading	240

CHAPTER 18 Self-assessment emotion tool: nonverbal measurement tool of user's emotional experience 243

Jianmin Wang, Yujia Liu, Yuxi Wang, Jinjing Mao, Tianyang Yue and Fang You

18.1	Introduction	243
18.2	Related works	244
	18.2.1 Emotion and emotional space	244
	18.2.2 Nonverbal emotion measurement tool	245
18.3	Design of self-assessment emotion tool	246
	18.3.1 Emotion set	246
	18.3.2 Emotion design of self-assessment emotion tool	247
18.4	Validation of self-assessment emotion tool images	250
	18.4.1 Pilot study	250
	18.4.2 Validation study	253
	18.4.3 Summary of results	255
18.5	Conclusions and discussion	257
	18.5.1 Methods of application	257
	18.5.2 Conclusions	257
	References	257
	Further reading	258

CHAPTER 19 Robot transparency and anthropomorphic attribute effects on human-robot interactions 259

Jianmin Wang, Yujia Liu, Tianyang Yue, Chengji Wang, Jinjing Mao, Yuxi Wang and Fang You

19.1	Introduction	260
19.2	Robot-to-human communication	260
	19.2.1 Communication transparency	261
	19.2.2 Anthropomorphism	261

19.3 Human-robot interaction outcomes 262
19.3.1 Safety 262
19.3.2 Usability 262
19.3.3 Workload 262
19.3.4 Trust 263
19.3.5 Affect 263
19.3.6 Current study 264
19.4 Methods 265
19.4.1 Participants 265
19.4.2 Design 265
19.4.3 Apparatus and materials 266
19.4.4 Procedure 267
19.5 Results 268
19.5.1 Task one: welcome 268
19.5.2 Task two: incoming call 268
19.5.3 Task three: chatting 273
19.6 Discussion 275
19.6.1 Transparency 275
19.6.2 Anthropomorphism 276
19.6.3 Human-robot interaction variable relationships 276
19.6.4 Implications for human-robot interaction design 276
19.6.5 Limitations of the current study 277
19.7 Conclusions and future directions 277
References 277
Further reading 278

CHAPTER 20 Design of proactive interaction of in-vehicle robots based transparency 281
Jianmin Wang, Tianyang Yue, Yujia Liu, Yuxi Wang, Chengji Wang, Fei Yan and Fang You
20.1 Introduction 281
20.2 Related works 282
20.2.1 Proactive interaction 283
20.2.2 Transparency 283
20.3 Transparency design of proactive interaction 284
20.3.1 Human-robot interface levels 285
20.3.2 Transparency design assumptions 285
20.4 Experiment 287
20.4.1 Participants 287
20.4.2 Design of experiment 287

20.4.3 Measurement 289
20.4.4 Apparatus and materials.. 289
20.4.5 Procedure 291
20.5 Results 292
20.5.1 Telephone task results... 292
20.5.2 Speeding task results... 295
20.6 Discussion... 297
20.7 Conclusions 299
References 299
Further reading 299

PART 5 Automated vehicles' external human interfaces design and evaluation

CHAPTER 21 Interaction design for environment information surrounding vehicles in parking scenarios 305

Jianmin Wang, Jingyi Pei and Wei Cui

21.1 Introduction and research methodology 306
21.1.1 Introduction 306
21.1.2 Research methodology.. 306
21.2 Design space of environment information surrounding vehicles........ 307
21.2.1 Future trends of environment information surrounding vehicles 307
21.2.2 Case analysis 308
21.3 User research of environment information demand in parking scenarios 310
21.3.1 Questionnaire survey on users' demand for information in parking scenarios 311
21.3.2 In-depth interview on users' demand for information in parking scenarios 311
21.3.3 Field research 312
21.4 Design practice of environment information surrounding vehicles in parking scenarios.. 313
21.4.1 User activity analysis... 313
21.4.2 Scenario analysis... 314
21.4.3 Display scheme design.. 315
21.5 Usability test of design 319
21.5.1 Basic information of test... 319
21.5.2 Comparison of text scheme and graphic scheme..................... 320
21.5.3 Comparison of display positions of environment information 322
21.6 Results and recommendations.. 323
21.6.1 Results 323

21.6.2 Recommendation...323
Further reading 323

CHAPTER 22 Unmanned vehicle external interaction design based on an automation acceptance model 325
Bihan Zhang and Fang You

22.1 Introduction 325
22.2 Automation acceptance model.. 326
22.3 Scenario-based pedestrian behavior research 327
    22.3.1 Design background 327
    22.3.2 In-depth interview and user research on pedestrian-unmanned vehicle interaction...327
    22.3.3 Behavioral characteristics and demand analysis of pedestrian-unmanned vehicles 328
22.4 Vehicle diplomacy mutual strategy based on automation acceptance model 329
    22.4.1 Construction of vehicle diplomacy mutual strategy model 329
    22.4.2 Anthropomorphic design promotes trust and emotion...............330
    22.4.3 Classification and visualization of information outside the vehicle..........331
    22.4.4 Multimodal design of external vehicle interaction 332
    22.4.5 Interactive prototype design based on vehicle diplomacy mutual strategy 332
22.5 Design test evaluation 334
22.6 Conclusion 334
Reference 334
Further reading 335

CHAPTER 23 Designing communication strategies of autonomous vehicles with pedestrians: an intercultural study 337
Mirjam Lanzer, Franziska Babel, Fei Yan, Bihan Zhang, Fang You, Jianmin Wang and Martin Baumann

23.1 Introduction 337
23.2 Related works 338
23.3 Method 339
    23.3.1 Sample 339
    23.3.2 Procedure 340
    23.3.3 Materials 341
    23.3.4 Questionnaires 342
23.4 Results 343
    23.4.1 Manipulation check 343

		23.4.2 Compliance	343
		23.4.3 Acceptance	345
		23.4.4 Trust in automation	346
	23.5	Discussion	347
		23.5.1 Practical implications	348
		23.5.2 Limitations	348
	23.6	Conclusion	349
		Further reading	349

PART 6 Intelligent cockpit HMI evaluation

CHAPTER 24 Augmented reality cognitive interface in enhancing human vehicle collaborative driving safety: a design perspective ... 357
Fang You, Yuwei Liang, Qianwen Fu and Jun Zhang

24.1	Introduction		358
24.2	Context and theoretical framework		359
	24.2.1	Human-vehicle team and cooperation mechanism	359
	24.2.2	Cognitive interface and augmented reality information elements	362
24.3	Design method for augmented reality human-machine interface		364
	24.3.1	Collaborative contextual analysis	365
	24.3.2	Cognitive information visualization	368
	24.3.3	Component adjustment and overlay	370
24.4	Empirical evaluation		371
	24.4.1	Hypotheses	372
	24.4.2	Experimental design	372
	24.4.3	Apparatus and materials	374
	24.4.4	Procedure	375
	24.4.5	Measures and analysis	375
24.5	Results		377
	24.5.1	Subjective scales	377
	24.5.2	Fixation metric	378
	24.5.3	Fixation heatmap	380
	24.5.4	Interviews	380
24.6	Discussion		381
	24.6.1	Visualization components and quantities	381
	24.6.2	Attention and cognitive demands	382
	24.6.3	Training and mental models	383
	24.6.4	Limitations	384
24.7	Conclusions		384
	References		384

CHAPTER 25 Behavioral indicators affecting driving performance in human-machine interface assessments with simulation 391
Yukun Xie, Tianyang Yue, Preben Hansen and Fang You
- 25.1 Introduction 391
- 25.2 Method 392
  - 25.2.1 Participants 392
  - 25.2.2 Experimental design 392
  - 25.2.3 Experimental procedure 394
  - 25.2.4 Data collection 394
- 25.3 Result 394
  - 25.3.1 Vehicle vertical indicators 394
  - 25.3.2 Vehicle horizontal indicators 394
- 25.4 Discussion 395
- 25.5 Conclusion 396
- Further reading 396

CHAPTER 26 Electrodermal activity measurement in a driving simulator 399
Fang You, Yukun Xie, Yaru Li and Yijun Xu
- 26.1 Introduction 399
- 26.2 Applications and advantages of electrodermal activity measurement 400
  - 26.2.1 Applications of electrodermal activity measurement in evaluation 400
  - 26.2.2 Advantages of experiments combined with electrodermal activity measurement 401
- 26.3 Teaching objectives and method exploration 401
  - 26.3.1 Teaching objectives 401
  - 26.3.2 Method exploration 402
- 26.4 Course experiment methods 403
  - 26.4.1 Experimental design 403
  - 26.4.2 Experimenter 403
  - 26.4.3 Design scheme 403
  - 26.4.4 Experimental environment and equipment 404
  - 26.4.5 Electrodermal activity measurement equipment 405
- 26.5 Analysis and discussion of experimental results 406
  - 26.5.1 Electrodermal activity and system availability scale 406
  - 26.5.2 Electrodermal activity and task behavior time 406
- 26.6 Conclusion 409
- References 409
- Further reading 409

CHAPTER 27 Using eye-tracking to help design HUD-based safety indicators for lane changes 411
Fang You, Jürgen Friedrich, Yang Li, Jianmin Wang and Ronald Schroeter
- 27.1 Introduction and related works 411
- 27.2 Methods 412
  - 27.2.1 Experimental setup 412
  - 27.2.2 Data collection 412
  - 27.2.3 Data analysis 413
- 27.3 Results 413
- 27.4 Discussion 414
- 27.5 Lessons learned 416
  - Further reading 417

CHAPTER 28 The situation awareness and usability research of different HUD HMI designs in driving while using adaptive cruise control 419
Jianmin Wang, Wenjuan Wang, Preben Hansen, Yang Li and Fang You
- 28.1 Introduction 420
  - 28.1.1 Adaptive cruise control 420
  - 28.1.2 Head-up display: windshield and augmented reality 420
  - 28.1.3 Situation awareness 421
  - 28.1.4 Usability test 421
- 28.2 Research propositions 422
- 28.3 Experimental method 422
  - 28.3.1 Participants 422
  - 28.3.2 Technical equipment 422
  - 28.3.3 Experimental design 423
  - 28.3.4 Experimental tasks 425
  - 28.3.5 Experimental process 425
- 28.4 Results and preliminary findings 425
  - 28.4.1 Situation awareness 426
  - 28.4.2 Usability 427
- 28.5 Discussion 428
- 28.6 Limitations 429
- 28.7 Conclusion 429
  - References 429
  - Further reading 429

Index 433

List of contributors

Franziska Babel
Department of Human Factors, Institute of Psychology and Education; Faculty of Engineering, Computer Science and Psychology, Ulm University, Germany

Martin Baumann
Department of Human Factors, Institute of Psychology and Education; Faculty of Engineering, Computer Science and Psychology, Ulm University, Germany

Yongkang Chen
College of Design and Innovation, Tongji University, Shanghai, P.R. China

Cuiqiong Cheng
Car Interaction Design Lab, College of Arts and Media, Tongji University, Shanghai, P.R. China

Wei Cui
College of Design and Innovation, Tongji University, Shanghai, P.R. China; Car Interaction Design Lab, College of Arts and Media, Tongji University, Shanghai, P.R. China

Huijun Deng
Car Interaction Design Lab, College of Arts and Media, Tongji University, Shanghai, P.R. China

Weixuan Fang
Car Interaction Design Lab, College of Arts and Media, Tongji University, Shanghai, P.R. China

Jürgen Friedrich
Center for Computing Technologies, University of Bremen, Bremen, Germany

Mengting Fu
Car Interaction Design Lab, College of Arts and Media, Tongji University, Shanghai, P.R. China

Qianwen Fu
College of Design and Innovation, Tongji University, Shanghai, P.R. China

Preben Hansen
Department of Computer and Systems Sciences, Stockholm University, Stockholm, Sweden

Siguang Huo
College of Design and Innovation, Tongji University, Shanghai, P.R. China

Fusheng Jia
College of Design and Innovation, Tongji University, Shanghai, P.R. China

Chenxi Jin
Car Interaction Design Lab, College of Arts and Media, Tongji University, Shanghai, P.R. China

Xin Jin
Car Interaction Design Lab, College of Arts and Media, Tongji University, Shanghai, P.R. China

Mirjam Lanzer
Department of Human Factors, Institute of Psychology and Education; Faculty of Engineering, Computer Science and Psychology, Ulm University, Germany

Liping Li
Intelligent Driving Experience Center, Baidu, Shenzhen, GD, P.R. China

Xiaomeng Li
Centre for Accident Research and Road Safety-Queensland (CARRS-Q), Institute of Health and Biomedical Innovation (IHBI), Queensland University of Technology (QUT), Kelvin Grove, Brisbane, Australia

Yang Li
Car Interaction Design Lab, College of Arts and Media, Tongji University, Shanghai, P.R. China; Karlsruher Institute of Technology, Baden-Württemberg, Germany

Yaru Li
Car Interaction Design Lab, College of Arts and Media, Tongji University, Shanghai, P.R. China

Yuwei Liang
Car Interaction Design Lab, College of Arts and Media, Tongji University, Shanghai, P.R. China

Zhenghe Lin
Car Interaction Design Lab, College of Arts and Media, Tongji University, Shanghai, P.R. China

Yujia Liu
Car Interaction Design Lab, College of Arts and Media, Tongji University, Shanghai, P.R. China; College of Design and Innovation, Tongji University, Shanghai, P.R. China

Jiawei Lu
College of Design and Innovation, Tongji University, Shanghai, P.R. China

Xiaojun Luo
Intelligent Driving Experience Center, Baidu, Shenzhen, GD, P.R. China

Liya Mai
College of Design and Innovation, Tongji University, Shanghai, P.R. China

Jinjing Mao
Car Interaction Design Lab, College of Arts and Media, Tongji University, Shanghai, P.R. China

Jingyi Pei
Car Interaction Design Lab, College of Arts and Media, Tongji University, Shanghai, P.R. China

Ronald Schroeter
Centre for Accident Research and Road Safety-Queensland (CARRS-Q), Queensland University of Technology, Brisbane, Australia

Lian Shen
Car Interaction Design Lab, College of Arts and Media, Tongji University, Shanghai, P.R. China

Chengji Wang
Car Interaction Design Lab, College of Arts and Media, Tongji University, Shanghai, P.R. China

Jianmin Wang
Car Interaction Design Lab, College of Arts and Media, Tongji University, Shanghai, P.R. China

Jifang Wang
Intelligent Driving Experience Center, Baidu, Shenzhen, GD, P.R. China

Qiaofeng Wang
Car Interaction Design Lab, College of Arts and Media, Tongji University, Shanghai, P.R. China

Wenjuan Wang
Car Interaction Design Lab, College of Arts and Media, Tongji University, Shanghai, P.R. China

Yujia Wang
Car Interaction Design Lab, College of Arts and Media, Tongji University, Shanghai, P.R. China

Yuxi Wang
Car Interaction Design Lab, College of Arts and Media, Tongji University, Shanghai, P.R. China

Yukun Xie
Car Interaction Design Lab, College of Arts and Media, Tongji University, Shanghai, P.R. China

Yijun Xu
Car Interaction Design Lab, College of Arts and Media, Tongji University, Shanghai, P.R. China

Fei Yan
Department of Human Factors, Institute of Psychology and Education; Faculty of Engineering, Computer Science and Psychology, Ulm University, Germany

Xu Yan
Car Interaction Design Lab, College of Arts and Media, Tongji University, Shanghai, P.R. China

Weiguang Yang
Car Interaction Design Lab, College of Arts and Media, Tongji University, Shanghai, P.R. China

Yifan Yang
Car Interaction Design Lab, College of Arts and Media, Tongji University, Shanghai, P.R. China

Fang You
Car Interaction Design Lab, College of Arts and Media, Tongji University, Shanghai, P.R. China

Tianyang Yue
Car Interaction Design Lab, College of Arts and Media, Tongji University, Shanghai, P.R. China

Bihan Zhang
Car Interaction Design Lab, College of Arts and Media, Tongji University, Shanghai, P.R. China

Jie Zhang
School of Design, The Hong Kong Polytechnic University, Hong Kong, P.R. China

Jinghui Zhang
Car Interaction Design Lab, College of Arts and Media, Tongji University, Shanghai, P.R. China

Jun Zhang
College of Design and Innovation, Tongji University, Shanghai, P.R. China; School of Design, The Hong Kong Polytechnic University, Hong Kong, P.R. China; Car Interaction Design Lab, College of Arts and Media, Tongji University, Shanghai, P.R. China

Xichan Zhu
School of Automotive Studies, Tongji University, Shanghai, P.R. China

Acknowledgments

The publication is funded by the 2023 Shanghai Municipal Financial Support Funds for Promoting the Development of the Cultural and Creative Industries. This book is also listed and funded by in the 14th Five-Year Textbook Plan of Tongji University. The research in the book is supported by Intelligent New Energy Vehicle Collaborative Innovation Center, Tongji University, based on the Fundamental Research Funds for the Central Universities; Shenzhen Science and Technology Program (GJHZ20220913142401002).

PART 1

Intelligent cockpit HMI information perception and understanding

Interface color design of intelligent vehicle central consoles

1

Fang You[1], Yaru Li[1], Preben Hansen[2], Liping Li[3], Mengting Fu[1], Yifan Yang[1], Xin Jin[1] and Jianmin Wang[1]

[1]*Car Interaction Design Lab, College of Arts and Media, Tongji University, Shanghai, P.R. China* [2]*Department of Computer and Systems Sciences, Stockholm University, Stockholm, Sweden* [3]*Intelligent Driving Experience Center, Baidu, Shenzhen, GD, P.R. China*

Chapter Outline

1.1 Introduction 3
1.2 Experimental design 4
    1.2.1 Preliminary preparations 4
    1.2.2 Simulator settings 5
    1.2.3 Driving scene settings 5
1.3 Experiment execution 6
1.4 Experimental results 6
    1.4.1 Data collection 6
    1.4.2 Subjective data 6
    1.4.3 Objective data 8
1.5 Discussion 9
Further reading 9

1.1 Introduction

With the increasing development of cloud platforms and 5G technology, smart cars are emerging as a pivotal direction for the world's future automobile industry. Given its convenience, the human-machine interface has become the primary interactive method for users. The vehicle-machine interface contains many elements, including text, icons, colors, etc.

The objective of this study is to explore how different contrasts influence users' assessment of information accuracy based on color brightness. To achieve this research objective, we will address the following research questions.

RQ1: How does the contrast value of different hues on the vehicle-machine interface affect driving performance and vehicle control under different lighting conditions (daytime diffuse reflection versus nighttime)?

RQ2: What criteria should be considered when selecting foreground and background color contrasts to achieve the best recognition when designing vehicle-machine interface?

Our contributions include three main aspects. First, we verified that the recommended contrast values of vehicle-machine interface under different lighting conditions comply with ISO 15008 standards. Second, through the analysis of subjective and objective data, we inferred that different contrasts have different effects on drivers' color perception usability. Finally, we extended the recommended range of color contrasts to vehicle interface design, thus establishing a forward-looking reference value in human-computer interaction.

1.2 Experimental design
1.2.1 Preliminary preparations

In the early stages, the vehicle interfaces of 22 models, including Mercedes-Benz A200L, were sorted out through network resources and 4S shop visits. Based on the theme colors identified, our analysis revealed a predominant concentration of blue, green, and red hues. Consequently these three colors were selected as the experimental objects. Subsequently, we selected five models with better assessments and proceeded to extract the RGB values of the three colors along with their backgrounds. Adjustments were made using online tools such as WebAIM's color contrast checker to derive experimental materials with different contrasts. We found that the hues became more diverse when applied to icons. Considering the influence of different color contrasts on driving performance, we applied foreground colors to the icons.

According to ISO 15008 guidelines, the minimum color contrast value is 3:1 during daytime diffuse lighting and 5:1 at night. Accordingly this experiment selected two sets of contrasts that were different in front and rear ranges. For daytime conditions, color contrast was divided into 1.5:1, 2:1, 3:1, 4:1, and 5:1, while for nighttime conditions, contrasts were set at 1.5:1, 3:1, 5:1, 7:1, and 9:1 (Fig. 1.1).

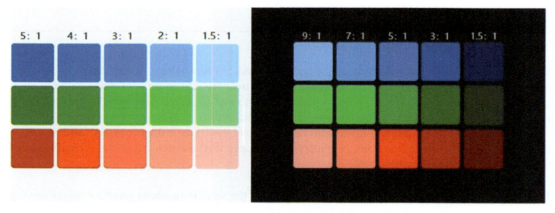

FIGURE 1.1

Schematic diagram of color contrast selection for daytime (*left*) and nighttime (*right*) conditions.

Our primary focus was the impact of brightness contrast on driving. Thus we used WestboroPhotonics WP6120E colorimetric brightness meter to examine different background areas. Our findings indicated that, excluding the influence of environmental illuminance, differences in contrast between foreground and background colors showed no difference under different screen brightness levels.

1.2.2 Simulator settings

To simulate a realistic driving environment and ensure a safer immersive experience for drivers, the experiment was carried out using a car driving simulator. The simulator used in this experiment comprised essential components such as seat, steering wheel, brake, accelerator, and three screens with a resolution of 7680*1440, creating a virtual road environment.

For light condition settings, we adhered to the SAE J1757/1:2007 specifications. Daytime conditions were simulated using Fleisch LED lights to provide diffuse ambient light. The ambient light intensity measured on the display surface was maintained at 5 klx ± 5%. Nighttime conditions, on the other hand, were simulated in a dark environment where the maximum illuminance of the measured object did not exceed 10 lx, with a relative tolerance of 5% (Fig. 1.2).

1.2.3 Driving scene settings

To distinguish between specific phenomena caused by secondary tasks and those inherent to driving tasks, specific driving scenarios need to be set in the driving simulator. To verify the effectiveness of driving scenarios, we interviewed six professional test drivers. Based on their evaluations, it was determined that the operation would be carried out only under safe road conditions, such as low traffic flow, and would not be used in nonsafe conditions, such as curves. Consequently, for the experiment, the road scene was configured as a straight urban road with a traffic flow of 2−3 vehicles per minute.

FIGURE 1.2

Overall view of the simulator under daytime lighting (*left*) conditions and nighttime lighting (*right*) conditions.

1.3 Experiment execution

The experiment is composed of 30 similar tasks. The first 15 tasks are performed under daytime lighting conditions, and the latter are performed under nighttime lighting conditions. The task is to travel stable and straight on urban road at a speed of 30 km/hour, and complete the identification and click operation of icons with different color contrasts. 31 subjects participated in the study (age $M = 33.4$, $SD = 5.277$) (Fig. 1.3).

1.4 Experimental results

1.4.1 Data collection

We used a multidirectional method to collect both subjective and objective data. Subjective data includes workload and usability indicators such as clarity, comfort, and satisfaction, while objective data includes task response time and vehicle speed standard deviation. We adopted a synthetic analysis method to obtain the appropriate recommended value of icon-background contrast under specific lighting conditions (Table 1.1).

1.4.2 Subjective data

Subjective data includes usability and workload. Usability evaluation involved scoring clarity, comfort, and satisfaction using Likert's 7-level scale, where 1–7 points represent gradually increasing usability. Workload, on the other hand, encompasses multiple dimensions, which include mental strength,

FIGURE 1.3

Experimental process.

Table 1.1 Spearman correlation coefficient: color contrast and subjective evaluation data.

		Clarity	Comfort	Satisfaction	Usability	Workload
Daytime	Mixed	0.406**	0.353**	0.386**	0.402**	−0.274**
	Red	0.516**	0.431**	0.490**	0.508**	−0.320**
	Blue	0.511**	0.475**	0.473**	0.502**	−0.360**
	Green	0.183*	0.158*	0.207**	0.195*	−0.145
Nighttime	Mixed	0.273**	0.270**	0.261**	0.275**	−0.186**
	Red	0.210**	0.205**	0.184**	0.208**	−0.118*
	Blue	0.270**	0.272**	0.260**	0.276**	−0.189**
	Green	0.259**	0.278**	0.270**	0.290**	−0.194**

**Note: $p < 0.001$
*$p < 0.01$

attention, physical burden, time pressure, and frustration required to complete the existing task. Participants scored the workload necessary to effectively maintain safe driving while browsing the screen on a scale ranging from 0 to 10 points with higher scores indicating greater workload (Fig. 1.4).

During daylight conditions, with hues not distinctly separated, the changing trend of subjective evaluation tends to show a gradual shift, with contrast ranging from 3:1 to 4:1. When considering the daytime color contrast of the vehicle-machine interface, contrast ratios of 3:1 and 4:1 can be selected for debugging purposes. This conclusion applies to red, green, and blue hues. Under the circumstance where the RGB value of the background color of the central control screen is (245, 245, 245), the contrast of different hues of the icons and their corresponding RGB values are shown in Table 1.2.

Under nighttime light conditions, when hues are not distinctly separated, subjective evaluations tend to remain relatively consistent across contrast ratios ranging between 3:1 and 7:1. Since they are both inflection points, a contrast ratio of 5:1 is recommended when considering the color contrast of the car-machine interface at night, which can be adjusted to higher than 3:1. for debugging purposes. Furthermore it is recommended to debug red at a contrast ratio of 3:1 at night, while blue and green are recommended to be debugged at 5:1. When the RGB value of the background color of the central control screen is (10,10,10), the contrast of the different hues of the icons and their corresponding RGB values are shown in Table 1.3.

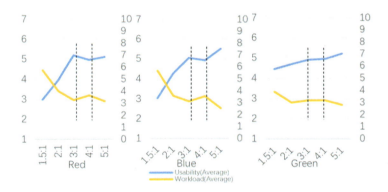

FIGURE 1.4

Color contrast evaluation: usability and workload (daylight).

Table 1.2 Recommended color contrast and corresponding RGB value during the day.

Hue	Contrast ratio	R	G	B
Red	3:1	239	94	78
Red	4:1	233	43	22
Blue	3:1	5	143	255
Blue	4:1	0	121	219
Green	3:1	30	164	68
Green	4:1	25	138	55

8 Chapter 1 Interface color design of intelligent vehicle central consoles

Table 1.3 Recommended color contrast and corresponding RGB value at night.

Hue	Contrast ratio	R	G	B
Red	3:1	189	0	22
Blue	5:1	5	128	240
Green	5:1	27	147	61

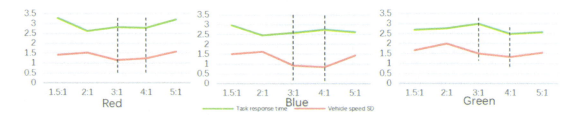

FIGURE 1.5

Color contrast: task response time and speed SD (daylight).

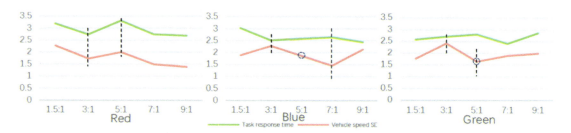

FIGURE 1.6

Color contrast: task response time and speed SD (nightlight).

1.4.3 Objective data

Objective data includes task response time and vehicle speed standard deviation. Task response time reflects how quickly drivers distinguish different color contrasts during driving. Because the task requires the drivers to maintain a stable speed of about 30 km/hour, the standard deviation of the vehicle speed is used as a metric for safe and attentive driving as well as the stability of driving control (Fig. 1.5).

Under daylight conditions, when the contrast ratio is between 3:1 and 4:1, the change trend in vehicle speed standard deviation is relatively gradual. At this time, participants' control of the vehicle is stable and task response time is relatively short. Combining subjective and objective data, when considering the color design of the vehicle interface in the daytime, a contrast value between 3:1 and 4:1 should be selected (Fig. 1.6).

Under nighttime light conditions, when the contrast of red is in the range of 3:1 to 5:1, task response time and speed standard deviation are on the rise. After the contrast exceeds 5:1, the objective data shows a downward trend. Under comprehensive consideration, designers should opt for a contrast ratio of 3:1 for red at night. For blue, the turning points of data changes are at 3:1 and 7:1. Combining subjective and objective data, 5:1 is deemed the appropriate contrast value for blue at night. When the contrast ratio of green is higher than 5:1, the standard deviation of vehicle speed tends to stabilize; thus green is recommended to be adjusted at 5:1 at night.

1.5 Discussion

The results of our study confirm the most recommended contrast values for vehicle interface design under different lighting conditions mentioned in ISO 15008. Furthermore these findings are extended to include the three most widely used hues in vehicle interfaces: red, blue, and green. Through a thorough analysis of subjective and objective data, the results obtained from the experiment can be extended to various applications related to the car's central control screen. Blue is a ubiquitous color in vehicle interfaces, often denoting settings, navigation, steering wheel, Bluetooth, wireless network connectivity, etc., generally indicating turning on or normal status display. Green indicates that the vehicle accessories are in good condition, and is usually used for icon colors such as telephone, voice, music, car lock, etc. Conversely red serves as a warning or reminder, signaling abnormal conditions such as alarms and tire pressure issues.

When selecting colors for the vehicle's central control screen design, hues within the visible wavelength range can be chosen based on the following steps: first, determine the RGB values of the foreground and background colors; next use online tools such as WebAIM's color contrast checker to enter the foreground and background color numbers and check whether the color contrast aligns with the experimental conclusions; finally, based on visual characteristics and functional requirements, designers can assign red to represent warnings, blue as the common color for vehicle interfaces, and green to indicate turning on. Designers can select color contrasts of vehicle-machine interface referencing the recommended values, aiming for a contrast of 3:1 to 4:1 for red, blue, and green during daytime conditions and 3:1 for red and 5:1 for blue and green during nighttime conditions.

Further reading

I.O. Standardization, F. Road vehicles — ergonomic aspects of transport information and control systems — specifications and test procedures for in-vehicle visual presentation. (2009).

S. Fabius, et al., From road distraction to safe driving: evaluating the effects of boredom and gamification on driving behavior, physiological arousal, and subjective experience, Computers Hum. Behavior. 10 (2017).

D.B. Vivek, Ergonomics in the Automotive Design Process, CRC Press, Boca Raton, 2012.

J. Hunsley, G.J. Meyer, The incremental validity of psychological testing and assessment: conceptual, methodological, and statistical issues, Psychol. Assess. 15 (4) (2003) 446−455.

P. Vermersch, Mental workload—its theory and measurement - Moray, N, Annee Psychologique 81 (2) (1981) 591−592.

CHAPTER 2

Icon design recommendations for central consoles of intelligent vehicles

Fang You[1], Yifan Yang[1], Mengting Fu[1], Jifang Wang[2], Xiaojun Luo[2], Liping Li[2], Preben Hansen[3] and Jianmin Wang[1]

[1]*Car Interaction Design Lab, College of Arts and Media, Tongji University, Shanghai, P.R. China*
[2]*Intelligent Driving Experience Center, Baidu, Shenzhen, GD, P.R. China* [3]*Department of Computer and Systems Sciences, Stockholm University, Stockholm, Sweden*

Chapter Outline

2.1 Introduction 11
2.2 Method 12
    2.2.1 Participants 12
    2.2.2 Experimental environment 12
    2.2.3 Task 13
    2.2.4 Procedure 13
    2.2.5 Questionnaires and data analysis 14
2.3 Results 15
    2.3.1 Influence of icon size on usability and workload 15
    2.3.2 Impact of icon thickness on usability and workload 16
2.4 Conclusion and discussion 16
Further reading 17

2.1 Introduction

With the development and application of technology in mobile computing and the Internet of Things (IoT) in vehicles, the automotive industry in China faces new challenges in designing innovative human-machine interfaces (HMIs) for the mass production of intelligent vehicles. A critical aspect of HMI is related to the interface design of the central console, where drivers experience different nondriving related tasks, potentially leading to distractions while driving and a high workload due to the increased amount of information displayed. As the central control screen is getting bigger and the in-vehicle infotainment expands to provide more entertainment and life services, the design of icons on the central console becomes a focal point of discussion in the current international standards and design guidelines. A common concern of the standards and design guidelines for in-vehicle interface design is the size of icons, prompting further research into the recommended values of icons' minimum size and exploration of line thickness.

With the rapid innovation in automotive HMI in China, many HMI products have already entered production and are available in the Chinese market. We investigated 22 central console interfaces of intelligent vehicles in the Chinese market. We then summarized their information architectures, and the main features of icons, especially in terms of size and line thickness. Subsequently, we invited six professional test drivers to subjectively score the central control screen interfaces, with several vehicle interfaces receiving high recognition in the subjective evaluations. Our findings revealed the minimum icon size of these interfaces to be 6 mm, with an average size of about 18 mm and a most used value of about 13 mm. Based on our investigation and literature review, we selected a reference range of the icon size (ranging from 6 to 18 mm) and line thickness proportional to the icon size (ranging from 5.88% to 17.64% of the icon size) for the experiment.

We provided the participants with a relatively realistic driving environment by using driving simulators. The benefits of using simulators include avoiding unpredictable safety issues concerning participants and saving resources. In this paper, an experiment is presented that will investigate the proposed conditions of displayed size and the line thickness of icons on the central control screen of intelligent vehicles using a driving simulator.

2.2 Method

2.2.1 Participants

A cohort of 17 participants (11 males and six females), aged between 25 and 44, were recruited for this study, with a mean age of 33.35 and a standard deviation of 5.97 years. Each participant had at least two years of driving experience, with the majority frequently or occasionally using central consoles while driving.

2.2.2 Experimental environment

Because the location of the devices can affect the reading and operation of icons, we simulated the position of the driver and the central control screen of a real car based on the ergonomics in the automotive design process. In terms of seating position, the seat height was standardized at 260 mm, with the horizontal distance from the steering wheel to the accelerator heel point set at 440 mm. The height of the steering wheel was standardized at 660 mm. As for the position of the central control screen, we set the height from the center of the screen to the ground as 710 mm, the transverse distance between the center of the screen and the center of the steering wheel as 390 mm, and the horizontal longitudinal distance as 150 mm. At the same time, we tried to ensure that the eyes of participants were 710 mm away from the center of the central control screen. A tablet computer was used to simulate the central control screen and visualize the test content, which ran on a Unity application designed to issue commands, such as initiating a trial and logging data generated by the content. An experimenter is responsible for controlling the tablet computer.

The daytime light, with an intensity of about 5 klx and a relative tolerance of 5%, was selected as the illumination condition in this test of vehicle visual display, in accordance with the specifications outlined in SAE J1757/1:2015 (Fig. 2.1).

2.2.3 Task

We employed the classic research model of in-vehicle secondary tasks. In this setup, the primary driving task of the participants was to maintain a straight path forward at a speed of about 30 km/hour, while their secondary task involved clicking on the icon specified by the main test moderator on the central console under adherence to safety protocols and stability of driving conditions.

The clicking tasks varied according to two independent variables: the size of the icons (6×6 mm, 9×9 mm, 12×12 mm, 15×15 mm, and 18×18 mm) and the line thickness of the icons represented by the proportion of line to icon size (5.88%, 11.78%, and 17.64% of the icon size). Each participant performed all 15 tasks combined in terms of size and line thickness (5 levels of size $\times$ 3 levels of line thickness = 15 tasks).

The icon contents consist of functions, such as navigation, music, camera, and buying movie tickets (Fig. 2.2). Icons were displayed in pure white (RGB: 255, 255, 255) against a background of pure black (RGB: 0, 0, 0) in the center of a 12.3-inch screen with a 16:9 ratio (Fig. 2.2).

2.2.4 Procedure

Participants first completed a consent form and a questionnaire to provide demographic information. Next, the experimenter described the purpose of the test, test content, and the tasks involved. Participants were then instructed on how to interact with the driving simulator and underwent several training trials to familiarize themselves with the secondary task. Once the participants were comfortable with the procedure, the formal experiment began.

In this experiment, participants received instructions for secondary tasks while driving smoothly at a speed of about 30 km/hour. Upon completion of each task, participants were asked to fill out a

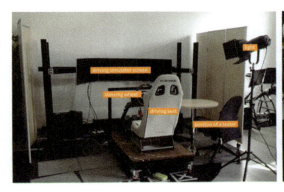

FIGURE 2.1

Test environment.

14 Chapter 2 Icon design recommendations for central consoles

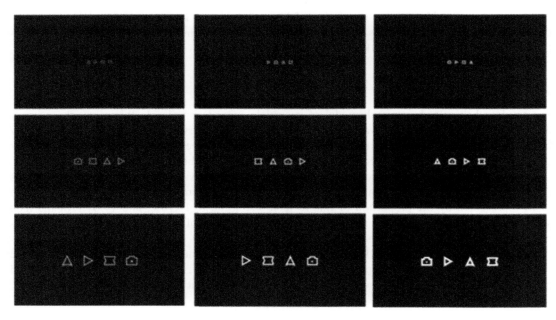

FIGURE 2.2

Test contents.

questionnaire about usability and a single global workload rating. Test contents were carried out in random order, but the tasks were released in a fixed order by the experimenter. The whole procedure lasted approximately 20 minutes.

2.2.5 Questionnaires and data analysis

In the experiment, we used a questionnaire to assess usability and an overview workload rating questionnaire. Generally, usability is evaluated in terms of efficiency, effectiveness, and satisfaction. Given that the experiment started with visual evaluation, we chose the clarity of the icon as indicative of efficiency and the understanding of the icon as representative of effectiveness. We then designed a subjective questionnaire to gauge clarity, understanding, and satisfaction using Likert's 7-point scale ($\alpha = 0.870$). We combined the averages of the subjective scores from three aspects to provide a comprehensive description of the usability of the icons. Regarding workload assessment, we opted for the global workload rating, which was consistent with NASA-TLX's approach. The global rating is a short form to rate how much workload was associated with each task on a scale of 0 to 10, where 0 signifies a very low workload and 10 denotes a very high workload.

Since the data did not conform to a normal distribution, nonparametric tests, including the Kruskal–Wallis test, were used. All data were analyzed and visualized using SPSS and Excel.

2.3 Results

2.3.1 Influence of icon size on usability and workload

In the experiment, when the size of the icons was taken as an independent variable, it was found that the size of the icons had a significant effect on usability ($p < 0.05$) (Table 2.1). As shown in Fig. 2.3, the usability of the icons increases as the icon gets larger. The average usability score of the 9 mm icon size is 14.5% higher than that of the 6 mm icon size. In addition, the average usability score for the 12 mm icon size is 5.9% higher than that of 9 mm icon size. As a result, we can find that there are clear differences in usability scores between the icon sizes of 6, 9, and 12 mm. As the size of icons exceeds 12 mm, the usability growth trend slows down or even decreases. Thus, it was observed that once the size of icons exceeds 12 mm, further increases in size have little impact on usability. However, the transition from 6 to 9 mm significantly enhances usability.

By calculation, we can see that the workload decreases as the size of icons increases ($p < 0.05$), which is contrary to the trend observed in usability values (Table 2.1 and Fig. 2.3). The average workload score of 9 mm icon size is 25.4% lower than that of the 6 mm icon size. In addition, the average workload score for the 12 mm icon size is 29.3% lower than that of the 9 mm icon size. As a result, we can find that there are clear differences in workload scores between the icon sizes of 6, 9, and 12 mm. In other words, when the size changes from 12 to 15 mm, the driver's workload decreases with a smoother trend (Fig. 2.3). Therefore, when the icon size is greater than 12 mm, the size of icons has little impact on the workload.

Table 2.1 Results of the Kruskal–Wallis test for icon size with respect to clarity, understanding, satisfaction, usability, and workload.

Dependent variables	Independent variables	Sig.
Clarity	Icon size	0.004
Understanding	Icon size	0.026
Satisfaction	Icon size	0.003
Usability	Icon size	0.000
Workload	Icon size	0.000

 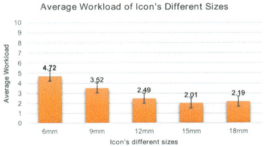

FIGURE 2.3

Average usability and workload scores across different icon sizes.

Table 2.2 Kruskal–Wallis test results of the icons' line thickness with clarity, under-standing, satisfaction, usability and workload.

Dependent variables	Independent variables	Sig.
Clarity	Icon line thickness	0.933
Understanding	Icon line thickness	0.762
Satisfaction	Icon line thickness	0.537
Usability	Icon line thickness	0.813
Workload	Icon line thickness	0.588

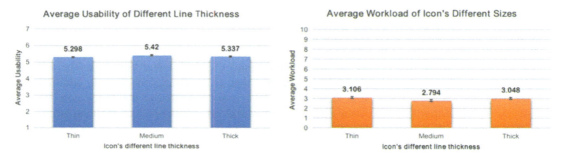

FIGURE 2.4

Average usability and workload of different icon sizes.

2.3.2 Impact of icon thickness on usability and workload

As for the test of the line thickness of icons, we still used usability and workload to describe the effect of different line thickness. Table 2.2 shows that there is no obvious trend in usability ($p > 0.05$) and workload ($p > 0.05$) as the line of icons becomes thicker. When the icon line is of medium thickness (11.78% of the icon size), the usability value is the highest and the workload is the lowest (Fig. 2.4). It can be seen icons with lines of medium thickness perform better. But the different thickness of lines has no significant effect on usability or workload.

2.4 Conclusion and discussion

In this study, we present an experiment to assess the impact of different displayed sizes and line thicknesses of icons on the central control screen of intelligent vehicles using a driving simulator. Our findings suggest that the minimum size of icons should be 9 mm, with the recommended icon size ranging between 12 and 18 mm for central console interface design. Although the differences in usability attributed to different line thicknesses were not statistically significant, we recommend a medium (11.78% of the icon size) line thickness. Further studies need to be conducted to investigate the line thickness in combination with icon size to uncover potential patterns and redefine design guidelines accordingly.

In addition, it is important to note that this experiment was conducted using a 12.3-inch screen with a 16:9 ratio. Thus caution should be exercised when extrapolating the experimental results to other screens of different sizes and proportions. Nevertheless, this experiment proposes a fundamental research methodology from the perspective of visual design and workload assessment for in-vehicle equipment in a dynamic environment. The study lays the groundwork for further exploration and detection of optimal design patterns using larger datasets in future studies.

Further reading

H. Alm, L. Nilsson, The effects of a mobile telephone task on driver behaviour in a car following situation, Accid. Anal. Prev. 27 (5) (1995) 707–715.

N. Beckers, et al. Comparing the demands of destination entry using google glass and the Samsung Galaxy S4, Proc. Human Factors Ergonom. Soc. Ann. Meet. 58(1) (2014) 2156–2160.

J.L. Campbell, J.B. Richman, C. Carney, J.D. Lee, In-Vehicle Display Icons and Other Information Elements. Volume I: Guidelines (2004).

ISO 15008, Road vehicles – ergonomic aspects of transport information and control systems – specifications and compliance procedures for in-vehicle visual presentation. International Organization for Standardization, Geneva, 2017.

SAE J1757/1, Standard metrology for vehicular displays. Vehicular Flat Panel Display Standards Committee, 2015.

S. Siegal, J. Castellan, Nonparametric Statistics for the Behavioural Sciences, McGraw Hill, Boston, 1988.

D. Vivek, Bhise.: Ergonomics in the Automotive Design Process, Taylor and Francis, 2011.

CHAPTER 3

Design guidelines for the size and length of Chinese characters displayed in the intelligent vehicle's central console interface

Fang You[1], Yifan Yang[1], Mengting Fu[1], Jun Zhang[2] and Jianmin Wang[1]

[1]Car Interaction Design Lab, College of Arts and Media, Tongji University, Shanghai, P.R. China [2]College of Design and Innovation, Tongji University, Shanghai, P.R. China

Chapter Outline

- 3.1 Introduction 19
- 3.2 Related works 20
  - 3.2.1 Classification of visual signals and human-computer interaction model 20
  - 3.2.2 Standards and guidelines about size and length of text 21
  - 3.2.3 Driving cognitive model 21
- 3.3 Method 22
  - 3.3.1 Participants 22
  - 3.3.2 Experimental environment 22
  - 3.3.3 Task and stimuli 23
  - 3.3.4 Procedure 24
  - 3.3.5 Questionnaires and data analysis 24
- 3.4 Results 25
  - 3.4.1 Reaction time 25
  - 3.4.2 Visual subjective rating 26
  - 3.4.3 Usability scores and workload 27
- 3.5 Guidelines and discussion 29
- 3.6 Limitations 31
- 3.7 Conclusions 31
- Further reading 31

3.1 Introduction

With the development and application of technology, the automotive industry faces new challenges in designing revolutionary human-machine interfaces (HMIs) for the mass production of intelligent vehicles. With this development, the complexity of in-vehicle interfaces has increased dramatically.

Central console interfaces now provide the driver with access to varied applications, from information to infotainment functions. Many of these applications use texts to display information on the central control screen, from live weather reports, and vehicle fault warnings to integrated navigation systems with address lists and the next direction of a trip.

It should be taken into account that these screens provide ever-richer information, and drivers use them to accomplish more nondriving tasks, so that drivers' attention may turn to the device with increasing frequency and duration, rather than to road conditions. Providing drivers with visual user interfaces whose information content can be easily read is important. If text characters are hard to read, user satisfaction will be low, and the risk of accidents may increase. Therefore the interface design of characters should be optimized to lessen visual demands and reaction time to complete nondriving tasks, thereby reducing workload and freeing up for more attentional resources.

From the visual point of view, the information layout of the interface will have an impact on the scanning time, and the information itself will also have an impact. Reading text with different amounts of information or words can affect the reading time and accuracy. From the perspective of font designers, factors that influence legibility can be divided into extrinsic and intrinsic. Extrinsic factors are physical considerations such as size, contrast, polarity, and color. These factors have received attention within the in-vehicle visual presentation and are covered by various standard documents or design guidelines. A significant factor that influences reading speed is the text size, which has been confirmed in automotive research. However, this research is based on English characters. There are few studies about the effect of the size and length of Chinese fonts on the in-vehicle interface. At the same time, they are assumed to have a great impact on driver behavior, so we focus on these two aspects of Chinese characters.

3.2 Related works

Previous research on the legibility of text focused on embedded reading, using metrics and tasks about reading entire paragraphs of text. However, modern reading behavior is fragmented with brief glances, whether looking at a smartphone or scanning at an in-vehicle screen. In the scene of driving, the displayed text on the in-vehicle screen has a significant impact on safe driving. Therefore existing studies have also included information on visual signal interaction, text design standards, and driving cognitive models.

3.2.1 Classification of visual signals and human-computer interaction model

In visual interaction, humans and computers exchange information through interfaces in the form of visual signals composed of text, icons, and colors. Visual signals are of primary importance in driving and can be divided into various sensory dimensions such as color, luminance, and contrast, as well as stimulus dimensions such as location, size, height, length, shape, spacing, and periodicity. Visual signals must be seen to be effective, and placing them in optimal locations in vehicles can help quickly detect visual signals and promote faster responses to them. In addition, the sensory and stimulus dimensions mentioned above can be combined to maximize the legibility and comprehensibility of messages.

In each step of the interaction, drivers keep offering instructions to the computer and receiving feedback, and the interfaces become the communication channel between humans and computers. The visual information provides its meaning for humans and the computer, respectively. To establish a successful human-computer interaction through visual signals, Bottoni et al. suggest that correct communication occurs if a pair of inverse morphisms can be established between user meaning and computer meaning.

3.2.2 Standards and guidelines about size and length of text

The optimum text presented in the driver-vehicle interface is legible under a large number of viewing distances, viewing angles, and environmental conditions. A lot of literature exists about reference values for text size.

ISO 15008 recommends that the character heights for in-vehicle display alphanumeric text should be at least 20 arc-minutes of visual angle, but 16 arc-minutes are also acceptable. Furthermore, it is one of the few standards that provide reference values for Chinese font sizes, which recommends the minimum or modification size of Chinese characters to be 24 pixels. However, studies on Chinese character sizes and related recommended values are very limited. HUAWEI company suggests that 5.3 mm is the minimum size of Chinese characters that can be recognized.

Referring to the length of characters per row, there are few standards providing reference values. However, the capacity of short-term memory is instructive to the length design, which indicates that it is easy for people to remember 5—7 items.

Before starting this research, we investigated the central console interfaces of 22 intelligent vehicles in the Chinese market and sorted out their information architectures, and the main features of Chinese characters, especially in terms of size and length. The results show that the font size of the first-level heading is 36—40 pt, and that of the second-level heading is 20—24 pt. In terms of length, the first-level heading is 4—6 characters, and the second-level heading is 10—16 characters.

After that, we invited six professional test drivers to subjectively score user satisfaction with central control screen interfaces, and some vehicle's interfaces received a high score of recognition. The minimum font size on these interfaces is 3 mm and the average number per row is about seven characters. Based on this literature and our statistical results, we selected the range of the two independent variables of the size and length of Chinese characters.

3.2.3 Driving cognitive model

There are conceptual cognitive models, which try to explain components and processing stages, including information perception while driving. Mathematical cognitive models explain cognitive principles through signal detection theory for a driver's information processing, information theory, and computational cognitive models, which focus on understanding the driving process and interaction with different cognitive patterns to reproduce driver behaviors.

Among the above cognitive models, we chose Endsley's Situation Awareness to explain the cognitive process in the test tasks. Endsley defines situation awareness as "knowing what is happening around you," which refers to the user's cognition of the surrounding environment and state. The model describes three levels of understanding: perception of elements in the current situation, comprehension of the current situation, and projection of future status. In other words, factors in the environment are perceived within a certain time and space, their meanings and relationships are understood, and their states in the near future are predicted by drivers.

The driver must draw a conclusion quickly from the deserved information and must understand the meaning or importance of the information. Therefore the time from perception to understanding is very important. In the process of driving, 90% of the driver's information comes from the human visual system. When engaging in nondriving tasks, the driver cannot withdraw their eyes off the road over 2 seconds. Young et al. summarized the decline of the cognitive ability and physical ability of the elderly through research on their use of interactive interfaces and also put forward requirements for the design of HMI, such as the guidelines for the amount of text, including the research on the optimal font size or display luminance. We explain the experiment tasks using this situation awareness model in the procedure section.

3.3 Method

We used the driving simulator to conduct the experiment. We required the experiment participants to complete a series of text information cognitive tasks while keeping driving safe. The font size and the length of the sentence were used as the variables of the experiment. We counted the time that took the participants to complete the task, and after a group of tasks were completed, we asked the participants to complete the usability scale and the workload scale.

3.3.1 Participants

We recruited a representative sample of experienced drivers to take part ($n = 30$), comprising 16 males and 14 females, with ages ranging from 25 to 44 years with a mean age of 33.2. All participants had more than 2 years of driving experience. Most of them had the experience of using central consoles frequently or occasionally while driving.

3.3.2 Experimental environment

The location of the in-vehicle environment and the devices can affect the reading and understanding of the text, so we designed the position of the driver and the central control screen of a real car based on the ergonomics of the automotive design process. Based on a seat height of 260 mm, the horizontal distance from the steering wheel to the accelerator heel point was set to 440 mm, and the height of the steering wheel was set to 660 mm. As for the position of the central control screen, we set the height from the screen center to the ground as 710 mm, the transverse distance between the screen center and the center of the steering wheel as 390 mm, and the horizontal longitudinal distance as 150 mm. At the same time, we tried to ensure that the eyes of participants were 710 mm away from the center of the central control screen, considering that it may vary with participants' height. A surface tablet computer was placed on the central control screen to display and visualize the test content, which ran on a Unity application configured to execute commands such as initiating a trial. The experimenter was responsible for controlling the tablet computer. In this test of vehicle visual display, the daytime light (about 5 klx with a relative tolerance of 5%) was selected as the illumination condition, as proposed by SAE J1757/1:2015 (Fig. 3.1).

3.3 Method   23

FIGURE 3.1
Test environment.

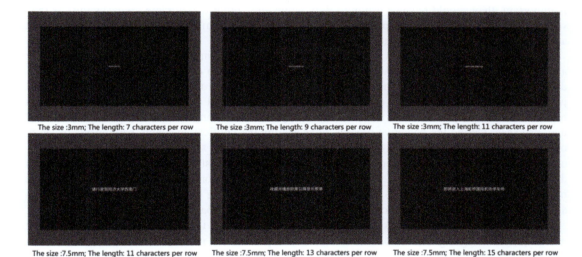

FIGURE 3.2
Test content.

3.3.3 Task and stimuli

In task design, driving straight forward at a speed of about 30 km/hour was the primary driving task and the nondriving task was to name the text on the central console under the premise of safety and stability. Naming tasks have been used in a lot of English and Chinese character experiments, in which participants identify and read the target stimulus. The processing of fonts was measured by reaction time from the appearance of the target stimulus to the response of the participants.

The stimuli of retelling tasks were Mandarin sentences written in simplified Chinese characters. The text contents consist of the functions of navigation, entertainment, vehicle settings, and driver assistance (Fig. 3.2). The stimuli varied according to two independent variables: the size of the character (3, 4.5, 6, and 7.5 mm) and the length or the number of characters per row (7, 9, 11, 13, and 15/row).

Each participant performed all 20 tasks combined in terms of size and length (4 levels of height * 5 levels of the number of characters per row = 20 tasks).

Chinese characters in Noto Sans Chinese Regular typeface were displayed in pure white (RGB: 255, 255, 255) against a background of pure black (RGB: 0, 0, 0) at the center of a 12.3-inch screen with a 16:9 ratio (Fig. 3.2).

3.3.4 Procedure

Participants read and signed an informed consent form and filled out a questionnaire covering demographic information and driving experience. They then came to the simulator and adjusted their seats so that they were comfortable. Next, the experimenter described the test purpose, test content, and test tasks to them, and they received instructions on how to interact with the driving simulator. It must be emphasized to the participants that while performance on the nondriving task is important, they should balance driving safety while attempting to complete the tasks, just as they were driving a real car. A brief practice session familiarized them with the simulator environment and task process before the formal experiment commenced.

In this experiment, participants received instructions for nondriving tasks while driving smoothly at a speed of about 30 km/hour. A notification tone was employed to cue the participants that the nondriving task was ready and that they should scan and try to keep in mind the meaning of the text that just appeared. They could let the experimenter know that they had completed the task by pressing a button on the steering wheel, and then the text disappeared prompting, the participants to repeat what they had seen. During the cognitive process, after hearing the notification sound, participants began to obtain visual information from the central control screen. They then selected a method to scan and recognize the text information based on the driving condition and the surrounding environment, to assess their ability to completely and quickly retell the information based on short-term memory. Participants made decisions regarding when to press the button to start retelling based on their individual abilities and experiences, subsequently executing the corresponding actions based on their decision.

After each task was completed, the participants were asked to fill out a questionnaire about usability and a single global workload rating. After repeating the process 20 times, the experiment concluded. Test contents were presented in a random order throughout. The entire test lasted for approximately 30 minutes.

3.3.5 Questionnaires and data analysis

In the experiment, we used a questionnaire about the visual subjective rating, a usability questionnaire, and a global workload rating questionnaire. The visual subjective rating questionnaire was a 7-point scale (on a scale of -3 to 3) used to describe participants' visual impressions of the text regarding the size and length of the characters. For the size of characters, -3 meant the participants thought the characters were too small, 0 meant the characters were just the appropriate size, and 3 meant the participants thought the characters were too big. Similarly, for the length of characters, -3 and 3 meant that the length of characters was too short and too long, respectively, while 0 was the length that was most appropriate. Generally, the usability questionnaire describes efficiency, effectiveness, and satisfaction. Due to the experiment starting with a visual evaluation, we chose the clarity of Chinese characters as the representative of efficiency and the understanding of Chinese characters as the representative of effectiveness. The chosen clarity and understanding are the operated items for the two dimensions. We then designed a usability questionnaire using Likert's 7-point scale (=0.814), and we combined the three aspects to comprehensively describe the usability of Chinese characters. For assessing workload, we utilized the global workload rating questionnaire, known for its consistency with NASA-TLX. The global rating was a short form to rate how much workload was associated with each type of task using a 0 to 10 scale, where 0 = very low workload, and 10 = very high workload.

Since the data did not meet the normal distribution, Spearman's rank correlation coefficient was used. All data were analyzed and visualized in SPSS and Excel.

3.4 Results

We analyzed objective and subjective data, including response time, visual satisfaction, usability, and workload. The results show that the font size and length have a certain influence on the reaction time. When the font size is less than 6 mm, the reaction time is positively correlated with the font size. When the font size is greater than 6 mm, the reaction time does not change significantly. When the font size is less than 13 characters, the response time is negatively correlated with the number of characters, while when the font size is more than 13 characters, the response time does not change significantly.

3.4.1 Reaction time

Longer reaction times may indicate increased processing demands and higher uncertainty or difficulty. Therefore shorter reaction times might reduce uncertainty and mitigate negative influence on drivers' behavior. A number of reaction time effects are shown in the data. The main effect of different sizes of characters on reaction time ($r = 0.125$, $p < 0.05$) was found, as shown in Fig. 3.3, in which the reactions for the larger Chinese characters are faster. The reaction time for a 4.5 mm character size is 6.97% shorter than that of 3 mm, and the reaction time for a 6 mm character size is 6.63% shorter than that of a 4.5 mm character size, while participants reacting to a 7.5 mm character is just 2.39% faster than reacting to 6 mm characters. We can find that there are clear differences in reaction time between the sizes of 3 mm, 4.5 mm, and 6 mm. As the size exceeds 6 mm, reaction time begins to decrease. As a result, when the size of Chinese characters is larger than 6 mm, the size has little impact on the reaction time.

As for the length of Chinese characters per row, the reaction time increases with the increasing number of Chinese characters per row ($r = 0.337$, $p < 0.05$). We can see in Fig. 3.4 that when the number of characters per row changes from 7 to 13, the reaction time tends to increase. Reaction times even decrease from 13 to 15 characters per row. Therefore the drivers use more attention resources as the length increases. When the number of Chinese characters is more than 13, the effect of the length on the reaction time and the driver's performance is unclear.

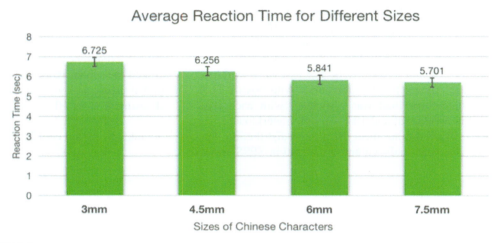

FIGURE 3.3

Average reaction time for different sizes of Chinese characters.

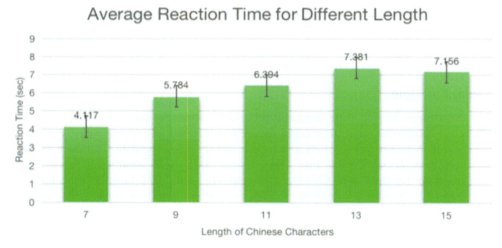

FIGURE 3.4

Average reaction time for different length of Chinese characters.

3.4.2 Visual subjective rating

After each task was completed, participants were immediately asked to score their visual perception of the Chinese characters in the task. When the size of Chinese characters is taken as an independent variable, it is found that the size of Chinese characters has an effect on the visual subjective rating and there is a strong correlation between the two variables ($r = 0.725$, $p < 0.01$). As shown in Fig. 3.5, the

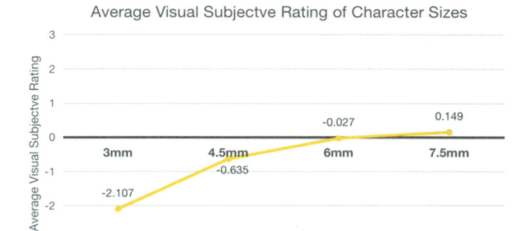

FIGURE 3.5

Average visual subjective rating of the sizes of Chinese characters.

participants thought the Chinese character sizes of 6 mm and 7.5 mm were both appropriate, but the size of 6 mm was better. At the same time, they believed that the 4.5 mm Chinese characters were a little small, while the 3 mm Chinese characters were particularly small, affecting the legibility of the characters.

The length of Chinese characters significantly impacts the visual subjective rating ($r = 0.603$, $p < 0.01$). As illustrated in Fig. 3.6, the participants rated 7 or 9 words per row as the closest score to 0, which means 7 or 9 words are the most visually appropriate length for them. As the number of characters per row increases, the visual subjective rating of the length increases. When the number of words per row was more than or equal to 13, the participants considered the text to be particularly long.

3.4.3 Usability scores and workload

In this section, usability and workload questionnaires are used to measure participants' perceptions of the sizes and lengths of different Chinese characters.

In this experiment, character size has a significant effect on usability ($r = 0.346$, $p < 0.01$) and workload ($r = 0.399$, $p < 0.01$). As the character size increases from 3 to 6 mm, the usability increases by 19.8% and 8.2%, respectively. When the characters change from 6 to 7.5 mm, the usability decreases by 1.6%. Regarding the workload, when character size increases from 3 to 6 mm, the workload decreases by 31.2% and 27.5%, respectively, while the workload scores of 7.5 mm characters are 5.3% higher than that of 6 mm characters (see Fig. 3.7). We can find that the character size of 6 mm is a turning point. When the character size is larger than or equal to 6 mm, the changing trend of usability and workload is smoother than the scoring trend of character

28   Chapter 3 Design guidelines for Chinese characters

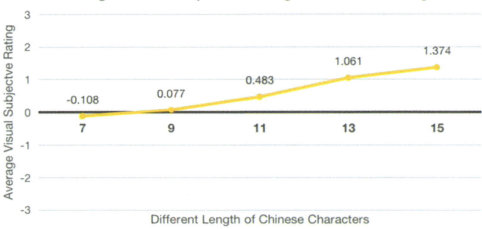

FIGURE 3.6

Average visual subjective rating of the length of Chinese characters.

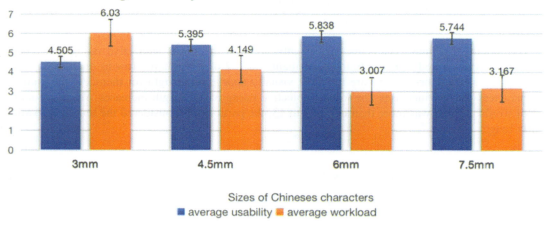

FIGURE 3.7

Average usability and workload for the sizes of Chinese characters.

sizes smaller than 6 mm. In summary, participants' perception of character size is getting better as the size increases, but this effect becomes smaller when the size exceeds 6 mm.

As shown in Fig. 3.8, as the number of words per row increases, the usability decreases, and the workload increases. When the number of words per row increases from 7 to 15, the usability scores decrease

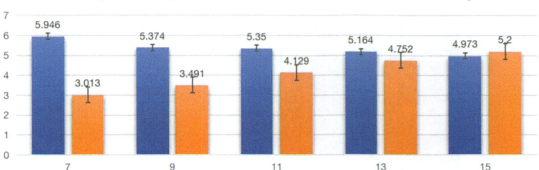

FIGURE 3.8

Average usability and workload for the length of Chinese characters.

by 9.6%, 0.4%, 3.5%, and 3.7%, in turn. In other words, 9 words/row is a turning point in usability scoring. When the number of words per row exceeds 9, the usability decreases in a smoother trend. At the same time, with the increase in length, the workload has been gradually increasing in a consistent trend. In general, when the number of words per row is 7 or 9, participants' perception of the text is better. Conversely, participants' perception of text gradually deteriorates as the length continues to increase.

3.5 Guidelines and discussion

The study explored the usability and drivers' workload influenced by different displayed sizes and lengths of Chinese characters in the intelligent vehicle's central control screen in a driving simulator. Results show that the recommended minimum character size is 4.5 mm and the recommended appropriate text size is 6 mm, with the premise of ensuring legibility. Participants also accept when the size of Chinese characters is larger than 6 mm. For character length, 7 or 9 words or less than 7 or 9/row are most appropriate, which also conforms to the law of short-term memory. At the same time, it is recommended that the maximum number of Chinese characters per row should not exceed 13 words.

Extending the findings from the experiment to specific design guidelines, designers can build the text hierarchy of the interface by choosing the different character sizes suggested. In the navigation scenario shown in Fig. 3.9, the recommended minimum character size of 4.5 mm can be used for the explanatory text, 6 mm characters can be used for the main list display, and 7.5 mm characters can be used for the title bar display. Designers can arrange the text size to highlight the different levels of text information according to design requirements. For the text length, the text in the title bar is better within 9 characters, and each row of text needs to be set within 13 characters in

30 Chapter 3 Design guidelines for Chinese characters

FIGURE 3.9

Examples of Chinese character interface design.

the pop-up window about warnings or notifications. This allows drivers to quickly understand information and reduce distractions for driving tasks.

To facilitate designers to directly use the results in the design phase, we provide reference values for the Chinese character size in points. The minimum recommended font value is 12.76 points, and the most appropriate recommended value is 17 points. Designers can refer to the design guidelines to improve and adjust interface design based on different scenarios.

In addition, because this experiment is carried out under a 12.3-inch screen with a 16:9 ratio, more considerations are required when using experimental results on other screens with different sizes and proportions. It can be further utilized to detect optimal design patterns using a larger data set.

3.6 Limitations

In this study, we studied the impact of the size and length of the characters on human cognition. However, there are several limitations to consider:

(1) There are essential differences between Chinese and English. While a single Chinese character has its own meaning, in English, a word composed of multiple characters can express meaning. This experiment did not consider the impact of this distinction. Perhaps following the English language conventions, such as grouping characters that make up a word, might also reduce comprehension time.

(2) Additionally, factors like color, layout, etc., can affect the degree of difficulty in people's perception. However, our experiment involved testing under a single color condition.

(3) Finally, our experiment was conducted using a simulator. Since the simulator lacks dynamic vibration, the experiment did not take into consideration whether the vibration would affect the recognition of text in the real vehicle.

3.7 Conclusions

This experiment proposes a basic research method from the perspective of visual design and workload for in-vehicle equipment in a dynamic environment. This is not limited to Chinese characters and can be expanded into other fonts. Starting from the size and length of the Chinese characters, we can further study the dimensions of icons, colors, and layouts in the central console interface design in the future. The proposed design guidelines for the intelligent vehicle's central console can provide a reference for the current innovative design wave of the global automotive HMI. Of course, we cannot rely too much on guidelines, as Tetzlaff et al., argued that we should reduce the dependence on the guide, treat the guide as a supplementary tool, and focus on the information that these tools cannot obtain from the interaction design. Our guide is just an exploration. It can also improve the driver's information processing for driving tasks and in-vehicle nondriving tasks, and expand on the current knowledge of user-centered design.

Further reading

J. Aasman, Modelling driver behaviour in soar. Available at: <http://repository.tudelft.nl/view/tno/uuid: f6ec9503-a255-4b58-b40a-3b79f97586dc/>.

N. Beckers, S. Schreiner, P. Bertrand, B. Mehler, B. Reimer, Comparing the demands of destination entry using Google Glass and the Samsung Galaxy S4 during simulated driving, Appl. Ergon. 58 (2017) 25–34.

V.D. Bhise, Ergonomics in the Automotive Design Process, Taylor and Francis; CRC Press, Boca Raton, FL, 2019.

C. Bigelow, S. Matteson, Font improvements in cockpit displays and their relevance to automobile safety (2011).

P. Bottoni, S. Levialdi, G. Păun, Successful visual human-computer interaction is undecidable, Inf. Process. Lett. 67 (1998) 13–19.

K.A. Brookhuis, D.D. Waard, W.H. Janssen, Behavioural impacts of advanced driver assistance systems: an overview, Eur. J. Transp. Infrastruct. Res. 1 (2001) 3.

H. Cai, P. Green, Range of character heights for vehicle displays as predicted by 22 equations, Proc. SID Vehicle. Disp. Symp. (2005).

J.L. Campbell, J.L. Brown, J.S. Graving, C.M. Richard, M.G. Lichty, T. Sanquist, J.L. Morgan, Human factors design guidance for driver-vehicle interfaces (No. Report No. DOT HS 812 360); National Highway Traffic Safety Administration: Washington, DC, USA, 2016.

J.L. Campbell, Z.R. Doerzaph, C.M. Richard, L.P. Bacon, Human factors design principles for the driver-vehicle interface (DVI), in: Proceedings of the Adjunct International Conference on Automotive User Interfaces and Interactive Vehicular Applications, Seattle,WA, USA, 17−19 September 2014, pp. 1−6.

C. Castro, Visual demands and driving, Hum. Factors Vis. Cogn. Perform. Driv. 45 (2008) 2−26.

Y.-N. Chang, C.-H. Hsu, J.-L. Tsai, C.-L. Chen, C.-Y. Lee, A psycholinguistic database for traditional Chinese character naming, Behav. Res. Meth. 48 (2015) 112−122.

R.K. Deppe, The human factor in road safety. Available at: <https://trid.trb.org/view/215844> (accessed 13.12.20).

J. Dobres, B. Reimer, B. Mehler, N. Chahine, D. Gould, A pilot study measuring the relative legibility of five simplified Chinese typefaces using psychophysical methods, in: Proceedings of the 6th International Conference on Communications and Broadband Networking-ICCBN 2018, Volume 14, Singapore, 24−26 February 2018, pp. 1−5.

M.R. Endsley, Toward a theory of situation awareness in dynamic systems, Hum. Factors J. Hum. Factors Ergon. Soc. 37 (1995) 32−64.

K. Fujikake, S. Hasegawa, M. Omori, H. Takada, M. Miyao, Readability of character size for car navigation systems, Trans. Petri Nets Other Models Concurr. XV 4558 (2007) 503−509.

Huawei, HUAWEI Hicar Ecological White Paper, Huawei, Shenzhen, China, 2019.

M.M. Kokar, M.R. Endsley, Situation awareness and cognitive modeling, IEEE Intell. Syst. 27 (2012) 91−96.

G.E. Legge, C.A. Bigelow, Does print size matter for reading? A review of findings from vision science and typography, J. Vis. 11 (2011) 8.

R. Li, Y.V. Chen, C. Sha, Z. Lu, Effects of interface layout on the usability of in-vehicle information systems and driving safety, Displays 49 (2017) 124−132.

J.H. Lim, O. Tsimhoni, Y. Liu, Investigation of driver performance with night vision and pedestrian detection systems—part I: empirical study on visual clutter and glance behavior, IEEE Trans. Intell. Transp. Syst. 11 (2010) 670−677.

Y. Liu, R. Feyen, O. Tsimhoni, Queueing network-model human processor (QN-MHP), ACMTrans. Comput. Interact. 13 (2006) 37−70.

M. Lu, K. Wevers, R. Van Der Heijden, Technical feasibility of advanced driver assistance systems (ADAS) for road traffic safety, Transp. Plan. Technol. 28 (2005) 167−187.

I.S. MacKenzie, Fitts' law as a research and design tool in human-computer interaction, Hum.−Comput. Interact. 7 (1992) 91−139.

G. Marchionini, E.A. Fox, Progress toward digital libraries: augmentation through integration, Inf. Process. Manag. 35 (1999) 219−225.

G.A. Miller, The magical number seven, plus or minus two: Some limits on our capacity for processing information, Psychol. Rev. 101 (1994) 343−352.

R.L. Moore, Paper 5: some human factors affecting the design of vehicles and roads, Inst. Highw. Eng. J. 16 (1969) 13−22.

S. O'Day, L. Tijerina, Legibility: back to the basics, SAE Int. J. Passeng. Cars-Mech. Syst 4 (2011) 591−604.

R. Ratcliff, G. McKoon, The diffusion decisionmodel: theory and data for two-choice decision tasks, Neural Comput. 20 (2008) 873−922.

C.E. Shannon, W. Weaver, The Mathematical Theory of Information, University of Illinois Press, Urbana, IL, USA, 1949.

B.G. Simons-Morton, F. Guo, S.G. Klauer, J.P. Ehsani, A.K. Pradhan, Keep your eyes on the road: young driver crash risk increases according to duration of distraction, J. Adolesc. Health 54 (2014) S61–S67.

W.P. Tanner Jr, J.A. Swets, A decision-making theory of visual detection, Psychol. Rev. 61 (1954) 401–409.

L. Tetzlaff, D.R. Schwartz, The use of guidelines in interface design, in: Proceedings of the ACM SIGCHI Conference on Human factors in computing systems, New Orleans, LA, USA, 27 April–2 May 1991.

C. Wu, Y. Liu, Queuing network modeling of driver workload and performance, IEEE Trans. Intell. Transp. Syst. 8 (2007) 528–537.

K. Young, S. Koppel, J. Charlton, Toward best practice in human machine interface design for older drivers: a review of current design guidelines, Accid. Anal. Prev. 106 (2017) 460–467.

CHAPTER 4

The research on basic visual design of the head-up display of an automobile based on driving cognition

Fang You, Jinghui Zhang, Jianmin Wang, Mengting Fu and Zhenghe Lin
Car Interaction Design Lab, College of Arts and Media, Tongji University, Shanghai, P.R. China

Chapter Outline

4.1 Introduction 35
4.2 Method 36
    4.2.1 Experimental environment 36
    4.2.2 Subjects 38
    4.2.3 Test contents and steps 38
4.3 Results 38
    4.3.1 Influence of weather conditions on reaction time 38
    4.3.2 Influence of age on response time 40
    4.3.3 Trends of workload 41
4.4 Discussion 42
Further reading 43

4.1 Introduction

In the process of human-vehicle-environment interaction, drivers assimilate a certain amount of onboard information through the visual interaction interface in the vehicle. This cognitive process is the basis of all driving behavior and driver performance. The driver observes the change in the interface visually and determines the type and meaning of the information through cognitive processing. Following this, the driver makes decisions and executes corresponding interactions based on the content of the information. Ultimately, the interaction yields feedback in the vehicle interaction system. Throughout the process, the driver unavoidably diverts attention from driving to engage in visual interpretation and cognitive thinking. If the visual interface is not designed properly, the driver will need to spend more cognitive resources to identify the information. In the process of identification and thinking, the driver's line of sight and attention shifts away from the primary driving tasks, potentially leading to unsafe driving distraction behaviors.

It is reasonable to measure mental workload with a subjective evaluation scale, given its inherent psychological structure. Hart and Staveland's description of the Task Load Index (TLX) mental workload measurement method, validated through empirical evidence, represents perhaps the closest approach to the essence of mental workload mining and could provide the most common, effective, and sensitive indicators for assessing mental workload.

The NASA-TLX provides a multidimensional scoring process that derives the overall score based on six psychological workload factors: mental load, physical burden, time stress, task performance, effort, and frustration. The TLX consists of two measurement evaluation processes: weight and rating. This experiment also used the NASA-TLX multidimensional rating scale for workload testing of different sizes.

Head-up display (HUD) can reduce the frequency and duration of the driver diverting their eyes off the road by projecting the required information in front of the driver. This enables the driver to steer easily and respond quickly to information provided by the road conditions and communication systems. It is of great significance to study the visual interaction interface of HUD based on the driving cognition and human-computer interaction. With the popularization and promotion of HUD, it is very important to explore and establish a set of effective HUD visual interaction principles and norms.

This paper investigates and categorizes the information architecture and information layout of existing automotive HUDs, extracting key design elements. According to the illumination standard provided by ISO 15008, the light conditions of daytime direct light (45,000 lx) were simulated, and the optical calibration was performed by the brightness meter. The experiment was carried out in the daytime and snowy scenarios, wherein the three main design elements of text, numbers, and icons were tested for different sizes of recognition responses. Besides, the participants were asked to evaluate the workload in six dimensions. The integrated subjective and objective test parameters provided a theoretical basis for the HUD interface design.

4.2 Method

4.2.1 Experimental environment

In this experiment, a HUD optical machine is used to put the test content on. The communication interface of the optical machine is connected to a tablet computer. A tester is responsible for controlling the tablet computer. ISO 15008 suggests that testing vehicle visual displays should encompass simulations of various lighting conditions, including night (maximum illumination less than 10 lx), evening light (about 250 lx), and daytime direct light (about 45,000 lx). The purpose of the experiment is to test the minimum size and comfort domain size range that the user thinks can be accepted in the HUD visual design, and provide a theoretical basis for the HUD interface design. Therefore this experiment selects the daytime direct sunlight with the smallest contrast between the simulated environment and the optical machine. The light was tested and optically calibrated using a luminance meter. In terms of weather, due to the greater reflection of the road surface during snowy days, the effect on the image of the light machine is more obvious. Therefore this experiment selects two weather conditions, sunny and snowy, to explore whether weather conditions influence the identification of the subject. The experiment uses a projector to project a picture containing weather and road conditions in front of the driving field of view, statically simulating environmental conditions (Figs. 4.1 and 4.2).

4.2 Method

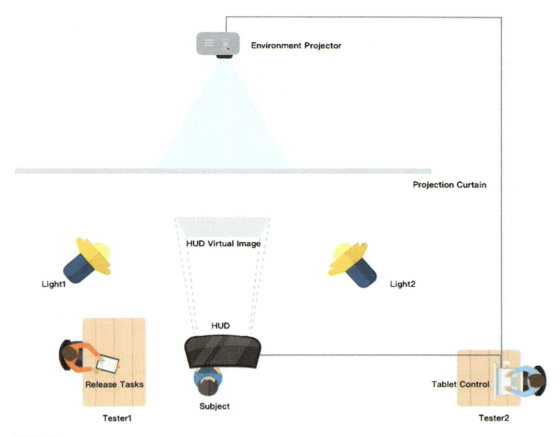

FIGURE 4.1

Test environment.

FIGURE 4.2

Test site.

4.2.2 Subjects

Thirty-six participants, aged between 21 and 46 years, participated in the experiment (mean age = 32.8, SD = 6.8 years). All subjects had at least 1 year of driving experience, were noncolor blind, and possessed a visual acuity (including corrected visual acuity) of 4.8 or above. Additionally each participant had a certain understanding of HUD before the experiment, although none of them had prior experience driving a car equipped with Windshield-HUD.

4.2.3 Test contents and steps

The size range exceeds the size of the text, numbers, and icons identified through market surveys. Specifically, text and number values are 24, 30, 36, and 48 px, and the icon size values are 50 × 24 px, 50 × 40 px, 68 × 70 px, and 71 × 88 px. Moreover, to avoid the learning effect, the text, numbers, and icons in the sunny and snowy scenes are deliberately varied. At the same time, all elements are displayed in white.

After the participant enters the test area, the tester asks the participant to review and sign the informed consent form, provide the basic information, and undergo a vision test. Following that, participants are directed to read a test documentation detailing the purpose, content, and tasks of the test. The experiment was carried out according to the test sequence of the sunny day scenario followed by the snowy day scenario. The three types of elements (text, numbers, and icons) are divided into groups of small, medium, large, and superlarge groups. The experiments were carried out in the order of Latin squares (Fig. 4.3).

The test used Unity to develop a display program, which underwent verification by the tester before the experiment. Subsequently, with the subject pressing the right button on the keyboard, the task commenced. Following a 2-second interval of a black screen, the HUD displayed the test content. Upon recognizing the content (number, text, or icon), subjects pressed the space key and then dictated what they observed. The reaction time (RT) of the subjects was then displayed on the tester's tablet computer interface. After recording the time, the tester informed the subjects to proceed with the next identification RT experiment. After the test of each group of three elements was completed, the participants were asked to complete the NASA-TLX questionnaire to score the workload value for that particular size group.

4.3 Results

4.3.1 Influence of weather conditions on reaction time

After calculating the average recognition RT of each element in both the sunny and snowy scenarios, the data from 36 people revealed that only the text, number, and icon in the small size group had 85.80%, 88.57%, and 91.43% completeness in the sunny scenario, and only the text, number and icon in the small size group had 97.14%, 91.41%, and 68.57% completeness in the snowy scenario respectively, and the other size groups can be fully recognized.

As shown in Fig. 4.4, the change of the overall recognition response time in sunny and snowy scenarios conforms to the general cognitive law, and the response time of icons is generally higher

4.3 Results 39

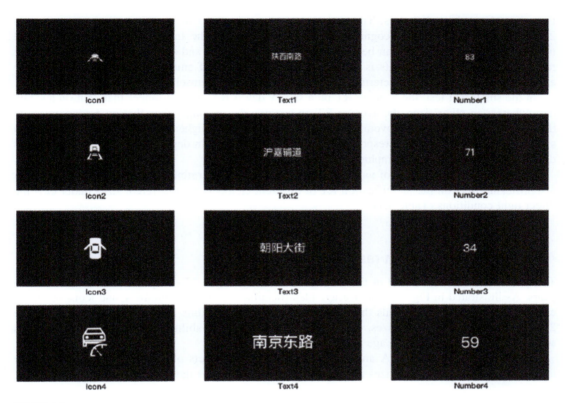

FIGURE 4.3

Test contents.

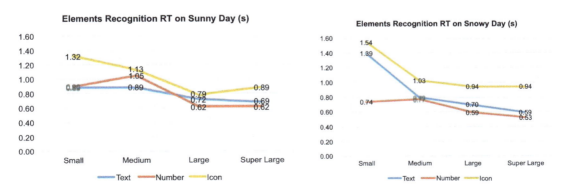

FIGURE 4.4

Average reaction time for elements on sunny and snowy days.

than that of words and figures. The average response time of icons with small sizes is close to 2 seconds, indicating that the cognitive resources occupied by the small icon are high and should not be used in the design. It can be observed that in both sunny and snowy scenarios, the response rate for medium-sized elements is generally higher than that of small-sized ones, which may be attributed to the inherent characteristics of the number shapes themselves.

In the small text test, the average RT on a snowy day is 0.4 seconds longer than that on a sunny day, whereas in other sizes, the average RT on a sunny day is longer than that on a snowy day. In the RT test of numbers, the overall response time is slightly higher on sunny days compared to snowy days. As for icons, the response time increases to a certain degree for 71 × 88 px size icons due to the complexity of the graphics.

Through one-way analysis of variance (ANOVA), it was determined that weather does not exert a significant effect on the identification RT of different elements with different sizes under daytime direct light conditions (Table 4.1).

4.3.2 Influence of age on response time

Thirty-six subjects were tested in this experiment, with 16 aged 21–30 years and 20 aged 31–46 years. As illustrated in Fig. 4.5, the average identification RT of texts, numbers, and icons of different sizes on a sunny day reveals that the 31–36 age group exhibits higher RT compared with the 21–30 age group across all sizes, indicating that the cognitive ability of the 21–30 age group is superior to those of the 31–46 age group across all sizes.

However, one-way ANOVA analysis indicates that the effects of different age groups on the identification RT of different sizes are not significant under sunny day conditions (Table 4.2).

Table 4.1 One-way ANOVA analysis of weather conditions.			
Size	Text	Number	Icon
Small	0.15	0.20	0.44
Medium	0.41	0.42	0.62
Large	0.78	0.53	0.29
Superlarge	0.21	0.08	0.76

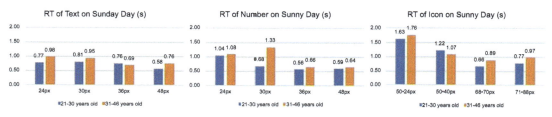

FIGURE 4.5

Recognition response time of elements for two age groups on a sunny day.

4.3 Results

Table 4.2 One-way ANOVA analysis of age groups on a sunny day.

Size	Text	Number	Icon
Small	0.37	0.40	0.22
Medium	0.99	0.35	0.76
Large	0.54	0.13	0.32
Superlarge	0.18	0.68	0.48

FIGURE 4.6

Recognition response time of elements for two age groups on a snowy day

Table 4.3 One-way ANOVA analysis of age groups on a snowy day.

Size	Text	Number	Icon
Small	0.37	0.40	0.22
Medium	0.99	0.35	0.76
Large	0.54	0.13	0.32
Superlarge	0.18	0.68	0.48

In the snowy day scenario, experiments involving different contents but identical sizes were carried out. As shown in Fig. 4.6, the average recognition RT of texts, numbers, and icons of different sizes showed that the 31–46 age group exhibited lower RT compared with the 21–30 age group across all sizes, except for the smallest size elements (words, numbers, and icons) where the trend was reversed. This indicates that for the smallest size elements, the identification of the 31–46 age group is more difficult compared with the 21–30 age group, although the difference between the two groups is minimal, with the older age group showing slightly better recognition. Similarly, the one-way ANOVA test revealed that the effect of different age groups on the identification response was not significant under snowy day conditions, as shown in Table 4.3.

4.3.3 Trends of workload

Based on the selection of scores and weights, the average workload value is calculated for each size category. It can be seen that the workload value gradually decreases as the size increases. On snowy days, the shift from small to medium size is large, indicating that within the size range of

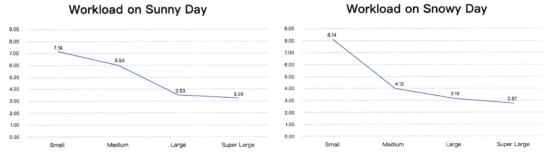

FIGURE 4.7

Workload trend of each size group.

this experiment, the primary driving force for workload reduction occurs in the transition from small to medium size. Conversely, in daytime sunny scenarios, the reduction in workload is more pronounced for medium to large sizes. In both weather conditions, the disparity in workload values between large and superlarge sizes is not significant (Fig. 4.7).

4.4 Discussion

When considering the dimension design of elements on HUD display terminals from the perspective of RT and workload, it becomes challenging to define the minimum age range among the four levels identified from market survey results. This difficulty may be attributed to the focus on indoor simulation static experiments rather than dynamic testing in actual driving scenarios. As for the comfort range of size, it is suggested that in HUD interface design, text sizes should be between 36 and 48 px, and digital sizes should range between 36 and 48 px. However, due to semantic considerations, the icon size should not be less than 68 × 70 px. At the same time, for icons with warning meanings, designers should exercise careful deliberation concerning size, color, layout, and dynamic effects.

By observing the influence of weather conditions on size identification, it is determined that weather does not exert a significant influence. However, when observing the influence of the two age groups on size identification, it is found that for smaller sizes, the older age group has a higher recognition response. Specifically, for the smallest size, the RT of the older group tended to be slower than that of the younger group. Conversely, at other sizes, the differences between the different age groups were not as pronounced.

In addition, given that this experiment was carried out under standard illumination conditions and with constraints on the brightness of the optical machine itself, the experimental results have certain limitations. However, it is worth noting that this experiment introduces a fundamental research methodology from the perspective of visual design and workload for new in-vehicle equipment. This aspect holds significant importance in the context of the study.

Further reading

H. Alm, L. Nilsson, The effects of a mobile telephone task on driver behaviour in a car following situation, Accid. Anal. Prev. 27 (5) (1995) 707–715.

L.M. Bergasa, J. Nuevo, M.Á. Sotelo, et al., Visual monitoring of driver inattention, in: Computational Intelligence in Automotive Applications, 2008.

S.G. Hart, NASA-task load index (NASA-TLX); 20 years later, Proc. Hum. Factors Ergon. Soc. Ann. Meet. 50 (4661) (2006) 904–908.

ISO 15008, Road vehicles – ergonomic aspects of transport information and control systems – specifications and compliance procedures for in-vehicle visual presentation (2003).

R. Schmidt, Unintended acceleration: human performance considerations, in B. Peacock, W. Karwowski (eds.), Proceedings of Automotive Ergonomics, Taylor and Francis, London, 1993, pp. 633–662.

CHAPTER 5

A novel cooperation-guided warning system utilizing augmented reality head-up display to enhance driver's perception of invisible dangers

Fang You[1], Jun Zhang[2,3], Jie Zhang[3], Lian Shen[1], Weixuan Fang[1], Wei Cui[2] and Jianmin Wang[1]

[1]Car Interaction Design Lab, College of Arts and Media, Tongji University, Shanghai, P.R. China [2]College of Design and Innovation, Tongji University, Shanghai, P.R. China [3]School of Design, The Hong Kong Polytechnic University, Hong Kong, P.R. China

Chapter Outline

- 5.1 Introduction 46
- 5.2 Context and theoretical framework 47
  - 5.2.1 Human-vehicle teaming and cooperative interface 47
  - 5.2.2 Situational awareness in cooperative driving 48
  - 5.2.3 Cognitive information design 50
  - 5.2.4 Automation surprises and trust 51
- 5.3 Designing HMI 52
  - 5.3.1 HMI for Case 1 52
  - 5.3.2 HMI for Case 2 54
- 5.4 Method and materials 55
  - 5.4.1 Case Study 1 55
  - 5.4.2 Case study 2 59
- 5.5 Results and analysis 61
  - 5.5.1 Results of Case 1 61
  - 5.5.2 Results of Case 2 64
- 5.6 Discussion 66
  - 5.6.1 Case 1 66
  - 5.6.2 Case 2 67
  - 5.6.3 Limitations 69
- 5.7 Conclusion 70
- References 70

Human-Machine Interface for Intelligent Vehicles. DOI: https://doi.org/10.1016/B978-0-443-23606-8.00017-8
© 2024 Tongji University Press Co., Ltd. Published by Elsevier Inc. All rights are reserved, including those for text and data mining, AI training, and similar technologies.

5.1 Introduction

Human beings are not fully capable of handling all the dangers presented by an uncertain environment.r Consequently, the functions of intelligent systems are continually evolving to supplement human insufficient abilities. Currently, an intelligent vehicle and a human driver can be seen as a joint driver-car system [1], in which the two elements need to cooperate to provide a safe and comfortable driving experience [2]. Imagine a scenario where the vehicle in front of you suddenly brakes due to an accident, which is a typical rear-end collision on expressways. Another scenario is when a car merges into your lane without using turn signals, resulting in a crash [3]. However, augmented reality head-up display (AR-HUD) can provide information annotation in real environments, which seems to be a solution to such problems. Therefore, this study investigates how AR technology assists human drivers in situations where they cannot obtain important information by observing the environment.

AR display technology has been integrated into this collaboration to enhance drivers' responses to various road and traffic scenarios [4,5]. AR serves as an advantageous interactive interface, enabling drivers to maintain sufficient situational awareness (SA) without diverting their attention from the road, which is pivotal for safe driving. The task of human drivers is to control the vehicles while remaining vigilant of changes in the environment. At the same time, intelligent systems sense the changes through devices, such as radars and cameras and provide cognitive assistance to drivers when detecting hazardous situations. AR can improve drivers' perception by conveying cognitive information from intelligent systems, encompassing their understanding and prediction of the current environment. The task of drivers during autonomous driving at level 2 (L2) is to observe changes in the environment and supervise the automated operation [6]. At this point, the task for the interface is to assist the human driver in observing the environment and supervising the intelligent system, which is usually achieved through a transparent interface.

Few studies have explored how an AR interface should be designed for situations where humans cannot perceive dangerous elements that affect driving safety. This gap may stem from AR's reliance on visual interfaces, which cannot redirect attention to unseen areas, resulting in minimal designs for blind spots. However, similar scenarios, such as, when the line of sight is blocked by the vehicle ahead or when other traffic participants do not use turn signals, can also lead to serious traffic accidents.

Therefore, the objective of this study is to design cooperative interfaces for dangerous scenarios in which environmental information may be missing. In this paper, a framework for human-machine cognitive cooperation is proposed, along with a design method for AR-HUD interfaces that enable intelligent systems to provide appropriate cooperation. To achieve these goals, the following research questions (RQ) need to be addressed:

RQ1: What cues are lacking in the real environment that requires the intelligent system and the driver to cooperate? How do they cooperate?

RQ2: What kind of information should the intelligent system provide when cooperation is necessary?

RQ3: How should the information be displayed and how can AR-HUD's strength be integrated into the design?

5.2 Context and theoretical framework

This section reviews relevant theories for this research and proposes a conceptual human-vehicle teaming framework and an information design method to guide the design of the HMI (Human-Machine Interface) in response to the RQs stated above.

5.2.1 Human-vehicle teaming and cooperative interface

With the advancement of vehicle intelligence, vehicles are transitioning from mere tools to becoming collaborative partners for humans. To be specific, humans and vehicles can cooperate at three levels of driving tasks [7]: operational, tactical, and strategic levels. For instance, at the operational level both horizontal and longitudinal control of vehicles is managed through human-vehicle input [8], current tasks, such as overtaking, leaving expressways, making turns, etc., are planned at the tactical level, and the selection of destination occurs at the strategic level. No matter at which level humans and vehicles cooperate, completing a particular driving task involves a cognitive cycle of perception and action, allowing for humans and vehicles to cooperate in this cycle as well.

Human-vehicle cooperation is therefore a continued process of communication, through which a common knowledge base and goals are created [9]. Such cooperation requires agents to share a situation model so that they both can understand the current or future environment or status of the process [8]. Changes in the human-vehicle relationship also change the human-machine interface, thus emphasizing the importance of cooperative interfaces as the evolving trend in intelligent vehicle interfaces.

Cooperation at L2 requires the human driver to pay attention to the vehicle's driving in addition to the environment [6,10]. Therefore, the driver needs to monitor the two variables simultaneously. Nevertheless, the traditional interface (center control) requires the driver to look away from the environment, which can be potentially dangerous. In this regard, the advantage of AR-HUD is that the driver does not need to take his or her eyes off the road. Besides, as an interactive mixed-reality interface, it allows the real and virtual environments, which show the same space, to be synchronized [4]. More importantly, all the decisions of the intelligent system are made depending on environmental changes, so AR enables the driver to directly observe the causal relationship between the vehicle's decisions and actions and the environment.

Reflecting on the question mentioned above: What cues are lacking in the real environment that require the intelligent system and the driver to cooperate? In potentially dangerous scenarios, some information cannot be obtained due to the limitation of the human driver's perception, but it can be sensed by the intelligent system and transmitted to the human driver through HMI, thereby realizing cognitive information support of the human and the machine (Fig. 5.1). This is the cooperation at the levels of perception and prediction. While there are few studies on the cooperation at the perception level in autonomous driving, because the reaction time is short and the driver has no time to guide the timely control of the vehicle [11], the information on the environmental perception level in autonomous driving helps with the driver's understanding of vehicle actions and avoids "automation surprises." This is also a kind of cooperation, which is discussed below. In addition, manual driving involves the largest amount of cooperation at the perception level, which can be seen from various ADAS functions. Thus, the information at the levels of perception and prediction promotes cooperation in both manual and autonomous driving.

48 Chapter 5 A novel cooperation-guided warning system

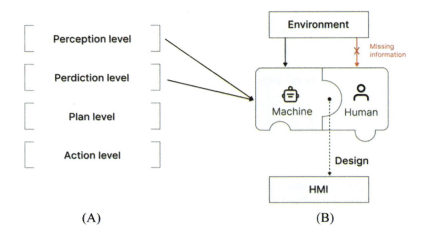

FIGURE 5.1

Developed cognitive information support for humans and machines through HMI. (A) Four levels of cooperation. (B) Cooperative interface for perceptual and prediction level compensation.

5.2.2 Situational awareness in cooperative driving

SA is "the perception of the elements in the environment within a volume of time and space, the comprehension of their meaning, and the projection of their status in the near future" [12]. For driving, the driver first needs to obtain the SA elements relevant to the driving task, understand the meaning of these elements, and predict their future changes. Failure of the driver to obtain the SA elements (perception), incorrect understanding of the SA elements (understanding), or failure to predict the next actions of related elements (prediction) may lead to accidents. Fortunately, with the advancement in technology, intelligent systems can also have SA.

While the SA theory proposed by Endsley (1988) is the mainstream at present, other SA theories also have their advantages. For example, in the SA theory proposed by Smith and Hancock [13] based on the schema theory [14] and the perceptual cycle model by Neisser [15], SA is believed to be a "generative process of knowledge creation and informed action taking," so it not only exists in the world and in humans but also exists in the interaction between humans and the environment. According to this theory, people react to and judge the environment based on their inherent mental models and schemata, which help with the projection of situations and events, drawing people's attention to the clues in the environment and guiding their final courses of action. When unexpected events beyond mental models occur in the environment, people carry out further external searches (look for other clues in the environment). The results of interaction, in turn, modify initial models and eventually form new schemata or mental models to guide further interaction. The significance of this theory is that it describes a dynamic process and a constantly renewing cognitive cycle.

In the existing literature, while the perceptual cycle theory has not been extended to the team level, some studies have pointed out the possibility of extending it to the team cognitive cycle. Stanton et al. [16] believed that the process of a periodic cognitive cycle was to develop individual SA, share SA with other team members, and modify the team and individual SA based on the SA of other team members. This directly extended the explanation of the cycle, in which other team members represented parts of the surrounding environment. SALAS et al. [17] pointed out that the understanding of information communicated in a team was influenced by the interpretations of other team members, which, in turn, were influenced by their own individual SA. The extended points of view are important for this research, meaning that humans can directly adopt the information provided by agents as they search further for clues. On top of that, if our interaction with the world is guided by our inherent schemata or mental models [15], a digital AR world can be construed as the schema of an agent, manifesting the agent's cognition (perception, understanding, and prediction) of the real world rather than the simulation of the physical world.

Based on the above understanding, a framework is proposed to guide the design of the AR-HUD cooperative interface (Fig. 5.2). In this framework, the human (on the left) and the intelligent system (on the right) carry out synchronized perceptual cycles in the same environment. When necessary information is lost in the current scenario or when the human is unable to perceive some necessary information, he or she explores the outer circle according to the perceptual cycle theory. In a smart vehicle with AR-HUD, some information in the real environment can be replaced with that from the intelligent system as the human explores outward. The human reacts when they receive the corresponding signal, and the reaction changes the real and virtual environments and trains the intelligent system in machine learning. This process is repeated again and again. Therefore, in this model, a single SA forms a team SA and then modifies itself [16]. It can be seen in this model that the schema of the agent's understanding of the environment is the part of HMI that needs to be designed, which is obtained by the human driver through AR-HUD.

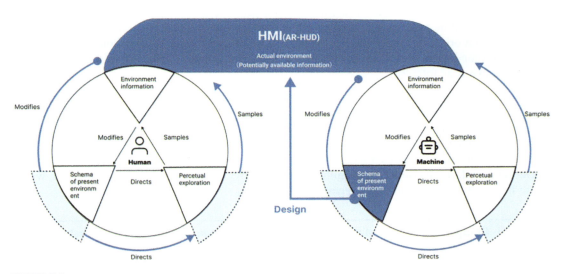

FIGURE 5.2

Presented conceptual human-vehicle teaming perceptual cycle framework derived from previous perceptual cycle [13,15]. Note that the schema of the agent's understanding of the environment is the part of HMI that needs to be designed, which is obtained by the human driver through AR-HUD.

5.2.3 Cognitive information design

Based on the above cooperation frameworks, the next step is to design the information on AR-HUDs. Fig. 5.2 emphasizes that the cognitive information of the intelligent system (i.e., the schema of its perception and understanding of the current environment) is a crucial part of the HMI design. The key challenge is to establish how the schemas in the intelligent system's mental model correspond to the information processing stage of human beings. To address this problem, the design method for AR-HUDs proposed by Debernard et al. [18] was drawn upon. Their approach used cognitive work analysis and common workspaces to categorize the information requirements of human drivers and determine the cooperative information displayed on AR-HUDs. However, their approach is not entirely applicable to the project, as they consider it a process of dealing with tasks, while the focus of the project is on the process of information cognition. Therefore, automation corresponds to the perceptual and processing processes of human beings.

This correspondence is supported by two theories. Parasuraman et al. [19] introduced the concept that automation can be employed in four categories of functions, namely information acquisition, information analysis, decision-making, and action implementation. This is a simplified correspondence of the human cognitive process to the input processing of automation. The second is Endsley's (1988) definition of SA, namely perception, comprehension, projection, and decision and action. The two theories were compared, and the results shown in Table 5.1 were obtained.

Fig. 5.3 shows the information design framework proposed, in which the information transferred can be designed to be multilevel to support the driver.

Environmental information, such as roads, street signs, signals, and other traffic participants, is situated in the middle and can be perceived by both humans and machines. Humans cannot enter the process of situation awareness when they do not perceive the relevant environmental information. AR-HUD can be perceived by humans when superimposed on real environmental information. The information originates from intelligent systems, which can furnish information across different phases corresponding to the phases in human cognition, in response to changes in the environment. The breakdown is as follows:

- **Information acquisition/perception:** Intelligent systems provide information to guide the human driver's attention.
- **Information analysis/comprehension and projection:** Intelligent systems provide insights into their understanding or prediction of the environment, enabling the human driver to make informed decisions rather than independently grasping the environment.
- **Decision-making/decision:** The intelligent system provides decision-making information, with the human driver following the suggestions.

Table 5.1 Comparison of information processing between humans and intelligent systems.

Human	Intelligent system	Explanation of detail
Perception	Information acquisition	Human/machine's perception of environmental cues
Comprehension	Information analysis	Human/machine's understanding of environmental information
Projection	Information analysis	Human/machine's prediction of environmental information
Decision	Decision-making	Human/machine's decision-making

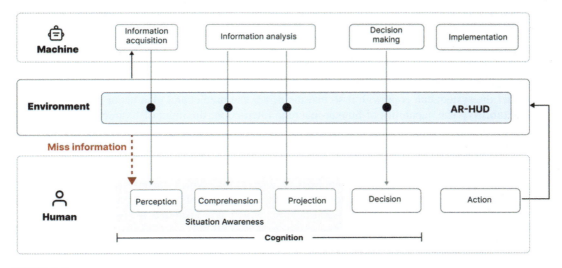

FIGURE 5.3

Proposed information design for cognitive cooperation. Note that the machines can provide information through AR-HUD in different phases, corresponding to the phases in human cognition according to changes in the environment.

For example, during the information acquisition stage of the machine, it can display instructions to remind humans of certain environmental information. When this information is not present in the scenario, the machine can provide information during the information analysis stage to aid in the understanding and prediction of the environment. In urgent situations, such as when the car needs to brake immediately, the machine can directly provide decision-making information. The level of information displayed is determined by the specific scenario and the degree of urgency.

5.2.4 Automation surprises and trust

The concept of automation surprises was first defined in aviation as the operator's failure to understand what automation is doing or why [20,21]. Dekker [22] provided the following definition: "The automation does something without immediate preceding crew input related to the automation's action, and in which that automation action is inconsistent with crew expectation." Christoffersen and Woods [23] categorized discrepant situations in automatic driving into the absence of an expected action and an unexpected action of vehicles. For example, a vehicle is expected to slow down in the case of a red light signal but does not do so (the absence of expected action). Another example is when a vehicle suddenly slows down on an empty road (an unexpected action). It can be seen that automation surprises mainly arise from different decisions that human drivers and automatic systems make in the same environment, and the reactions of vehicles are different from the expectations of human mental models. To reduce such automation surprises,

Christoffersen and Woods [23] also mentioned that the HMI design should exhibit the following characteristics:

- **Observability:** The driver's ability to understand the status of the system and whether the system detects the required information.
- **Predictability:** HMI is supposed to help the driver understand the current environment and whether the system can cope with it so that the driver can predict and plan his or her next actions.
- **Directability:** The capability of human-machine interaction; the input of bidirectional commands between the human and the machine to affect the actions of each other.
- **Timelines:** Relevant information or warning messages are supposed to be provided for the driver in advance, allowing sufficient time for the driver to react.

In addition, automation surprises can lead to uncertainty, which means that our beliefs and representations of the world may not accurately predict future events in our environment [24]. This can have an impact on human trust in automation, which is considered a key factor in human-machine cooperation efficiency and driving safety in automatic driving [25]. According to Lee and See [26], human trust in an agent is "the attitude that an agent will help achieve an individual's goals in a situation characterized by uncertainty and vulnerability." It can be seen that trust means that a human believes that their partner is able to help them out and achieve goals in the case of uncertainty [27]. Automation surprises, by contrast, create more uncertainty.

In this study, when the autonomous vehicle suddenly slows down and the driver cannot promptly discern the clues leading to this action from the environment in time, the human driver becomes perplexed about the vehicle's maneuver. Some studies have shown that visually presenting information regarding the reliability or uncertainty of the system can help the driver form appropriate trust [28,29]. Therefore, the HMI design demonstrates the vehicle's understanding of the intentions of other traffic participants and its decision-making process, aiming to eliminate the influence of automation surprises and enhance the driver's trust in automation.

5.3 Designing HMI

In this section, two dangerous scenarios are selected where environmental information is missing for HMI design (Fig. 5.4). One scenario is where environmental information is obscured by other objects, making it difficult for human drivers to perceive, and is referred to as Case 1. The other scenario is where other vehicles do not transmit intent information and is referred to as Case 2.

5.3.1 HMI for Case 1

In the Case 1 scenario (Fig. 5.4B), Car A, which is traveling straight ahead, collides with Car B, which suddenly brakes sharply in front of our ego-car (a manually driven vehicle), as Car A does not have enough time to avoid the collision. In this situation, the driver cannot observe the road conditions ahead because Car A is blocking the view. Therefore, before Car A's emergency brake,

5.3 Designing HMI 53

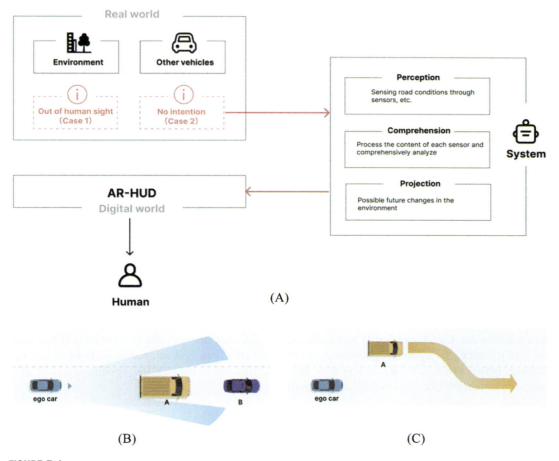

FIGURE 5.4

Dangerous scenarios with missing information. (A) Two kinds of scenarios of missing information (out of human sight or no intention signal) that can be obtained by intelligent system. (B) The vision of the ego-car is blocked by the vehicle in front, and the situation further ahead cannot be seen. (C) The vehicle ahead of the ego-ocar does not turn on the turn signal, so the communication of intention information is lacking.

the driver assumes that everything is normal ahead of the vehicle, and has no psychological expectation of avoiding a collision.

Fig. 5.5 illustrates the scenario of Case 1 and the two HMI design schemes [HMI1 (world-fixed) and HMI2 (screen-fixed)]. The AR information design involves two stages: the information acquisition stage (to detect anomalies in the preceding scene) and the information analysis stage (to determine whether the anomaly poses a danger). Specifically, when the intelligent system measures the

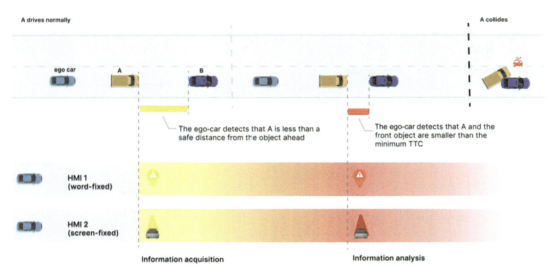

FIGURE 5.5

Case 1 scenario and HMI design. Note that HMI1 and HMI2 represent two different designs, but both include information in the perception and understanding stages.

distance between Car A and Car B and finds that the distance between the two vehicles is less than the safe distance, the system highlights it as abnormal (the icon is displayed in yellow). When the distance between the two cars is less than the TTC (time-to-collision), the system judges it as a dangerous situation (the icon is displayed in red).

5.3.2 HMI for Case 2

In the Case 2 scenario (Fig. 5.4), Car A is traveling straight in the left lane in front of our ego-car, driving in our lane without turning on the turn signal. At this point, the driver understands that Car A intends to go straight without changing lanes. However, the intelligent system believes that Car A intends to change lanes because its sensors have detected changes in the truck's behavior, such as slowing down and moving closer to the right lane, which occur only when a lane change is needed. Therefore, the automation system decides to reduce the speed. The problem is that automated systems initiate deceleration plans when people do not receive information about the intentions of other participants, such as turn signal lamps, creating automation surprises. As environmental clues are not sufficient for human drivers to explain machine behavior reasonably, this affects human trust in automatic systems.

Fig. 5.6 illustrates the scenario of Case 2 and two HMI design schemes. The information design of AR includes three stages: information acquisition stage, analysis stage (predicting vehicle lane-changing behavior), and decision-making stage (making a deceleration decision for the car). Specifically, the intelligent system collects the movement behavior of Car A and determines its intention to change lanes (the icon displays the prediction of the vehicle's future movement trajectory). When Car A triggers the deceleration decision, the interface displays deceleration information.

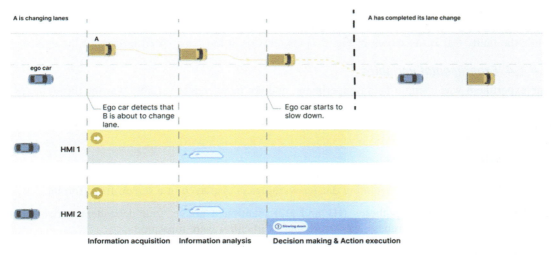

FIGURE 5.6

Case 2 scenario and HMI design. Note that HMI1 and HMI2 represent two different designs. Compared with HMI1, HMI2 also contains decision-making and action execution information.

5.4 Method and materials
5.4.1 Case Study 1
5.4.1.1 Objectives and hypotheses

As mentioned above in Section 5.3.1, there will be a risk of collision with a car at the scene of Case 1 (Fig. 5.5). Therefore, the purpose is to design AR information for potential dangers beyond human perception and assess whether it is helpful for drivers to understand the danger of visual blind areas. Therefore, the following hypotheses are made:

H1: AR interface can improve driver's situation awareness.

H2: AR interface can reduce driver's reaction time.

5.4.1.2 Participants
A total of 20 licensed drivers with normal or corrected vision were recruited for the study. The participants had an average age of 32.67 years (SD = 6.12) and an average of 4.52 years (SD = 3.67) of driving experience. All participants provided written consent and were compensated accordingly.

5.4.1.3 Experimental equipment and materials
This experiment was conducted using a driving simulator. Unity 3D Software1 was utilized to create a 3D highway simulation scene, and various road elements, such as green belts and road signs,

were incorporated to simulate the highway environment. In this scenario, the driver is required to drive the ego-car in the middle lane of a three-lane highway at a designated speed of 100 km/h, while the faulty Car B (as shown in Fig. 5.7) also travels at the same speed in front of the ego-car.

5.4.1.4 Experimental variables

The experiment was a single-factor experiment, and the independent variable was the different AR-HUD interface design schemes (world-fixed and screen-fixed). Table 5.2 presents the design details and demonstration graphs of the three augmented conditions. Fig. 5.8 displays the final comparison of interface schemes for Case 1.

1. **NO-HMI:** No AR information.
2. **HMI1:** AR information containing danger warnings is displayed in a world-fixed format.
3. **HMI2:** AR information containing danger warnings is displayed in a screen-fixed format.

Considering the two hypotheses (H1 and H2) mentioned above, four dependent variables of reaction time, situation awareness, workload, and availability were measured.

1. **Reaction time:** The reaction time was defined as the time for participants to make risk-off actions after the appearance of the HMI scheme.
2. **Usability:** The after-scenario questionnaire (ASQ) [30] is a widely used task-based assessment questionnaire that assesses three key aspects: ease of task completion, time required for task completion, and satisfaction with supporting information. The scale is set from 1 to 7 for options, with higher scores representing higher performance levels. The ASQ can be used for similar usability studies.
3. **Workload:** The driving activity load index (DALI) [31] is a revision of the NASA-TLX for driving tasks and contains six work dimensions: effort of attention, visual demand, auditory demand, temporal demand, interference, and situational stress. The scale options are set from 1 to 10, with higher scores representing higher degrees.

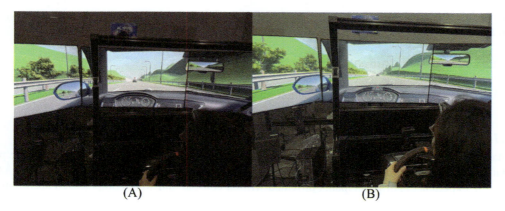

(A)   (B)

FIGURE 5.7

Experimental setup for Case 1. (A) Scheme 1. (B) Scheme 2. Note that each subject had to experience three experimental conditions on the simulator.

Table 5.2 Designing details and demonstration graphs of three augmented conditions.

Design	Condition	Detail	Information			
			Information acquisition	Information analysis	Decision-making	Action execution
	NO AR-HUD (NO HMI)					
	AR-HUD (HMI1)	When it is detected that the vehicle ahead and the object in front are less than the safe distance, a yellow warning will appear, and it will turn red when it is less than TTC				
	AR-HUD(HMI2)	When it is detected that the vehicle ahead and the object in front are less than the safe distance, a yellow warning will appear, and it will turn red when it is less than TTC				

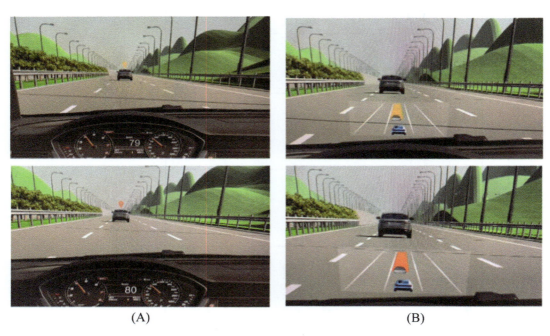

FIGURE 5.8

Comparison of interface schemes of Case 1. (A) HMI1 (world fixed). (B) HMI2 (screen fixed).

Table 5.3 Questions related to three levels of situational awareness (Case 1).

Level of situational awareness	Questions
Level 1	1. Do you see an abnormal signal in front of the vehicle ahead?
Level 2	2. Are you aware that there is danger in front of the vehicle ahead?
Level 3	3. Can you predict that the vehicle ahead is about to change lanes or brake?
	4. Do you realize that you should take some emergency measures to avoid danger?

4. **Situation awareness:** The situation awareness global assessment technique (SAGAT) [32] can assess the level of SA by asking participants relevant SA needs questions that address the three levels of SA (i.e., driver perception, understanding, and prediction) (Table 5.3).

5.4.1.5 Procedure

Before the experiment, participants were required to read and sign a knowledge notice and complete a demographic questionnaire. They were then invited to sit in the simulator and proficiently simulate the operation of driving. During the formal experiment, the presentation of the three schemes was randomized using a Latin square, and participants were asked to maintain a speed of 100 km/hour (within the highway speed limit range of 80–120 km/hour) in the middle lane, following Car A with the same speed, and perform hedging operations when danger was perceived. Upon completion of each scheme, participants were asked to complete the relevant scale. Semi-structured interviews with participants were conducted after the experiment. The simulator has built-in functions that record vehicle data, participants' pedal reaction time, speed changes, acceleration changes, etc.

5.4.2 Case study 2
5.4.2.1 Objectives and hypotheses
As mentioned in Section 5.3.2, there is a risk of automation surprises in the Case 2 scenario (see Fig. 5.6). The purpose of this experiment is to investigate the effect of HMI design on people's trust in autonomous vehicles and their ability to understand the intention of a vehicle changing lanes. Therefore, the following hypotheses are proposed:

H3: HMI design containing predictive information can increase driver understanding of the vehicle's intention to change lanes.

H4: HMI design containing predictive information can increase driver understanding and trust in automated vehicle behavior.

H5: HMI design containing decision information can increase driver understanding of the vehicle's intention to change lanes more than an HMI with predictive information.

5.4.2.2 Participants
A total of 25 licensed drivers with normal or corrected vision were recruited. The participants had an average age of 30.28 years (SD = 6.32) and an average of 5.88 years of driving experience (SD = 3.40). They all had experience with L1 or L2 autonomous vehicles and provided written acknowledgment of their participation in the study. Participants were compensated for their time and effort.

5.4.2.3 Experimental equipment and materials
In this study, participants were shown a first-person perspective video of driving to simulate the experience of being in an autonomous vehicle (Fig. 5.9). Each video included 3–4 vehicles in the side lane, and the vehicles changing lanes were different in each video. Therefore, participants did not know which car would change lanes. The video was played on a computer, and participants watched it in front of the computer while completing the scale. The video included a built-in timing function, allowing us to record participants' reaction time when they pressed a keyboard key according to the experimental requirements.

Video materials were utilized to simulate the first-person driving experience of the autopilot. A video (15 seconds) was created using Unity software, in which the ego-car was driven on the middle of a three-lane city road at a speed of 40 km/hour (city road speed limit 60 km/hour). Various road elements, such as buildings, green belts, bus stops, road signs, etc., were added to create a natural and realistic driving experience for the participants. An event was also included in this scene, where a van located in front of the left of the ego-car changed lanes to the middle of the three-lane road without using turn signals, as the video played.

Chapter 5 A novel cooperation-guided warning system

FIGURE 5.9

Video materials and comparison of interface schemes of Case 2. (A) Scheme 1. (B) Scheme 2. Note that the interface on the left contains AR analog turn signals and lane-changing prediction information of the vehicle in front, and the interface on the right adds the vehicle deceleration icons.

5.4.2.4 Experimental variables

The experiment was a single-factor experiment with the independent variable being the AR-HUD interface design, and each participant underwent three different AR-HUD display schemes. Table 5.5 shows the design details and demonstration graphs of the three augmented conditions. Fig. 5.9 shows the final comparison of the interface schemes for Case 2.

1. **NO-HMI:** Absence of AR information.
2. **HMI1:** Display AR analog turn signals and lane-changing prediction information of the vehicle in front.
3. **HMI2:** Display AR analog turn signals, lane-changing prediction information, and vehicle deceleration icons.

Considering the three hypotheses (H3, H4, and H5) mentioned above, four dependent variables were measured: reaction time, situation awareness, trust, and usability.

- **Reaction time:** In the experiment, reaction time was defined as the time from when the participant began watching the video to the time when they realized the lane-changing behavior of the vehicle ahead. Participants were asked to press the "M" key on the keyboard upon perceiving that the vehicle in front was intending to change lanes while watching the video, and the reaction time of 25 participants under three conditions was recorded.
- **Usability:** System usability scale [33] is a widely used subjective questionnaire for assessing system usability. In this study, the focus is on whether the design elements of AR would affect users' perception of the environment (occlusion of the line of sight) and whether users can fully understand the meaning of UI elements. The 10 declarative sentences form a description of the three dimensions of effectiveness, efficiency, and satisfaction, and participants are asked to rate how well they agree with these sentences. The scale is set from 1 to 5 for options, with higher scores indicating higher performance and system usability.

Table 5.4 Questions related to three levels of situational awareness (Case 2).

Level of situational awareness	Questions
Level 1 Level 2 Level 3	1. Can you see the change of turn signal of the vehicle in front of you? 2. What is the intention of the front car? 3. Are you aware that the autonomous vehicle is slowing down? By what? 4. Can you predict when the car in front will change lanes? 5. Can you predict where the car in front will change lanes? By what? 6. Are you aware of the danger to yourself from the car in front of you?

- **Trust:** Participants' trust in autopilot systems is measured using the internationally used Trust Scale [34], which contains 12 potential factors affecting trust between humans and automated systems. The scale is set from 1 to 7 for options, with higher scores representing higher levels of trust. Because interface design can affect users' trust and calibration of trust in automated systems [35], the UI design in this paper includes the cognitive processes and decision-making information of the machine, which enables the driver to quickly compare the visualized internal state with the changing environment. Research has shown that trust is increased when humans are able to compare the environment with the information provided by the agent [36].
- **Situation awareness:** Situation awareness global assessment technique (SAGAT) [32] can assess the participants' level of SA by asking relevant SA needs questions that cover the three levels of SA (i.e., the driver's perception, understanding, and prediction) (Table 5.4).

5.4.2.5 Procedure

Before the experiment, participants read and signed a notice of knowledge, and completed a demographic questionnaire. Then, participants were invited to sit in front of a computer and adjust their sitting position so that they could look at the screen. They were instructed to imagine themselves as the driver while watching the videos. To familiarize the participants with the experimental scenario, they underwent a practical trial in which all road settings were the same as in the formal experiment, except for the unpredictable lane-changing events of the vehicles ahead. Participants were instructed to press the designated key on the keyboard when they realized the change of the vehicle in front during the following formal experiment. After the practical trial, each participant completed three formal experiments (watched three videos), and the three videos were presented in random order. After watching each video, participants were asked to complete the SA scale, the trust scale, and the system usability scale (except for scenario 1, where no availability scale was required). After the completion of all experiments, semistructured interviews were conducted to ask participants about their feelings and attitudes toward different experimental schemes.

5.5 Results and analysis

5.5.1 Results of Case 1

In this experiment, Pingouin 2 [37] in Python was used to check the data for normality and perform various tests (including the Shapiro-Wilk test [38], nonparametric pairwise tests based on the

Mann-Whitney U test [39], repeated measures one-way ANOVA). Scipy 3 [40] was adopted to process and smooth the indicators related to reaction time; while plotnine 4 [41] was used to visualize results.

The Shapiro-Wilk test [38] results showed the measured values of each index were not normally distributed. Hence, the nonparametric pairwise tests based on Mann-Whitney U test [39] were adopted to compare the specific effects of the three schemes.

5.5.1.1 Situational awareness

The quantitative comparisons of three schemes for SA are presented in Fig. 5.10A, where a higher answer accuracy rate indicates better SA. As shown in the figure, the average and standard deviation of answer accuracy for the three schemes (i.e., HMI1, HMI2, and NO-HMI) were (M = 0.900, SD = 0.235), (M = 0.887, SD = 0.262), and (M = 0.662, SD = 0.399), respectively. The results of the Mann-Whitney U test indicate that the p-values of HMI1-HMI2, HMI1-(NO-HMI), and HMI2-(NO-HMI) were >0.05, <0.05, and <0.01, respectively. These results suggest that HMI1 significantly improved SA compared with HMI2 and NO-HMI.

5.5.1.2 Usability

The usability of the system was evaluated using the ASQ, where a higher score indicates a higher usability of the interface. Fig. 5.10B shows that the participants rated the usability of the AR-HUD interfaces significantly higher than that of no interface. The results, analyzed using the Mann-Whitney U test, revealed that HMI1 (M = 6.33, SD = 0.725) had the highest usability score, which

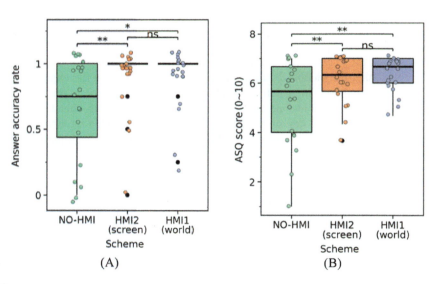

FIGURE 5.10

Scheme comparisons of Case 1. (A) answer accuracy rate. (B) ASQ score of usability. Ns (no significance) = >0.05; *p = ≤0.05; ** p = ≤0.01. Note that HMI1 performed the best in both dependent variables.

was significantly different from NO-HMI (M = 5.183, SD = 1.725), with a p-value of less than 0.001. Similarly, HMI2 (M = 6.123, SD = 0.993) also had a significantly higher usability score than NO-HMI, with a p-value of less than 0.001. Although HMI1 was rated higher than HMI2, the result was not significant, with a p-value greater than 0.05.

5.5.1.3 Workload

To assess the subjective workload, the DALI (driving activity load index) was used, which includes the effort of attention, interference, situational stress, temporal demand, and visual requirements, as shown in Fig. 5.11. Auditory demand was excluded because the interfaces had no sound. The global mean analysis revealed no significant differences among the three schemes. The data for each category option was analyzed using the Mann-Whitney U test.

In terms of effort of attention, although the cases with AR interfaces (HMI1 and HMI2) were better than the situation without interface (NO-HMI), the three schemes did not show significant differences, suggesting that the attentional effort is similar in all three situations.

For interference, HMI1 (M = 27, SD = 22.27) had the smallest results and was significantly different from NO-HMI (M = 42.5, SD = 29), with a p-value of less than 0.05. Moreover, the HMI1 scheme was superior to the HMI2 scheme (M = 39.5, SD = 26.05), with a p-value of less than 0.05, indicating that the driver had the least interference with the normal driving task when encountering the risk avoidance scene with the HMI1 scheme.

Regarding situational stress, the HMI1 scheme performed the best and was significantly better than NO-HMI. At the same time, HMI1 was less stressful than HMI2, but the results were not significant. For temporal demand, HMI1 (M = 32.5, SD = 21.24) performed the best and was significantly different from NO-HMI (M = 46.50, SD = 28.15), with a p-value of less than 0.01, but not significantly different from HMI2 (M = 55, SD = 28.75), with a p-value greater than 0.05. There was no significant difference in the three conditions in terms of visual requirements, and HMI1 was slightly better than HMI2 with significant differences.

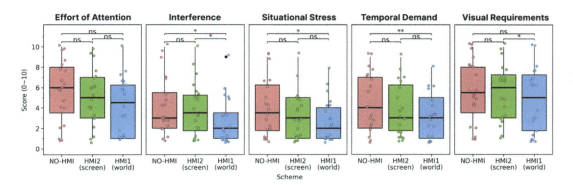

FIGURE 5.11

Rating of the DALI scale. Ns (no significance) = >0.05; *p = ≤0.05; ** p = ≤0.01. Note that the DALI was used to include effort of attention, interference, situational stress, temporal demand, and visual requirements (auditory demand was excluded because the interfaces had no sound).

5.5.1.4 Reaction time

After conducting the experiments, vehicle data of 20 drivers was collected using a simulated platform. Two indicators related to the reaction time of the three schemes, namely acceleration and braking, and speed, were recorded and compared. The average indicators around 5 seconds before/after the appearance time of the interface scheme (HMI) were computed and smoothed using a locally weighted scatterplot smoothing (LOWESS) method [42,43]. The results are shown in Fig. 5.12, where the position of 0 on the X-axis indicates the scheme's appearance time. Here, the reaction time is defined as the time when the driver actively applies the brake or releases the accelerator after the interface scheme appears.

Fig. 5.12A shows the mean change in the rate at which participants lift off the accelerator or brake when the interface appears for the three conditions. It can be observed that the driver decelerates approximately 2.5 seconds faster with the interface (HMI1 and HMI2) than without the interface (NO-HMI). A similar trend can be observed from the curve of velocity change in Fig. 5.12B, indicating that the interface had a positive impact on participants' reaction speed.

5.5.2 Results of Case 2

To test hypotheses H3, H4, and H5, descriptive analysis, manipulation checks, and one-way repeated measures ANOVA were performed.

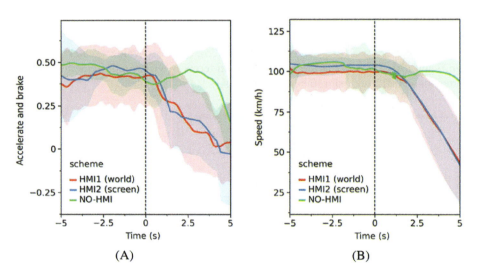

FIGURE 5.12

Indicator comparisons related to reaction time of three schemes. (A) accelerate and brake. (B) speed. The position of 0 on the X-axis in each subfigure indicated the appearance time of the interface scheme. Note that the driver with the interface (HMI1 and HMI2) was faster than that without an interface (NO-HMI) to release the accelerator in (A), and HMI1 and HMI2 were also faster than NO-HMI to decelerate in (B).

5.5.2.1 Reaction time

Fig. 5.13A shows the participants' reaction time results under the three experimental conditions. The reaction time was defined as the moment when the participant perceived that Car A had the intention to change lanes. The shorter the reaction time, the more effective the solution.

The results of repeated measures ANOVA showed that there was a significant difference among the three schemes F (1.644, 39.45) = 18.29, $p < 0.0001$. Post hoc analyses using Tukey's multiple comparison tests indicated that the reaction time of HMI1 (M = 6.582, SD = 1.899) was the fastest, which was significantly different from that of NO-HMI (M = 8.508, SD = 1.965), $p < 0.0001$. HMI1 was also slightly better than HMI2 (M = 6.806, SD = 2.082), but the difference was not statistically significant, $p = 0.7016$. HMI2, however, outperformed NO-HMI significantly, with a $p < 0.001$.

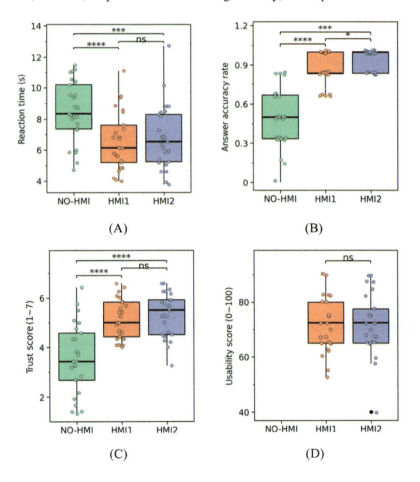

FIGURE 5.13

Scheme comparisons for Case 2. (A) Reaction time. (B) Answer accuracy rate. (C) Trust score. (D) usability score. ns (no significance) = >0.05; *p = ≤0.05; ** p = ≤0.01.

5.5.2.2 Situational awareness

Fig. 5.13B shows the correct rate of the SAGAT questionnaire in three conditions. The results analyzed by repeated measures ANOVA revealed that there were significant differences among the three groups $F\ (1.448, 34.76) = 82.46$, $p < 0.0001$, and the order is HMI2 > HMI1 > NO-HMI. Both HMI1 (M = 0.8660, SD = 0.1266) and HMI2 (M = 0.945, SD = 0.089) schemes had higher accuracy than NO-HMI (M = 0.4844, SD = 0.2276). Specifically, HMI2 had the highest accuracy rate, demonstrating a significant difference compared with HMI1 ($p < 0.05$) and an extremely significant difference compared with NO-HMI, $p < 0.001$. The accuracy of HMI1 was also higher than that of NO-HMI and a significant difference can be observed, with $p < 0.0001$. It can be concluded that the HMI2 was the best at supporting the driver's understanding of the situation.

5.5.2.3 Trust

Fig. 5.13C shows the results of the trust scale of three conditions, where a higher score indicates a greater level of trust. In general, the trust degree of the automatic driving system with the AR-HUD interfaces (HMI1 and HMI2) was higher than that of the system without the interface (NO-HMI). Specifically, subjects rated HMI2 (M = 5.311, SD = 0.896) the highest. HMI2 exhibited a significant difference from NO-HMI (M = 3.556, SD = 1.429), with $p < 0.0001$. However, it did not significantly differ from HMI1 (M = 5.136, SD = 0.806), with $p > 0.05$. In addition, HMI1 was also more trusted by the subjects compared with the NO-HMI situation, and there was an extremely significant difference, with $p < 0.0001$. Overall, the ranking of participants' trust in the system was HMI2 > HMI1 > NO-HMI.

5.5.2.4 Usability

The paired t-test results regarding the usability shown in Fig. 5.13D indicate that there was no significant difference ($t\ (24) = 0.1054$, $p = 0.9170$) between HMI1 (M = 72.20, SD = 10.57) and HMI2 (M = 71.90, SD = 11.46).

5.6 Discussion

During the driving process, the environment may lack cues that are perceivable by humans but can impact driving safety. This study aims to investigate how vehicles can offer drivers the necessary cognitive cooperation via AR-HUDs to improve their perception and cognition. This cooperation is deemed significant for both manual and autonomous driving. The subsequent section will discuss two cases.

5.6.1 Case 1

In manual driving, the driver's field of view is limited, and the information in the blind spots can cause danger. This study proposes the concept of human-machine cooperation to solve this problem, wherein the intelligent system transfers information about dangers in the blind spots to the driver through AR, thereby providing support to the driver. The information about cognitive cooperation on this support was designed according to the approach proposed in this study (Table 5.1

and Fig. 5.3). Cognitive information was divided into two stages, namely perceiving and understanding, and was placed in two different ways (world-fixed and screen-fixed).

The measurement results of SA showed a significant improvement in participants' perception of the danger ahead with the help of the information displayed on the AR interface (Table 5.3). During the interviews, all participants said that the signs on the vehicle ahead enabled them to be alert to possible hazards in advance. Therefore, H1 was supported. In addition, the data on reaction time sufficiently supported H2. The results showed that when the interface appeared to indicate that there had been an accident involving the vehicle ahead, the drivers who were alerted by the interface released the accelerator pedal or stepped on the brake 2.5 seconds earlier (Fig. 5.12). It suggests that the appearance of the interface reminds drivers of the possible danger and makes them more alert.

Although the total scores of the three experimental situations were not significantly different in terms of workload (DALI), each subcategory showed a lower workload with the presence of an AR interface. In particular, HMI1 (world-fixed) achieved the best performance in each subcategory. The results showed that HMI1 performed significantly better than NO-HMI in terms of interference, situational stress, and temporal demand and significantly better than HMI2 in terms of interference and visual requirements, suggesting that the world-fix mode generally has better effects. Over 70% of the interviewees said that the driving signal of the vehicle ahead was the only criterion to judge the situation ahead during high-speed driving. When the HMI1 (world-fixed) model is on, the information is displayed on the vehicle ahead, which can be easily seen. Half of the participants in the interviews argued that the design of HMI2 (screen-fixed) might not be easy to observe, probably because the visual horizon would be higher at 100 km/hour than at lower speeds. Other participants argued that the sudden appearance of an HMI1 (world-fixed) interface would be a bit abrupt. These two points should be taken into account in the design of AR-HUDs.

The results showed that both the world-fixed and screen-fixed approaches performed better than the NO-HMI approach in terms of usability, but there was no significant difference between HMI1 and HMI2. This suggests that participants are positive about this innovative design of AR cognitive information and can familiarize themselves with it quickly. These results suggest that vehicles equipped with both HMI1 and HMI2 can support drivers by providing missing information through AR-HUDs. In general, HMI1 performs better than HMI2. Therefore, the world-fixed mode is worth considering in the design of AR interfaces.

5.6.2 Case 2

Trust is pivotal in autonomous driving. When the autonomous driving system does something that does not meet the driver's expectations, the automation surprise phenomenon can undermine humans' trust in the vehicle. In Case 2, whether an intelligent system could decrease automation surprise and improve trust by displaying other the intentions of other vehicles through an AR-HUD was explored. The design method proposed by this study was also adopted in the design of this cooperation interface that displays different stages (i.e., information acquisition, information analysis, decision-making, and action implementation) to measure the impact on human drivers (Table 5.1 and Fig. 5.3).

First, it was found that the use of AR interfaces (Table 5.5) enabled drivers to perceive the intention of the driver in the car ahead to change lanes more quickly (by approximately 2 seconds),

Table 5.5 Designing details and demonstration graphs of three augmented conditions.

Design	Condition	Detail	Information acquisition	Information analysis	Decision-making	Action execution
	NO AR-HUD (NO HMI)					
	AR-HUD (HMI1)	A simulated turn signal and lane change prediction of vehicle ahead.	⬆	◧		
	AR-HUD (HMI2)	A simulated turn signal, lane change prediction of vehicle ahead, and vehicle deceleration warning.	⬆	◧	Slowing down	Slowing down

as evidenced by the results of the reaction time (Fig. 5.13A). Meanwhile, the measurement results of SA suggested that AR information enhanced the participants' understanding of vehicles' intentions. Therefore, H3 was supported. Second, the scores on trust supported H4. It can be seen that both HMI approaches improved the participants' trust in and understanding of the current performance of the autopilot system, and both were highly significant compared with the NO-HMI scenario, indicating that the automation surprise phenomenon was mitigated. In the posthoc interviews, more than 60% of participants were very positive about the information designed to predict the positions of vehicles in HMI1, saying that "the changing arrows on the ground are very clear and reassuring." However, unlike the case of HMI1, the information related to decision and action added to HMI2 did not significantly improve drivers' understanding of other drivers' intentions to change lanes, and H5 was not supported. This may be because the information provided by the machine about perception and prediction in this scenario is sufficient for human drivers to understand the slowdown behavior of the car, so the information related to decisions and actions and further explanation are not necessary. Over half of the participants in the interviews thought the information about "deceleration" was not important because they could feel the car slowing down. Eight participants said that the information indicating deceleration displayed at the bottom of the AR-HUD might affect their line of sight and suggested that the positions of the displayed information be considered in the design of AR-HUD interfaces. Lastly, there was no significant difference between HMI1 and HMI2 in terms of usability.

In summary, human-machine cooperation in autonomous driving requires a clear understanding of the behaviors and intentions of both humans and machines. AR technology enables simultaneous displays of the impacts of changing environments on system cognition in response to these changes. With the help of this technology, human drivers can monitor both variables (i.e., the environment and the intelligent system) in a more direct manner. This study proposes the idea of using AR technology to enable intelligent systems to transfer information about other traffic participants' intentions, which can improve drivers' understanding of changes in the environment and other vehicles' intentions, and enhance their trust in intelligent systems.

5.6.3 Limitations

This study has several limitations that should be considered. First, in Case 1, only one speed (100 km/hour) was examined when studying the transfer of information about dangerous situations ahead. However, speed may have an impact on human visual perception in high-speed environments, and different speeds should be considered in future studies. Second, the video experiment in Case 2 did not consider tactile feedback, which may be important for drivers to physically perceive the state of autonomous driving. Moreover, the distance to the car merging into the same lane was not considered, which could impact the reminder mechanism on the interface. Future studies should explore similar scenarios involving perceptual deficits, such as drivers' intentions at intersections and when allowing other cars to pass. Finally, this study only concentrated on the support provided by intelligent systems to human beings and did not examine human assistance for the systems. Therefore, there is still room for improvement in the model.

5.7 Conclusion

The present study examines the support offered by intelligent systems to human information cognition. This study is unique because it explores some common dangerous situations that are rarely considered due to the lack of necessary information in the environment. It is argued that the scarcity of relevant studies is not because the scenarios are atypical but rather due to limitations in display technology. Therefore, the use of AR technology on human-machine cooperative interfaces was considered to display important information on the interfaces for human drivers, ensuring their access to SA.

The contributions of this study are twofold. Theoretically, two frameworks were established: one that explains the mechanism of agents' cognitive cooperation in a human-machine cooperative driving team, and the other that guides the information design for interfaces. Two case studies were conducted to demonstrate that the cooperation frameworks could well explain the role of AR-HUDs in cooperation and the mechanisms of information design. Additionally, the cooperative interfaces designed through the method proposed in this study for the design of cognitive interfaces showed great effects.

In practical terms, the scenarios discussed in this study are believed to apply to many dangerous situations. For instance, in Case 1, the traffic conditions ahead that cannot be observed may also include anything that can trigger abnormal behavior in Vehicle A, such as a sudden appearance of an animal or an obstacle in a construction zone. Case 2 can also be extended to identify the intentions of various traffic participants. Although some studies have shown that human beings outperform intelligent systems in identifying implicit intentions, such as the expressions and gestures of passersby, intelligent systems perform better in perceiving changes in positions. Furthermore, from the perspective of vehicle behavior, it has been found that the driving intentions of human beings can be accurately recognized shortly after the action begins [44]. It is believed that ADAS systems have the cognitive ability to accurately predict some vehicle behaviors. Additionally, when the car is in autopilot mode, the human driver may be engaged in nondriving tasks and may ignore the understanding of environmental information. In such situations, information about the perception of the environment provided by the intelligent system is necessary.

The next plan is to explore more scenarios that lack environmental information and to consider the influence of more variables, such as speed and distance between two vehicles before a collision, on interface design. It is also suggested that for the design of AR-HUD interfaces, it is necessary to consider that the information provided is related to agents' cognition of the objective world and aims to provide adequate cognitive cooperation for human drivers. Additionally, at this stage, the model only focuses on the support of intelligent systems for human cooperation, and two-way cooperation will be studied in the future.

References

[1] E. Hollnagel, A function-centred approach to joint driver-vehicle system design, Cognition, Technol. & Work. 8 (3) (2006) 169–173.

[2] O. Carsten, M.H. Martens, How can humans understand their automated cars? hmi principles, problems and solutions, Cognition, Technol. & Work. 21 (1) (2019) 3–20.

References

[3] R. Ponziani. Turn signal usage rate results: a comprehensive field study of 12,000 observed turning vehicles. Technical report, SAE, 2012.

[4] M.L. Rusch, M.C. Schall, J.D. Lee, J.D. Dawson, M. Rizzo, Augmented reality cues to assist older drivers with gap estimation for left-turns, Accid. Anal. Prev. 71 (2014) 210–221.

[5] G. Moussa, E. Radwan, K. Hussain, Augmented reality vehicle system: left turn maneuver study, Transport. Res. Part. C: Emerg. Technol. 21 (1) (2012) 1–16.

[6] S.A.E. International, SAE international releases updated visual chart for its "levels of driving automation" standard for self-driving vehicles, 2018.

[7] J.A. Michon, A critical view of driver behavior models: what do we know, what should we do? In Human Behavior and Traffic Safety, Springer, 1985, pp. 485–524.

[8] F.O. Flemisch, K. Bengler, H. Bubb, H. Winner, R. Bruder, Towards cooperative guidance and control of highly automated vehicles: H-mode and conduct-by-wire, Ergonomics 57 (3) (2014) 343–360.

[9] L. Biester. Cooperative automation in automobiles. Humboldt University of Berlin, Faculty of Mathematics and Natural Sciences II, 2009.

[10] N. Merat, T. Louw, R. Madigan, M. Wilbrink, A. Schieben, What externally presented information do vrus require when interacting with fully automated road transport systems in shared space? Accid. Anal. & Prev. 118 (2018) 244–252.

[11] C. Wang, M. Kruger, C.B. Wiebel-Herboth, C. B. "Watch out!": Prediction-level intervention for automated driving. In 12th International Conference on Automotive User Interfaces and Interactive Vehicular Applications (pp. 169–180). Association for Computing Machinery, 2020.

[12] M.R. Endsley. Design and evaluation for situation awareness enhancement. In Proceedings of the Human Factors Society Annual Meeting (Vol. 32, pp. 97–101). Sage Publications, 1988.

[13] K. Smith, P.A. Hancock, Situation awareness is adaptive, externally directed consciousness, Hum. Factors 37 (1) (1995) 137–148.

[14] K.L. Plant, N.A. Stanton, The explanatory power of schema theory: theoretical foundations and future applications in ergonomics, Ergonomics 56 (1) (2013) 1–15.

[15] U. Neisser, Cognition and Reality: Principles and Implications of Cognitive Psychology, W H Freeman/ Times Books/ Henry Holt & Co, 1976.

[16] N.A. Stanton, P.M. Salmon, G.H. Walker, E. Salas, and, P.A. Hancock, State ofscience: situation awareness in individuals, teams and systems, Ergonomics 60 (4) (2017) 449–466.

[17] E. Salas, C. Prince, D. Baker, L. Shrestha. Situation awareness in team performance: Implications for measurement and training: Situation awareness. Hum. Factors 37 (1) (1995) 123–136.

[18] S. Debernard, C. Chauvin, R. Pokam, S. Langlois. Designing human-machine interface for autonomous vehicles. IFAC-PapersOnLine 49 (19) (2016) 609–614.

[19] R. Parasuraman, T.B. Sheridan, and, C.D. Wickens, A model for types and levels of human interaction with automation, IEEE Trans. Syst., Man, Cybernet.- Part. A: Syst. Hum. 30 (3) (2000) 286–297.

[20] N. Sarter, C. Wickens, R. Mumaw, S. Kimball, R. Marsh, M. Nikolic, et al., Modern flight deck automation: pilots' mental model and monitoring patterns and performance, Proc. Int. Symp. Aviat. Psychol., Dayton (2003).

[21] N.B. Sarter, D.D. Woods, How in the world did we ever get into that mode? mode error and awareness in supervisory control, Hum. Factors 37 (1) (1995) 5–19.

[22] S. Dekker, Report of the flight crew human factors investigation conducted for the Dutch safety board into the accident of TK1951, Boeing 737–800 near Amsterdam Schiphol Airport, Lund University, School of Aviation, February 25, 2009.

[23] K. Christoffersen, D. Woods, How to make automated systems team players, Adv. Hum. Perform. Cogn. Eng. Res 2 (2002) 1–12.

[24] F. Mushtaq, A.R. Bland, A. Schaefer, Uncertainty and cognitive control, Front. Psychol. 2 (2011) 249.

[25] P.A. Hancock, I. Nourbakhsh, J. Stewart, On the future of transportation in an era of automated and autonomous vehicles, Proc. Natl Acad. Sci. 116 (16) (2019) 7684–7691.
[26] J.D. Lee, K.A. See, Trust in automation: designing for appropriate reliance, Hum. Factors 46 (1) (2004) 50–80.
[27] F. Yan, L. Weber, A. Luedtke, Classifying driver's uncertainty about the distance gap at lane changing for developing trustworthy assistance systems, 2015 IEEE Intelligent Vehicles Symposium (IV), IEEE, 2015, pp. 1276–1281.
[28] T. Helldin, G. Falkman, M. Riveiro, S. Davidsson, Presenting system uncertainty in automotive UIS for supporting trust calibration in autonomous driving, in Proc. 5th International Conference Automotive User Interfaces and Interactive Vehicular Applications, 2013, pp. 210–217.
[29] A. Kunze, S.J. Summerskill, R. Marshall, A.J. Filtness, Automation transparency: implications of uncertainty communication for human-automation interaction and interfaces, Ergonomics 62 (3) (2019) 345–360.
[30] J.R. Lewis, Psychometric evaluation of an after-scenario questionnaire for computer usability studies: the ASQ, ACM Sigchi Bull. 23 (1) (1991) 78–81.
[31] A. Pauzié, A method to assess the driver mental workload: the driving activity load index (dali), IET Intell. Transp. Syst. 2 (4) (2008) 315–322.
[32] M.R. Endsley. Measurement of situation awareness in dynamic systems. Hum. Factors 37 (1) (1995) 65–84.
[33] A. Bangor, P. Kortum, J. Miller, Determining what individual sus scores mean: adding an adjective rating scale, J. Usability Stud. 4 (3) (2009) 114–123.
[34] J.-Y. Jian, A.M. Bisantz, C.G. Drury, Foundations for an empirically determined scale of trust in automated systems, Int. J. Cognit. Ergonomics 4 (1) (2000) 53–71.
[35] J. Kraus, D. Scholz, D. Stiegemeier, M. Baumann, The more you know: trust dynamics and calibration in highly automated driving and the effects of take-overs, system malfunction, and system transparency, Hum. Factors 62 (5) (2020) 718–736.
[36] M.R. Endsley, From here to autonomy: lessons learned from human-automation research, Hum. Factors 59 (1) (2017) 5–27.
[37] R. Vallat, Pingouin: statistics in Python, J. Open. Source Softw. 3 (31) (2018) 1026.
[38] S.S. Shapiro, M.B. Wilk, An analysis of variance test for normality (complete samples), Biometrika 52 (3/4) (1965) 591–611.
[39] A. Vargha, H.D. Delaney, A critique and improvement of the cl common language effect size statistics of McGraw and Wong, J. Educ. Behav. Stat. 25 (2) (2000) 101–132.
[40] P. Virtanen, R. Gommers, T.E. Oliphant, M. Haberland, T. Reddy, D. Cournapeau, et al., Scipy 1.0: fundamental algorithms for scientific computing in Python, Nat. Methods 17 (3) (2020) 261–272.
[41] J. Zhang, Beautiful Data Visualization of Python- How to Make Professional Charts, Publishing House of Electronic Industry, 2020.
[42] S. Seabold, J. Perktold, Statsmodels: Econometric and statistical modeling with Python, in: Proc. 9th Python in Science Conference, Volume 57, Austin, TX, 2010, pp. 10-25080.
[43] W.S. Cleveland, Robust locally weighted regression and smoothing scatterplots, J. Am. Stat. Assoc. 74 (368) (1979) 829–836.
[44] A. Pentland, A. Liu, Modeling and prediction of human behavior, Neur. Comput. 11 (1) (1999) 229–242.

PART 2
Intelligent cockpit HMI information decision and control

CHAPTER 6

Acting like a human: teaching an autonomous vehicle to deal with traffic encounters

Jianmin Wang[1], Jiawei Lu[2], Fang You[1] and Yujia Wang[1]

[1]*Car Interaction Design Lab, College of Arts and Media, Tongji University, Shanghai, P.R. China* [2]*College of Design and Innovation, Tongji University, Shanghai, P.R. China*

Chapter Outline

6.1 Introduction 75
6.2 Socially incapable autonomous vehicles 75
6.3 Method 76
6.4 Results and analysis 76
6.5 Discussion 78
6.6 Conclusion 79
Further reading 79

6.1 Introduction

During the transition from manual driving to autonomous driving, we anticipate a prolonged coexistence of manned vehicles and autonomous vehicles (AVs) running together on the roads. This raises a pertinent social question: Can human drivers and AVs coexist harmoniously on the roads? Initial indications suggest otherwise. This paper intends to explore a direction for developing socially acceptable AVs. We conducted a study involving 18 participants and asked them to drive in a university campus wearing Tobii Glasses to record their experiences. Traffic encounters during the experiment were extracted and analyzed. Our objective is to reveal human strategies and behaviors for dealing with traffic encounters, which can be leveraged by AVs. The results indicate the importance of increasing AVs' understanding of all stakeholders on the road, including drivers, other road users, and the environmental context.

6.2 Socially incapable autonomous vehicles

As all road users are required to move in a limited space, encounters between various road users become inevitable, leading to potential conflicts. Negotiation often becomes necessary to avoid conflicts and achieve mutually beneficial outcomes. Though road users' ability to communicate (mainly verbal communication) may be weakened by the confines (iron cages) of their vehicles, nonverbal communication still plays a crucial role in negotiation, which includes eye contact, body gestures, facial expressions, and so on. Human drivers possess the ability to predict the intentions of other road users by observing subtle changes in speed, direction, and nonverbal cues; they can even discern the mood and character of fellow drivers through these observations.

Brown and Laurier analyzed a number of videos of AVs' road test sourced from YouTube, unfolding the problem clearly—AVs may struggle to interpret human emotions and intentions accurately. Conversely, AV's actions and intentions may not be correctly interpreted by human drivers. problem, social capabilities, However, while improving AVs' ability to convey intentions could address part of the problem, we believe that a more comprehensive solution involves endowing AVs with social capabilities, enabling them to understand and be understood by other road users, akin to human behavior. Hence, we conducted experiments to discover how human road users deal with traffic encounters, seeking a humanized way of road socializing that AVs could learn from.

6.3 Method

This paper focuses on interaction-prone encounters where social intelligence is frequently involved. Participants were instructed to drive an SAIC Tiguan on a university campus while wearing a pair of Tobii Pro Glasses 2. This allowed us to obtain video data along with the participants' areas of interest, helping us to understand the intentions of the participants. In addition, considering the traffic on campus is rather simple than that off campus, we asked the participants to execute lane-changing tasks to generate more traffic encounters. A total of 18 participants with different career backgrounds were screened and recruited for the experiment, which included 9 males and 9 females. All participants owned a valid driver's license, with driving experience ranging from 1 to 13 years.

6.4 Results and analysis

From the original video recorded by Tobii Glasses, 12 encounter clips were extracted. We adopted a video analysis approach from Brown and Laurier's study on social interaction on the road. The actions and intentions of both the driver and other road users are listed in a time sequence. We present 3 of those encounters to discuss our main findings as follows.

Case 1. The presence of a bus being larger and heavier than a sedan, posed a potentially greater risk. Consequently, in this encounter, the driver chose to slow down (almost to a stop) immediately after seeing the bus. It can be seen from Fig. 6.1A that the participant maintained a considerable distance from the bus while interpreting the bus driver's intention. It took approximately 8 seconds for the participant to realize the bus driver was parking the bus. It was difficult because the bus was moving very slowly. Faced with the unpredictability of a large bus, it was reasonable to maintain such a long distance for safety. This case indicates that psychological factors should be taken into consideration when calculating safe distance. While physical safe distance could only protect

(A) 00:09 The participant observes a bus is ahead of the test car at a distance. Then he/she slows down (almost stop), starts to interpret.

(C) 00:18 The participant says "Let's wait for that bus parked".

(B) 00:17 The participant finally interprets that the bus driver is parking the bus.

(D) 00:19 The bus is parked. The participant continues to drive.

FIGURE 6.1

Illustration of Case 1 in time sequence.

(A) 00:06 From left rearview mirror, the participant detects a cyclist behind him/her.

(B) 00:11 The cyclist changes lane. The participant says 'good' and changes lane.

(C) 00:06–00:10 The cyclist is still in target lane. The participant sighs.

FIGURE 6.2

Illustration of Case 2 in time sequence.

the driver from physical harm, ensuring the driver's mental comfort is also equally important from a human-centered perspective.

Case 2. From Fig. 6.2A, it can be seen that the cyclist was positioned behind the test car. In this scenario, the participant could have initiated lane-changing tasks directly without considering the cyclist. However, they hesitated and waited. Despite appearing slightly frustrated at not being

78 Chapter 6 Autonomous vehicle acting like a human

able to commence the lane change, the participant still waited. Rather than disregarding the facilitator's left-turning instruction, the participant chose to avoid imposing stress on the cyclist. This decision illustrates that the participant's actions were motivated not only by safety concerns but also by consideration for others' feelings.

Case 3. Despite the participant turning on the left-turn signal to indicate their intention to change lanes, pedestrians continued walking straight without acknowledging the participant, as shown in Fig. 6.3C. Although there was no verbal or gestural communication, the pedestrians' intentions were clearly conveyed by their movements—they showed no inclination to yield to the participant. Despite pedestrians walking on the motor lane was against traffic regulations, the participant patiently waited, without showing any signs of frustration or anger. The reason could be inferred that the encounter happened on a university campus where pedestrians typically have priority over motors on the road. This indicates the significance of considering the environmental context when encountering such situations.

6.5 Discussion

Humans should be viewed as partners involved in the driving task together with AV rather than as masters served by AVs, even though humans may not need to take action in most situations. Understanding each other forms the foundation of effective teamwork. The academic community has been actively working on improving the collaboration between humans and robots, recognizing the importance of harmony between them. Whether cognitive or behavioral, human models have been developed as inputs for machines to understand human behavior. This approach could also be applied to human–AV collaboration, given that AVs are a kind of machine. Wei and Dolan

(A) 00:04 The participant notices two pedestrians on target lane.

(B) 00:08 The participant switches on left–turn signal and turns slightly left to inform the pedestrians his/her intention to change lane.

(C) 00:08–00:11 The pedestrians still walk straight, showing no intention to avoid.

(D) 00:12 The pedestrians pass. The participant starts to change lane.

FIGURE 6.3

Illustration of Case 3 in time sequence.

proposed a multilevel collaborative driving framework for AVs focusing on effectiveness and safety. However, the framework established on the consideration of effectiveness and safety does not fully address human–AV social conflicts. In our point of view, it is necessary to consider both physical safety and mental comfort from a human-centered perspective.

From the cases discussed above, it is evident that considering human psychological factors is crucial in the decision-making and action planning of AVs. Case 1 indicates the necessity of calculating psychologically safe distance from vehicles or other objects ahead. Case 2 demonstrates that a driver's decision-making process in a traffic encounter is influenced by the presence of other road users. In our experiment, participants completed lane-changing tasks while trying not to offend other road users, emphasizing the significance of AVs understanding all road users, not just the driver. By cultivating empathy towards all road users involved, AVs can adopt a more humanized approach to their actions.

In addition, it is important to recognize the impact of context on a driver's decision-making process during traffic encounters. In Case 4, for instance, the participant patiently waited for pedestrians to pass, acknowledging the higher priority of pedestrians on the road within a university campus setting. While walking on the motor lane goes against traffic regulations, it is socially acceptable in this context. Developing socially acceptable AVs requires the ability to perceive contextual cues, such as environmental factors. Moreover, studies suggest that apart from environmental context, cultural context also plays an important role. Further study is needed to investigate how cultural context would impact driver's behavior in traffic encounters.

6.6 Conclusion

This paper provides valuable insights for developing socially acceptable AVs by observing human strategies for dealing with traffic encounters. The results of our experiment indicate that AVs should be empowered with the ability to understand human partners (the driver and the other road users) as well as contextual cues. This would enable AVs to seamlessly integrate into the social fabric of the road. A limitation of this study is worth noting: the relatively small number of traffic encounters due to the interaction-limited nature of university campus contexts, despite participants being instructed to perform lane changes. Additionally, our suggestions are predicated on the assumption that human drivers would treat humanized AVs fairly akin to manned vehicles, which requires to be verified in further studies.

Further reading

B. Brown, E. Laurier, The trouble with autopilots: assisted and autonomous driving on the social road, in: Proc. 2017 CHI Conference on Human Factors in Computing Systems, ACM, Denver, 2017, pp. 416–429.

B. Brown, The social life of autonomous cars, Computer 50 (2017) 92–96.

L. Fletcher, L. Petersson, A. Zelinsky, Driver assistance systems based on vision in and out of vehicles, in: Proc. Intelligent Vehicles Symposium, IEEE, Columbus, 2003, pp. 322–327.

L. Hiatt, C. Narber, E. Bekele, S. Khemlani, G. Trafton, Human modeling for human–robot collaboration, The Int. J. Robot. Res. 36 (2017) 580–596.

T. Nordfjærn, S. Jørgensen, T. Rundmo, A cross-cultural comparison of road traffic risk perceptions, attitudes towards traffic safety and driver behaviour, J. Risk Res. 14 (2011) 657–684.

T. Özkan, T. Lajunen, J. Chliaoutakis, D. Parker, H. Summala, Cross-cultural differences in driving behaviours: a comparison of six countries, Transport. Res. Part F: Traffic Psychol. Behav. 9 (2006) 227–242.

T. Özkan, T. Lajunen, D. Parker, N. Sümer, H. Summala, Aggressive driving among British, Dutch, Finnish and Turkish drivers, Int. J. Crashworthiness. 16 (2011) 233–238.

J. Wei, J. Dolan. A multi-level collaborative driving framework for autonomous vehicles, in: RO-MAN 2009 - The 18th IEEE International Symposium on Robot and Human Interactive Communication, IEEE, Toyama, 2009, pp. 40–45.

Research on transparency design based on shared situation awareness in semiautomatic driving

CHAPTER 7

Fang You[1], Huijun Deng[1], Preben Hansen[2] and Jun Zhang[3]

[1]Car Interaction Design Lab, College of Arts and Media, Tongji University, Shanghai, P.R. China [2]Department of Computer and Systems Sciences, Stockholm University, Stockholm, Sweden [3]College of Design and Innovation, Tongji University, Shanghai, P.R. China

Chapter Outline

- 7.1 Introduction ... 82
- 7.2 Related works ... 82
  - 7.2.1 Shared situation representation ... 82
  - 7.2.2 Situation awareness ... 83
  - 7.2.3 Lyons's task model of transparency ... 83
  - 7.2.4 Relationship between transparency, performance, and trust ... 83
  - 7.2.5 Human-machine interface requirements of system transparency ... 83
  - 7.2.6 Handovers in automated driving ... 84
- 7.3 Study ... 84
  - 7.3.1 Scenario ... 84
  - 7.3.2 Framework ... 85
  - 7.3.3 Transparency design ... 86
- 7.4 Experiment ... 87
  - 7.4.1 Participants ... 88
  - 7.4.2 Prototype ... 89
  - 7.4.3 Procedure ... 90
  - 7.4.4 Apparatus and simulated environment ... 90
  - 7.4.5 Materials ... 90
- 7.5 Results ... 91
  - 7.5.1 Transparency ... 91
  - 7.5.2 Situation awareness ... 92
  - 7.5.3 Usability ... 92
  - 7.5.4 Workload ... 93
- 7.6 Discussion ... 94
- 7.7 Conclusions ... 95
- Reference ... 96
- Further reading ... 96

Human-Machine Interface for Intelligent Vehicles. DOI: https://doi.org/10.1016/B978-0-443-23606-8.00008-7
© 2024 Tongji University Press Co., Ltd. Published by Elsevier Inc. All rights are reserved, including those for text and data mining, AI training, and similar technologies.

Chapter 7 Research on transparency design

7.1 Introduction

The taxonomy of automation is essential when conducting automation research. The degree of automation can be defined as the fraction of automated functions from the overall functions of an installation or system. In an automated, or semiautomatic system, some or frequent human intervention is required. In L2 automated driving, the intelligent driving system provides driving assistance for steering wheel operation, acceleration, and deceleration operation, and other driving actions are performed by human drivers. In this process, drivers need to constantly supervise the information of the interface even if their feet are off the pedals and they are not steering. However, they must steer, brake, or accelerate as needed to maintain safety.

In L2 automated driving, the human driver cooperates with the virtual driver (intelligent driving system) to execute driving tasks together. Through the human-machine interface (HMI), the system can provide the human driver with environmental information. One way to provide drivers with information about system status is system transparency.

Lyons's research indicated that the transparency between the intelligent system and the human is a mechanism to promote the effective interaction between the human and the intelligent system. The intelligent system conveys its understanding of the current task to the driver to promote shared situation awareness. To improve the driver's understanding of the intelligent system and its perception level of the surroundings in assisted driving, this paper proposes a continuous change transparency design method according to the task stage of the scenario and compares it with the static transparency design method.

Based on the task model of transparency and situation awareness theory, we investigated which stage of situation awareness has a better transparency design effect as well as the impact of adding the situation awareness stage of transparency design on usability, driver's workload, and overall situation awareness. Thus an effective transparency design method is proposed to accurately display the current situation awareness information of the intelligent system.

7.2 Related works

Current research focuses on shared situation awareness and transparency. Related works encompass the content of shared situation awareness in HMI, the three levels of situation awareness, the three principles of the task model of transparency, the relationship between transparency level and work performance, the HMI requirements of the system transparency, and handovers in autonomous driving.

7.2.1 Shared situation representation

Christoffersen et al. [1] proposed that to promote team cooperation, the collaboration interface needs to provide a shared situation representation. This human-machine shared situation representation includes information about the machine's status, plan, objectives, and activities, as well as information about the current task status, environment, and situation. Therefore in human-machine cooperation, the automation system informs the driver of environmental information, decision-making, and subsequent behavior through the HMI, which can promote team cooperation.

7.2.2 Situation awareness

Situation awareness is a term used to describe the level of the operator's perception and understanding of the system environment. Accordingly, situation awareness is divided into three stages: SA1: perception, SA2: comprehension, and SA3: projection. The first stage (SA1: perception) involves the operator to perceive their own state, external environment, and information changes. The second stage (SA2: comprehension) is for the operator to identify, interpret, and evaluate the perceived information. The third stage (SA3: projection) is for the operator to comprehensively judge the scene attributes, states, and changes, and predict the upcoming actions or possible changes and situations.

7.2.3 Lyons's task model of transparency

Lyons summarized the transparency elements of the intelligent system in four models, which include the intention model, task model, analysis model, and environment model.

Debernard proposed three principles for the task model under automated driving conditions. According to him in autonomous mode:

1. The driver must be informed that the system will control the vehicle by following accepted driving practices and traffic laws (predictability of the behavior of the vehicle). Furthermore, the driver must be able to detect the actions (e.g., lane change) being performed by the vehicle and understand them.
2. The driver must be able to perceive the intention of the system (the maneuver it intends to carry out), why, how, and when this maneuver will be carried out. In the case of a lane change, the decision, as well as its cause, should be displayed (in pointing out, e.g., a slow vehicle in front of the ego vehicle).
3. The driver should know each maneuver that could possibly interrupt the current one. This information will help him/her avoid being surprised or frightened by what is happening.

7.2.4 Relationship between transparency, performance, and trust

Mercado et al. found that when interacting with the intelligent planning agent, the operator's performance and trust in the agent increase with the increase in the agent's transparency level. Parasuraman proposed that system transparency plays an important auxiliary role in explaining the behavior and decision-making of the intelligent system and can improve the driver's trust in the system and team performance. The degree of trust may decrease if there is inadequate transparency, and the driver's own observation of the intelligent system's behavior is insufficient to correctly understand what the vehicle is doing.

7.2.5 Human-machine interface requirements of system transparency

Beggiato et al. showed that users prefer to have information about the surrounding traffic for lane change maneuvers, current target speed with an explanation for free driving and speed limit scenarios, and information on route options, delays, and reasons for the congestion-related scenarios.

Pokam et al. defined HMI according to the information processing functions: information acquisition, information analysis, decision-making, and action execution. They found that the "information acquisition" and the "action execution" functions seem to be essential, and a level of transparency should consist of showing to the human agent what the technical agent sees and what it is doing.

7.2.6 Handovers in automated driving

In L2 automated driving, when the automated system faces a situation that is beyond its functional capabilities, the driver must take back control of the automated system immediately, a process known as handover.

During the handover process, the automated system has to transmit environmental information to the driver through the HMI, so that the driver can quickly acquire situation awareness, such as spatial awareness, identity awareness, temporal awareness, goal awareness, and system awareness.

Based on these studies on transparency and situation awareness, we selected a driving safety-related scenario. We created transparency designs for the intelligent system's comprehension of environmental information during human-computer interaction, according to Lyons's task model, and focused on the event development stage. This ensures that the driver can promptly understand the current environmental conditions and the system's analysis. We identified which stage of situation awareness has a better transparency design effect. Our proposal involves refining transparency design methods through experimental comparative analysis to enhance human-machine cooperation in improving user experience and driving safety.

7.3 Study

7.3.1 Scenario

According to Lyons's task model of transparency, we studied the mode of human-computer cooperation in security scenarios. The ego vehicle was in the L2 automated driving state and ran on a one-way three-lane road (as shown in Fig. 7.1, the ego vehicle is white, and other vehicles are yellow).

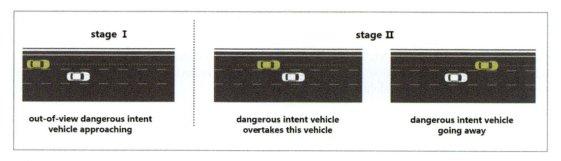

FIGURE 7.1

Driving scenario where a dangerous intent vehicle is approaching and overtaking the ego vehicle.

There was a vehicle (the yellow one) approaching in the left lane, and the vehicle had a dangerous intention to drive in the lane line. The driver of the ego vehicle needs to constantly supervise the driver support features and environment information through the interface, and they should respond accurately and promptly to the driving control switching request issued by the ego vehicle.

In L2 automated driving, the driver is primarily responsible for safety. Therefore the driver needs to take over for risk avoidance in dangerous situations. In this scenario, when the intelligent system of the ego vehicle (the white one) detected that the yellow vehicle crossed the line, the system judged that there might be a risk of collision with the ego car. It then informed the driver of the environmental monitoring information in real time so that the driver could take over the vehicle if the yellow vehicle was about to collide with the ego vehicle.

The scenario of this task (Fig. 7.1) was mainly divided into two stages, stage I: vehicles with dangerous intentions approaching out of sight, and stage II: vehicles with dangerous intentions approaching and moving away from view. In this scenario, the dangerous vehicle behind was driving on the line, and the speed was fast, which presented a potential risk of colliding with the white vehicle. Therefore the intelligent system of the ego vehicle needed to pay attention to the vehicle with dangerous intentions and inform the ego vehicle driver of the situation.

7.3.2 Framework

Combined with the theoretical background of shared situation representation, the task model of transparency and situation awareness, we analyzed the information in this interactive scenario according to the three stages of input, processing, and output. Then we made a transparency design of the environmental information in the scenario according to the situation awareness stage to compare the effects of the transparency design. We proposed a task process transparency design model based on shared situation awareness.

We hypothesized that transparency design can enhance users' perceptions of the environment, improve driving safety, and lessen workload. We further assumed that the dynamic transparency design for information can achieve greater usability and perception than the static transparency design by not only displaying the information state but also communicating the system's prediction of the state of the environment.

As shown in Fig. 7.2, the intelligent system collects data from sensors about the environment during the input stage, such as vehicle distance, vehicle speed, vehicle state, lane offset, and so on, and then analyzes the data, which is the first stage in situation awareness, the SA1 stage (perception). Then, through the transparency design, the intelligent machine provides the driver with fundamental information about their present condition, goals, intentions, and plans, assisting the driver in understanding the intelligent machine's present activities and plans.

In the process phase, the intelligent system understands and analyzes the information, which is the second stage in situation awareness, the SA2 stage (comprehension), and displays the understanding process through the transparency design to inform the driver of its understanding of the environment information and the task. In this phase, the information is status information (whether the approaching vehicle is dangerous, the degree of danger of the approaching vehicle, etc.), so that the driver can understand the actions performed by the vehicle and the intention of the system.

In the output phase, the intelligent system makes predictions based on the continuously collected information, which is the third stage in situation awareness, the SA3 stage (projection). These predictions are then displayed through the transparency design to inform the driver to avoid being surprised

Chapter 7 Research on transparency design

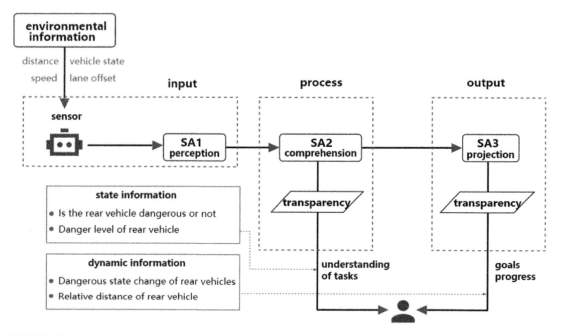

FIGURE 7.2

Task process transparency design model based on shared situation awareness.

or frightened by unexpected situations, such as vehicles automatically changing lanes. This phase involves dynamic information, such as changes in the dangerous state of rear vehicles or changes in the relative distance between rear vehicles and the ego vehicle. Thus the driver knows what environmental factors may be interrupting the current driving behavior and when to take over. This information will help drivers avoid being surprised or frightened by unexpected situations.

In the process phase of this model, the intelligent system displays the environmental state information through discontinuous changing interface elements, such as whether the rear vehicle is dangerous and the degree of danger of the rear vehicle. In the output phase, the intelligent system displays the dynamic information of the environment through continuously changing interface elements, such as changes in the dangerous state of the rear vehicle or changes in the relative distance between the rear vehicle and the ego vehicle, which causes the driver to pay attention to the change in environmental information at all times to deal with the occurrence of accidents.

7.3.3 Transparency design

According to the model, we designed three interface schemes for the scenario to verify which stage of situation awareness had a better transparency design effect, and to explore how increasing transparency design will affect the scheme's usability, driver's workload, and situation awareness.

As shown in Fig. 7.3, in the original state interface, the black vehicle in the middle lane is the ego vehicle, and the gray vehicle in the left lane is the vehicle with dangerous intentions. The gray vehicle gradually approached the ego vehicle, overtook the vehicle, and then drove out of sight.

Through this original interface, the driver could know that a vehicle passed by the ego vehicle in the left lane, but they could not know the safety information factors and potential risks identified by the intelligent system.

As shown in Fig. 7.4, in Design1, the content of the SA2 stage (understanding) of the intelligent system is displayed through transparency design, that is, the danger of vehicles coming from behind is expressed through the red color. In Design2, in addition to showing the danger information in SA2 through color, the color is also displayed with dynamic transparency design to express the content of SA3, that is, the vehicle informs the driver of the judged environmental safety state. Compared with Design2, Design3 has a more transparency design of information in the input stage (Fig. 7.4), which informs the driver of the surrounding environment as well as provides vehicle information and other basic information, so that the driver can understand the driving environment.

7.4 Experiment

We experimented with these three design schemes to assess the influence of transparency design in various situation awareness stages on the deployment of this task process transparency design

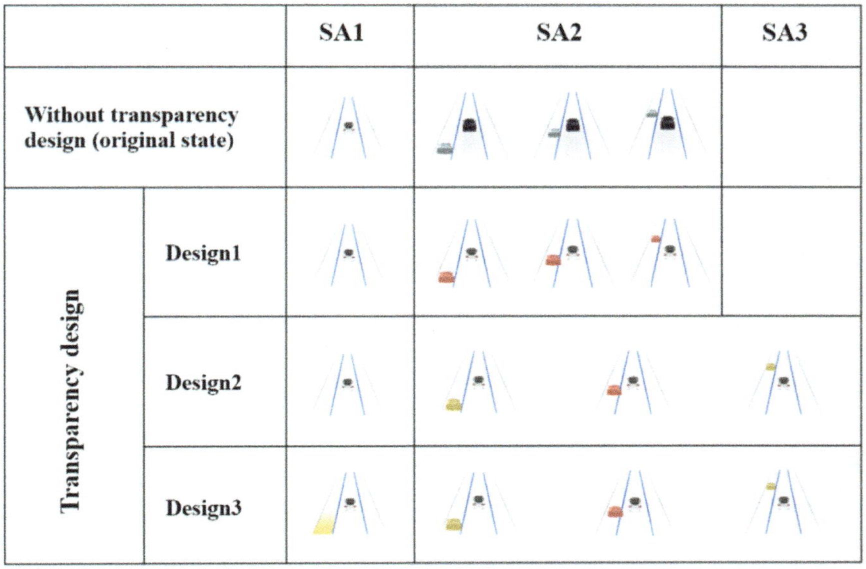

FIGURE 7.3

The original state of interface and three design schemes. The original interface can only show the change in distance between the rear vehicle and the ego vehicle. Design1 shows a change in distance and a red danger indication; Design2 shows the change in distance and the change of color indicating the danger level; and Design3 provides a forewarning of the rear vehicle based on Design2.

88 Chapter 7 Research on transparency design

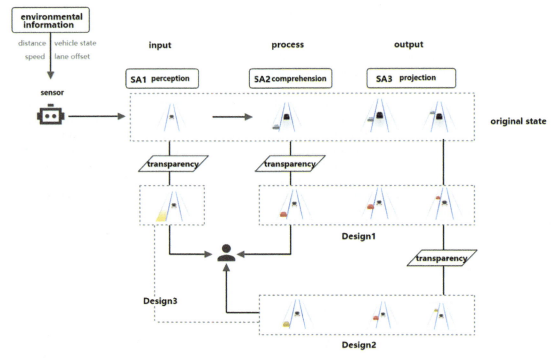

FIGURE 7.4

The analysis process of the three designs. Design1 enhances the original interface with the SA2 stage's transparency design, Design2 enhances Design1 with the SA3 stage's transparency design, and Design3 enhances Design2 with the SA1 stage's transparency design.

model based on shared situation awareness. Therefore we carried out a preliminary experimental verification. We recruited participants to drive on the driving simulator. When participants completed the specified tasks as required, we evaluated their workload, situation awareness, and the usability of the schemes through questionnaires.

7.4.1 Participants

We recruited 16 participants, 8 men, and 8 women, with an average age of 32.1 years (SD = 4.61) and an average driving experience of 7.3 years (SD = 3.11). All participants were skilled in driving, had not participated in any automated driving experiment, had not used this automated system before, had no cognitive or visual impairment, and were in good mental condition during the test. The detailed information is shown in Table 7.1.

7.4 Experiment

Table 7.1 Information of participants.

Category	Option	Proportion of total population (%)
Gender	Male	50
	Female	50
Age	25–30	37.5
	30–35	31.3
	35–40	31.3
Driving experience (year)	<5	25
	5–10	56.3
	>10	18.8
Driving frequency (day)	1–3/month	12.5
	1–3/week	6.3
	4–6/week	12.5
	Every day	68.8

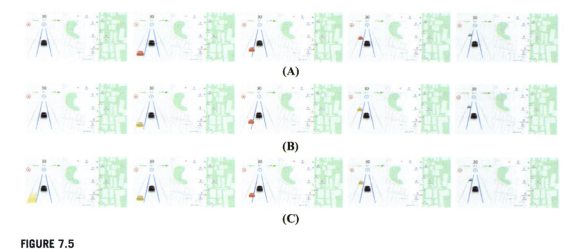

FIGURE 7.5

Experimental prototype of the three designs. (A) Design1. (B) Design2. (C) Design3.

7.4.2 Prototype

This experimental material used Tesla's interface as the prototype to carry out design and research, as shown in Fig. 7.5.

In Design1, a continuous red car was displayed on the interface to signal the danger of an approaching vehicle when a rear vehicle with dangerous intentions entered the field of view of the rearview mirror of the ego vehicle.

In Design2, the color change of the car on the interface displayed the danger level of the approaching car when a rear vehicle with dangerous intentions entered the field of vision of the rearview mirror. The danger level was then lowered from red to yellow and then to gray as this rear vehicle sped away. Design3 added the advance indication of the approach of the rear vehicle based on Design2. The intelligent system detected an approaching vehicle in the left lane and alerted the driver at the interface with yellow gradient color blocks.

7.4.3 Procedure

Before the experiment, the participants filled out the basic personal information form and the informed consent form of the experiment. The testers described the purpose and task of the experiment to the participants. After the participants understood and became familiar with the simulator, the testers wore the eye tracker for the participants and conducted calibration. After completion, the experiment officially began.

In the simulator, the vehicle was in an automated driving state, maintaining a constant speed of around 30 km/hour while moving steadily in the middle lane. This ensured that the vehicle was in a stable and safe driving state. The task started when the prompt sound ("di") started. At this time, the participant needed to observe the environmental information prompting method on the screen about the approaching vehicle with dangerous intentions. When the dangerous intent vehicle disappeared from view, the testers announced the end of the task, instructed the participant to complete the questionnaires, and then interviewed the participant about the three designs.

7.4.4 Apparatus and simulated environment

The study was conducted in a driving simulator (Fig. 7.6) that our laboratory created. The software system of this experimental platform was based on the development engine Unity. It could simulate a variety of driving scenarios, such as inner-city, highway, and suburbs. Three high-definition screens were placed in front of the driver, which displayed simulated driving scenarios that covered a 120-degree visual field of view. Participants drove in a vehicle mockup that implemented a fully equipped vehicle interface. The simulated environment of this experiment was a one-way three-lane road in the city during the day. The participants drove in the middle lane, and the experimental vehicle passed in the left lane.

7.4.5 Materials

All participants filled out questionnaires with the subjective scores of the three designs. After each experiment, we handed the three questionnaires to them. We selected the following questionnaires to measure these designs from the dimensions of availability, workload, and situation awareness.

7.4.5.1 After-scenario questionnaire

The after-scenario questionnaire (ASQ) was published by Lewis (1993). There are three items in total that measure the satisfaction of users in three aspects: task difficulty, completion efficiency, and help information. ASQ projects are scored from 1 (strongly agree) to 7 (strongly disagree). The ASQ score is the average score of the three items.

FIGURE 7.6

Driving simulator used in the study.

7.4.5.2 NASA-TLX
The driver workload after each scenario was measured with the NASA-TLX (NASA Task Load Index). This questionnaire consists of six dimensions (mental demand, physical demand, temporal demand, performance, effort, and frustration linked to completing a specific task). Participants were required to rate their perceived workload on these six Likert scales, ranging from 0 to 100.

7.4.5.3 Situation awareness global assessment technique
The situation awareness global assessment technique (SAGAT) is based on a comprehensive assessment of the operator's situation awareness needs and is used to assess all elements of situation awareness. Participants are required to answer questions about their understanding and evaluation of the environment after a single test. One point is given for each correct answer.

7.5 Results
Table 7.2 contains the experimental data.

7.5.1 Transparency
When comparing Design3 with Design2, Design3 demonstrated higher usability than Design2. From the transparency perspective, the transparency of SA1 (perception) in Design3 was higher than that in Design2, which allowed the driver to know the status information of a car coming from behind.

Table 7.2 Questionnaire results for the three designs (n = 16).

		NASA	ASQ	SAGAT
Design1	M	3.32	5.73	0.74
	SD	1.15	0.54	0.14
Design2	M	3.35	6.02	0.71
	SD	1.29	0.56	0.13
Design3	M	3.54	6.15	0.71
	SD	1.35	0.63	0.09

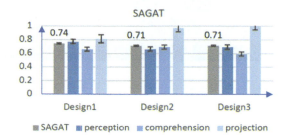

FIGURE 7.7

Scores from the situation awareness global assessment technique (SGAT) questionnaire. The first column of each design is the SAGAT score, and the next three columns are its three-dimensional scores.

When comparing Design2 with Design1, Design2 demonstrated higher usability than Design1. From the transparency perspective, Design2 had more transparency of SA3 (projection) than Design1. Through the dynamic change of color, the driver knew the dynamic change in the risk degree of the incoming vehicle.

7.5.2 Situation awareness

As shown in Fig. 7.7, Design1 enhanced participants' information perception, but their comprehension and projection were weak. Design1 performed relatively well in situation awareness and scored 0.74, while the other two designs scored 0.71. Design3 had the highest projection score since it performed better at forecasting the rear cars. We can see from comparing the "understanding" scores of Design1 and Design2 that the color change effectively assisted drivers in comprehending the situation.

7.5.3 Usability

Overall, the usability of Design3 was better (Fig. 7.8), which means that the driver was more effective in completing the task under this scheme. At the same time, Design2 scored higher in efficiency and satisfaction, indicating that users need direct and accurate information supply.

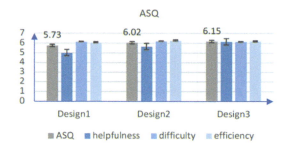

FIGURE 7.8

Scores from the after-scenario questionnaire (ASQ) questionnaire. The first column of each design is the ASQ score, and the next three columns are its three-dimensional scores.

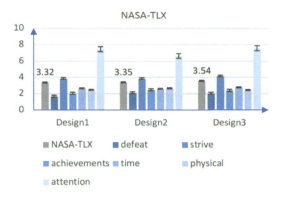

FIGURE 7.9

Scores from the NASA-TLX questionnaire. The first column of each design is the NASA-TLX score, and the next six columns are its six-dimensional scores.

7.5.4 Workload

The results of the NASA-TLX scale (Fig. 7.9) showed that although Design1 could improve participants' information perception, it also increased the attention workload.

According to the correlation analysis of the scale data of the three designs in Table 7.3, there was a significant difference between NASA-TLX and after-scenario questionnaire (ASQ) ($p = .011$), which showed a negative correlation ($r = -0.364^*$). The NASA score had a greater impact on the efficiency in ASQ ($r = -0.487^{**}$, $p = 0$).

According to the scale data (Fig. 7.8), increasing the transparency design stage greatly improved the interface's usability, especially in terms of helpfulness. The three designs showed significant differences in helpfulness ($F = 4.442$, $p = .017^*$). Among them, the helpfulness score of Design3 was higher than that of Design1, which means increasing the transparency design according to situation awareness can significantly improve the interface's usability and user satisfaction. Moreover,

Table 7.3 Results of Pearson correlation analysis.

		NASA	ASQ	SAGAT
NASA	r	1		
	p			
ASQ	r	−0.364[a]	1	
	p	.011		
SAGAT	r	−0.020	−0.056	1
	p	.890	.707	

[a] $p < .05$.

Table 7.4 Results of remarkable otherness analysis.

Helpfulness			
Projection	r		0.488**
	p*		0

Note: *$p < .05$, **$p < .01$.

through correlation analysis (Table 7.4), helpfulness was demonstrated to have a strong correlation with SA3 (projection) ($r = 0.488**$, $p = 0$). This implies that increasing the transparency design of the SA3 stage (projection) can greatly benefit users by improving their satisfaction with the interface and improving the interface's usability. Therefore the transparency design of the SA3 stage (projection) can significantly improve the usability of the interface and improve user satisfaction.

According to the interview, 12 of the 16 participants preferred Design3 (75%). In the comparison between Design1, Design2, and Design3, the participants indicated that the change of color conveyed a change in safety level (yellow indicated vigilance, red indicated danger) and attracted their attention. If the color did not change, they would have to pay long-term attention, which could cause stress, exhaustion, or neglect of the change. Comparing Design2 and Design3, the participants said that the color block reminded them in advance, attracted their attention, and provided additional environmental information. In addition to this, based on user feedback, increasing the transparency design resulted in a higher workload, possibly owing to the increased amount of information.

7.6 Discussion

The main purpose of this paper was to study the effect of interface transparency design according to the situation awareness stage and determine which situation awareness stage had a better transparency design effect.

We studied the impacts of transparency design of the situation awareness stages on usability, driver workload, and situation awareness in the scenario of dangerous vehicles approaching. According to the three stages of situation awareness and the task model of transparency, we proposed three transparency design schemes. We used the dynamic change of color to show the current environmental information, that is, to increase the driver's situation awareness in the comprehension and projection stages. The results showed that increasing the transparency design of situation awareness significantly improved the interface's usability and improved user satisfaction. Therefore Design3 outperformed Design2 and Design2 outperformed Design1. Moreover, the transparency design of the SA3 stage (projection) significantly improved the helpfulness of the prompt information on the interface, thus improving the interface's usability.

Transparency between intelligent systems and humans is a mechanism to promote effective interaction between humans and intelligent systems. Against the background of automated driving in the future, the transparency design of automobile HMI can effectively improve drivers' understanding of environmental information, thus strengthening cooperative driving. The system obtains the environmental information through the sensor, then analyzes and makes predictions, and prompts the driver through the HMI, which can broaden the driver's information perception and foresee potentially dangerous situations.

Although increasing the situation awareness stage of transparency design can significantly improve usability, the workload increases accordingly, especially in the attention aspect. This observation, combined with the analysis of the interviews, suggests that increased transparency design provides more information, demanding increased driver attention to the environment, and subsequently increasing workload. However, in more complicated situations, this may prove to be advantageous. For instance, heightened attention can prevent drivers from excessively trusting the automated driving system and disregarding crucial environmental cues, thereby averting potentially hazardous driving behaviors.

There was minimal variance in the total SAGAT scores among the three design schemes. Design3 exhibited a lower score in the comprehension dimension but a higher score in the projection dimension. This divergence may be attributed to Design3 incorporating prompt information in the stage when the rear vehicle is outside the driver's field of view, a feature absent in the first two designs. Although this increases the comprehension difficulty for the driver, it provides more premonitory cues of danger.

7.7 Conclusions

Drawing from typical security scenarios and adhering to the transparency requirements outlined in the task model, we explored the three stages of situation awareness and used three design schemes as examples to study the transparency design of human-machine team cooperation interface in semiautomated driving.

Overall, the research showed that increasing the transparency design of situation awareness can significantly improve the usability of the interface and improve user satisfaction. The transparency design of the SA3 stage (projection) can significantly improve the helpfulness of the prompt information on the interface, thus improving the interface's usability. The transparency design of shared situation awareness can accurately convey the perceived information of the intelligent system to help the driver understand the dynamic environment, which has a very important impact on human-machine cooperation. Changes in transparency will increase the workload, which may be because of the increase in information.

This experiment was carried out in the driving simulator, which was safer than a real vehicle experiment. Compared with the animation demonstration experiment, it was more immersive, more closely resembled a real driving scene, and the results obtained were more effective.

Reference

[1] K. Christoffersen, D.D. Woods, E. Salas, How to make automated systems team players, Advances in Human Performance & Cognitive Engineering Research, Volume 2, Emerald Group Publishing: Bradford, UK, 2001, pp. 1–12. Available from: https://doi.org/10.1016/S1479-3601(02)02003-9.

Further reading

V.A. Banks, K.L. Plant, N.A. Stanton, Driver error or designer error: using the perceptual cycle model to explore the circumstances surrounding the fatal Tesla crash on 7th May 2016, Saf. Sci. 108 (2018) 278–285. Available from: https://doi.org/10.1016/j.ssci.2017.12.023.

M. Beggiato, K. Schleinitz, J. Krems, What would drivers like to know during automated driving? Information needs at different levels of automation, in: Proceedings of the 7th Conference on Driver Assistance, Munich, Germany, 25–26 November 2015. <https://doi.org/10.13140/RG0.2.1.2462.6007>.

D.A. Drexler, Á. Takács, T.D. Nagy, T. Haidegger, Handover process of autonomous vehicles—technology and applica-tion challenges, Acta Polytech. Hung. 16 (2019) 235–255.

M.R. Endsley, Direct measurement of situation awareness: validity and use of SAGAT, Situation Awareness Analysis and Measurement, Lawrence Erlbaum Associates Publishers: Mahwah, NJ, USA, 2000. Available from: https://doi.org/10.4324/9781315087924-9.

S.G. Hart, L.E. Staveland, Development of NASA-TLX (task load index): results of empirical and theoretical research, Adv. Psychol. 52 (1988) 139–183. Available from: https://doi.org/10.1016/S0166-4115(08)62386-9.

J.D. Lee, K.A. See, Trust in automation: designing for appropriate reliance, Hum. Factors 46 (2004) 50–80. Available from: https://doi.org/10.1518/hfes.46.1.50_30392.

J.R. Lewis, Psychometric evaluation of an after-scenario questionnaire for computer usability studies: the ASQ, ACM SIGCHI Bull. 23 (1991) 78–81. Available from: https://doi.org/10.1145/122672.122692.

Y. Liu, J. Wang, W. Wang, X. Zhang, Experiment research on HMI Usability test environment based on driving simulator, Trans. Beijing Inst. Technol. 40 (2020) 949–955. Available from: https://doi.org/10.15918/j.tbit1001-0645.2019.176.

J.B. Lyons, Being transparent about transparency: a model for human-robot interaction, in: Proceedings of the AAAI Spring Symposium: Trust and Autonomous System, Palo Alto, CA, USA, 25–27 March 2013, pp. 189–204.

J.E. Mercado, M.A. Rupp, J.Y. Chen, M.J. Barnes, D.J. Barber, K. Procci, Intelligent agent transparency in human-agent teaming for multi-UxV management, Hum. Factors Ergon. Soc. 58 (2016) 401–415. Available from: https://doi.org/10.1177/0018720815621206.

S.Y. Nof, Automation: what it means to us around the world, in: S. Nof (Ed.), Springer Handbook of Automation, Springer: Berlin/Heidelberg, Germany, 2009, pp. 13–52. Available from: https://doi.org/10.1007/978-3-540-78831-7-3.

Further reading

R. Parasuraman, T.B. Sheridan, C.D. Wickens, A model for types and levels of human interaction with automation, IEEE Trans. Syst. Man. Cybern.-Part A Syst. Hum. 30 (2000) 286−297. Available from: https://doi.org/10.1109/3468.844354.

R. Pokam, S. Debernard, C. Chauvin, S. Langlois, Principles of transparency for autonomous vehicles: first results of an ex-periment with an augmented reality human−machine interface, Cogn. Technol. Work. 21 (2019) 643−656. Available from: https://doi.org/10.1007/s10111-019-00552-9.

CHAPTER 8

Human-machine interface design based on transparency in autonomous driving scenes

Jianmin Wang[1], Qiaofeng Wang[1] and Jun Zhang[2]

[1]Car Interaction Design Lab, College of Arts and Media, Tongji University, Shanghai, P.R. China
[2]College of Design and Innovation, Tongji University, Shanghai, P.R. China

Chapter Outline

8.1 Introduction 99
8.2 Theoretical basis 100
    8.2.1 Situational awareness and transparency 100
    8.2.2 Trust 102
    8.2.3 Human-machine interface design case study 102
8.3 Human-machine interface design 104
    8.3.1 Scenario analysis 104
    8.3.2 Interface element design 104
8.4 Experimental evaluation 108
    8.4.1 Participants 108
    8.4.2 Experimental environment 108
    8.4.3 Experimental process 109
8.5 Analysis of experimental results 110
    8.5.1 Scale data 110
    8.5.2 Correlation analysis 111
    8.5.3 Interview data 112
    8.5.4 Experimental conclusions 112
8.6 Conclusion 114
References 114
Further reading 114

8.1 Introduction

With the development of autonomous driving technology, smart cars equipped with advanced driving assistance systems (ADAS) have been industrialized [1]. Automation brings about three changes in human interaction: changes in feedback, changes in tasks and task structure, and changes in the operator's cognitive and emotional response to automation. These changes have introduced new challenges in the interaction between human drivers and ADAS technology. In traditional manual driving, the driver is required to complete all driving tasks. The driver will make driving decisions based on his own perception of the environment. However, in automatic driving, when the partially autonomous driving function is turned on (such as L2 and L3), ADAS handles some or all of the control and cognitive tasks. However, the process and results of this control and cognition necessitate human supervision. Thus the driver is responsible for supervising the vehicle's operation and taking over when necessary.

This level of attention can be achieved through human-machine interaction (HMI), making the design of interaction between humans and automation or intelligent systems the foremost concern. Many factors must be considered when designing information in HMIs, such as situational awareness, workload, trust, and so on. For human drivers to better understand machine cognition and behavior, HMI interface information needs to convey more accurate machine cognitive information to drivers. Appropriate transparency design of information has become the focus of interface design for intelligent systems, especially in self-driving vehicles.

Extensive research is necessary to explore how to incorporate transparency into interface design in autonomous vehicles. Based on the above background, this paper takes the incoming car as the research scenario and designs the information in the central control screen with different transparency levels. It is expected that these designs can effectively improve the driver's perception of the driving environment and influence the human driver's perception of automation and trust in the system.

The first section of this chapter provides an overview of important concepts such as situational awareness, transparency, and trust, as well as desktop research on autonomous vehicle interface design by important domestic car manufacturers. The second section conducts a cognitive needs analysis focusing on scenarios involving a car approaching from behind in autonomous driving, aimed at identifying the driver's information requirements. In the third section, the interface design process is undertaken based on insights from the Tesla Model 3 central control interface combined with the transparency (situational awareness-based agent transparency, SAT) model. Finally, this leads to the formulation of three levels of transparency in HMI design solutions.

8.2 Theoretical basis

8.2.1 Situational awareness and transparency

Situational awareness refers to a person's ability to perceive changes in the surrounding environment: what is happening currently, what it implies, and what to do. According to the definition of Endsley, it is divided into three levels: SA1 is the individual's perception of task-related elements in the surrounding environment; SA2 involves the interpretation of data at the SA1 level, which helps the individual understand the current task and understanding of goals; SA3 involves the prediction of the future state of elements in the system or environment. Situational awareness can be held by humans or by intelligent systems. The ADAS system on modern vehicles is an intelligent system with situational awareness capabilities. Such systems are transforming into autonomous agents with a high degree of autonomy (L3–L5 level autonomous driving). Before the realization of advanced autonomous driving, the current autonomous driving function was a multiagent (human and ADAS) joint driving activity, which implies that the driving task is completed by the human driver and ADAS. Therefore humans and ADAS work as a team to control the vehicle and sense the environment.

In the process of communication, achieving mutual understanding between parties depends on effectively conveying three levels of situational awareness information. Transparency emerges as a viable solution in this endeavor. Transparency refers to the ability of the interface to inform the operator of the purpose of the system and reason about future plans, thereby enhancing people's understanding and trust in the machine. Better transparency design should have a positive impact on drivers in driving scenarios because transparency plays an important supporting role in explaining the behavior and decision-making of intelligent systems. It improves operator trust in intelligent systems and team performance [2].

Using Endsley's situational awareness model, Chen and colleagues developed a SAT model that displays situational awareness information in layers to facilitate understanding for human partners. The SAT model consists of three separate layers, each representing information that needs to be conveyed when interacting with the driver. Among them, the information at the first level (SAT 1) is mainly used to help the driver perceive the current actions and plans of the system; the information at the second level (SAT 2) is used for the driver to better understand the current behavior of the system. Three levels (SAT 3) of information are used for driver predictions for future outcomes. When the driver turns on autonomous driving, ADAS monitors the surrounding environment at all times and transmits these three levels of situational awareness information to the human driver through the HMI interface. The human driver will use the information on the HMI interface and the information he or she gets from the environment. Information is used to comprehensively determine whether the vehicle needs to be taken over. The process is shown in Fig. 8.1.

Based on the above theoretical basis, this paper selects the scene of a car approaching from behind in autonomous driving. In this context, to satisfy the human driver's understanding of the machine's cognitive situation, the research team used the SAT model to transform the agent's situational awareness information about the car approaching from behind into HMI designs with

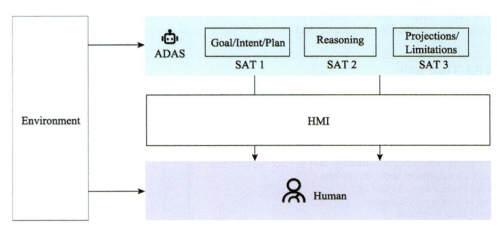

FIGURE 8.1

Transparency model based on situational awareness in autonomous driving.

different levels of transparency. The study investigates the impact of transparency interfaces in intelligent systems on human-machine collaboration.

8.2.2 Trust

Successful human-machine collaboration requires the driver to maintain a moderate level of trust in the automated system through trust calibration. If the driver does not have trust in the automated system, then even if the automation performance is good, the driver will not use it. On the contrary, if the driver has too much trust in the automated system, he will use it incorrectly due to over-reliance on the automated system. Low trust or too much trust can lead to accidents. HMI therefore plays a potential role in guiding drivers to acquire their appropriate trust in real driving conditions, which is crucial to ensuring driving safety and comfort.

Currently, there are two ways to measure trust. One is based on the vehicle automation trust model proposed by Muir [3], which divides trust into three dimensions: predictability, reliability, and loyalty. Predictability implies that the driver can predict the output of the system well, reliability implies that the driver can confidently let the automated system participate in the task and rely on the system, and loyalty implies that the driver uses automation After the system has completed its task, you want to continue using the system. The other is the trust scale developed by Jian et al. [4] which contains 12 potential factors influencing trust between people and automated systems (see Table 8.1).

Research by Chen and his team [5] has shown that people's trust in the agent is related to the transparency of the interface. The higher the level of transparency of the interface, the higher the person's trust in the system. However, there is no further description of how transparency affects trust. This paper provides additional insights by exploring the relationship between transparency and trust, addressing the gap in existing literature.

8.2.3 Human-machine interface design case study

In recent years, autonomous driving technology has developed rapidly. Many domestic and foreign mainstream car companies have released mass-produced vehicles equipped with assisted driving

Table 8.1 Trust scale.

Question number and description	1	2	3	4	5	6	6	8	9
1. The system is confusing									
2. The system operates in a covert manner									
3. I doubt the system's intentions, actions, or outputs									
4. I am wary of the system									
5. The system's behavior is harmful or will produce harmful results									
6. I feel very confident about this system									
7. This system makes me feel safe									
8. This system is very complete									
9. The system is reliable									
10. The system is trustworthy									
11. I completely believe in the system									
12. I am familiar with the system									

8.2 Theoretical basis

functions. They have their own HMIs for the typical scenario of a "car approaching from behind" in autonomous driving. plan. NIO ES8, Xiaopeng P7, and IM L7 have added information to display the current vehicle status on the instrument screen. In the HMI design, eye-catching car models are used to represent the vehicle, and gray car models are used to display surrounding vehicles and pedestrians. To a certain extent, it helps the driver to control the current vehicle's driving status without turning his head to observe the left and right rearview mirrors, but this solution still has room for optimization. Take the scene of a car approaching from behind, as an example. First, the car model does not visually distinguish between before and after overtaking. If there is a car approaching from behind, the driver may not be able to quickly perceive the vehicle behind. Second, if the car is approaching from behind, in a multilane driving environment where vehicles are approaching from both sides with the intention of merging, the automatic driving system adopts the braking action. However, the driver is unable to understand the emergency priority of the two vehicles because of the inability to make a judgment on the reason why the system made this decision. Finally, because other types of information, such as navigation and music cards, are also designed on the instrument panel, the driver will focus on this information when he cannot predict unknown dangers.

To solve the above problems, HMI solutions can increase the visual level of information from the perspective of transparency design, thereby improving the driver's ability to perceive information, helping them understand system behavior, and predicting possible future dangers. Table 8.2

Table 8.2 Summary of human-machine interface research for autonomous vehicles.

Vehicle name	HMI solution	Problems	Solution direction
Tesla Model 3		Instrument information accounts for a smaller proportion compared with navigation, and the sense of presence is slightly lower. It is difficult for the driver to detect information about cars approaching from behind	Visually attract the driver's attention when a car approaches from behind and improve the driver's ability to perceive the information of the car approaching from behind
NIO ES8		HMI cannot visually understand whether surrounding vehicles pose a threat to the vehicle	Keep the driver's visual focus on dangerous vehicles and help them understand system behavior
IM L7		The location of this car is at the bottom of the screen, and information about cars approaching from behind appears later	Display in advance when the oncoming vehicle from behind has not yet approached the vehicle
Xiaopeng P7		There is a lot of other information about the vehicle, which easily attracts the driver's attention. By the time the driver notices the danger, it is already too late.	Inform the driver of the danger information predicted by the system in advance

summarizes the four mass-produced vehicles currently equipped with assisted driving systems and summarizes their HMI design solutions in the scenario of vehicles approaching from behind. The problems and solutions are summarized.

8.3 Human-machine interface design

8.3.1 Scenario analysis

In this chapter, the vehicle is in autonomous driving mode. At this time, a car approaches from behind and swiftly overtakes the autonomous vehicle. The scene is shown in Fig. 8.3. The driving scene can be divided into three stages. The first stage is when the distance between the oncoming car and the autonomous vehicle is greater than 10 m (before overtaking). The second stage is when the distance between the oncoming car and the autonomous vehicle is 0–10 m (during overtaking), and the third stage is when the vehicle approaching from behind overtakes the vehicle (after overtaking), where (1) represents the vehicle traveling at a constant speed, and (2) represents the vehicle coming from behind, traveling at a constant speed faster than the vehicle.

There are often strong interactions between vehicles traveling on the road. It can be observed in daily traffic that in the first stage of overtaking, the driver often observes the driving environment behind through the left and right rearview mirrors. Sometimes the driver will slow down slightly or increase the lateral distance between the vehicles to smooth the way for the vehicle behind. Overtaking is accomplished, but in autonomous mode, the driver may take over the vehicle out of fear that the system will not be able to understand and perform this behavior. As shown in Fig. 8.4, the situational awareness information requirements at different stages during the entire overtaking process and the transparency levels of the situational awareness are described. First, when the rear vehicle is far away from the vehicle, the HMI displays SAT level 3 prediction information to warn the driver of vehicles behind or with dangerous intentions. When the driver's attention is drawn, the HMI displays SAT level 1 perception information, indicating the relative position of the rear vehicle to the autonomous vehicle. Second, when the position of the rear vehicle and the autonomous vehicle are closer, the HMI needs to inform the driver of the ADAS' understanding of whether the current driving situation is dangerous, while continuing to display the SAT level 1 perception information. This constitutes level 2 information. Finally, when the vehicle behind successfully overtakes, the HMI displays SAT level 1 perception, which is used to inform the driver of the current vehicle condition (Figs. 8.2 and 8.3).

8.3.2 Interface element design

The design of this chapter is based on the central control interface of Tesla Model 3. The design goal is to enhance the driver's awareness of events through the central control interface: Does the driver know what is happening behind him? Is the prompt information outside the Tesla screen range sufficient? How do you prompt the driver to perceive the driving environment? Furthermore it seeks to enhance users' trust in the information conveyed through the HMI.

Based on the above analysis, the interface elements were designed for the SAT level. The SAT 3 prediction information before overtaking is represented by a box with a yellow gradient.

8.3 Human-machine interface design   105

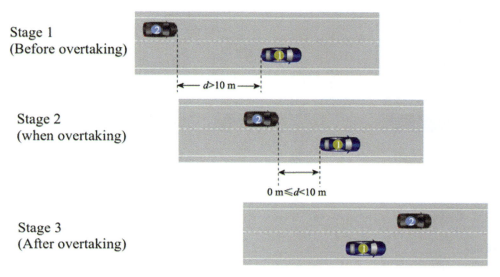

FIGURE 8.2

Front, middle, and backstages in the scene of a vehicle approaching from behind.

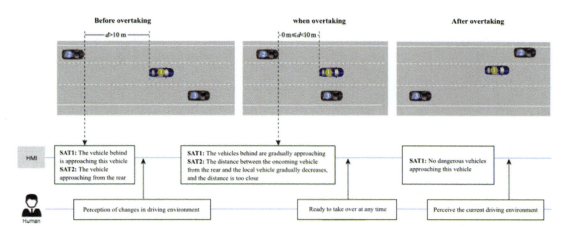

FIGURE 8.3

Interactive scenes based on transparency during overtaking.

For SAT 1 perception information, the car model is used to map the actual vehicle situation in real time; for SAT 2 understanding information during the overtaking process, yellow and red are used to display the degree of danger; for perception information after overtaking, a gray car model is used to map the actual vehicle condition in real time, as shown in Fig. 8.4.

106 Chapter 8 Human-machine interface design based on transparency

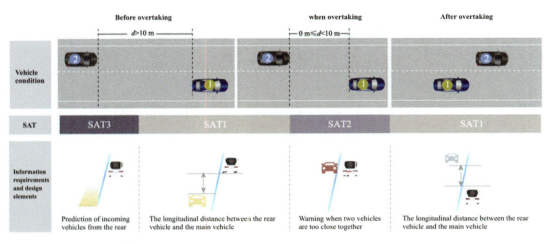

FIGURE 8.4

Information requirements and interface design elements based on transparency.

Table 8.3 Human-machine interface design scheme.

	SAT 1 (goals/intentions/plans)	SAT 2 (understand)	SAT 3 (Prejudgment)
HMI 0	N	N	N
HMI 1	Y	N	N
HMI 2	Y	Y	N
HMI 3	Y	Y	Y

Combining the three levels of the SAT model, three HMIs are defined. The information in each HMI reflects the information that needs to be transmitted at different SAT levels, as shown in Table 8.3, where HMI 0 is the Model 3 user interface, used as contrast, "N" implies that the information of a certain level is not displayed, and "Y" implies that the information of this level is displayed.

HMI 0 does not have any level of information in the SAT model. HMI 1 only displays information related to SAT 1, using the position of the vehicle behind the yellow car relative to the autonomous vehicle, as shown in Fig. 8.5; HMI 2 displays the information of SAT 1 and SAT 2. In the original scheme, the colors yellow, red, and gray are used to indicate the distance and degree of danger between the rear vehicle and the autonomous vehicle. When the rear vehicle is too close to the autonomous vehicle and is potentially dangerous, a red warning is used, as shown in Fig. 8.6; HMI 3 displays the information of SAT 1, SAT 2, and SAT 3. Based on HMI 2, it uses yellow color blocks to display the prediction information of the vehicle coming from behind (see Fig. 8.7).

8.3 Human-machine interface design   107

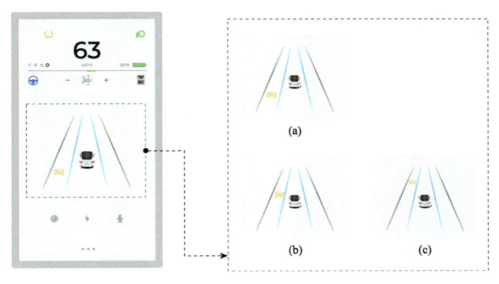

FIGURE 8.5

Human-machine interface 1 (displays SAT 1 information only).

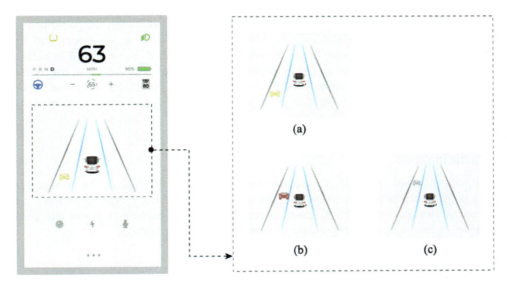

FIGURE 8.6

Human-machine interface 2 (displays SAT 1 + SAT 2 information).

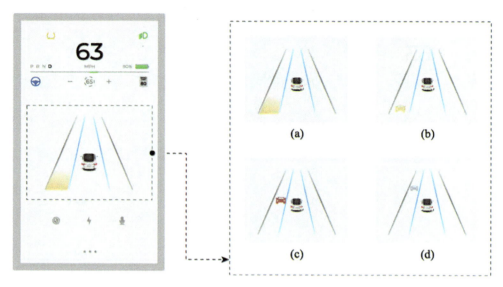

FIGURE 8.7

Human-machine interface 3 (displays SAT 1 + SAT 2 + SAT 3 information).

8.4 Experimental evaluation

A relatively real driving environment was built using a driving simulator; 16 participants were invited to evaluate these three HMIs using the HMI of Tesla Model 3 as a control (HMI 0). The HMIs were displayed in Latin square order, alternating between HMI 2 and HMI 3 to balance the learning effect. The subjective data uses the currently more common SAGAT scale, trust scale, and NASA-TLX scale to measure the impact of different HMI solutions on driver situational awareness, trust, and workload. At the same time, after each drive, drivers are let to evaluate the overall experience of the HMI solution.

8.4.1 Participants

The experiment involved 16 participants, comprising 8 males and 8 females. The average age was 32.1 years (with a standard deviation of 4.61), and the average driving experience was 7.3 years (with a standard deviation of 3.11). All participants held at least a bachelor's degree and possessed a certain level of knowledge and experience in autonomous driving systems.

8.4.2 Experimental environment

The experiment was completed on the driving simulator, as shown in Fig. 8.8. The test road was modeled in Unity. The lane width, lane line size, and lane line category were all national standard

FIGURE 8.8

Experimental environment.

sizes to maximize the authenticity of the driving scene in the driving simulation. In this experiment, a straight one-way two-lane road on a flat road was selected, and all simulated driving test tasks were completed on this road. In addition, a 10.8-inch (1 inch $\approx$ 3.33 cm) tablet was used to display the central control screen HMI interface involved in the task, and a 15.6-inch monitor was used to display the scene changes in the rearview mirror during the task.

8.4.3 Experimental process

Before commencing the experimental process, participants were briefed on the sequence of activities and the objectives of the study.

1. Before the test, participants filled out a personal information questionnaire, providing details such as name, age, driving experience, education, occupation, driving frequency, and so on.

2. Driving training was conducted, during which participants completed a 5-minute test drive on the simulator. They were familiarized with some basic simulator operations, such as turning on autonomous driving, taking over the vehicle, and so on. This step aimed to make participants accustomed to the simulator environment.

3. While conducting driving training, participants were instructed to turn on autonomous driving during the entire test and observe the surrounding driving environment using the central control HMI, as well as the left and right rearview mirrors. Participants were encouraged to monitor the autonomous driving system based on their information requirements.

4. During the test, each participant completed four driving sessions, each time presenting a different HMI. Since HMI 0 is a blank control, HMI 0 appeared on the central control screen during the first drive. For the next three drives, HMI 1, HMI 2, and HMI 3 were presented to participants in a Latin square order. The experiment lasted approximately 1 hour/participant.

5. After each drive, participants were required to fill out a questionnaire on trust, situational awareness, and workload. These responses were used to measure the impact of different levels of HMI transparency on driver trust in the system, driver awareness of their surroundings, and workload.

8.5 Analysis of experimental results
8.5.1 Scale data
8.5.1.1 Trust

When the screen displays HMI 0, the trust scale score is used as a control to study whether the transparency-based design will affect the driver's trust in the system. When the screen displays HMI 1, HMI 2, and HMI 3, the trust scale score is used to study how the three levels in the transparency model affect driver trust in the system. The results are shown in Fig. 8.9. The results show that when HMI 0 is on the screen, the driver's trust score in the system is 3.89 (SD value is 0.35), and when HMI 1 is on the screen, the driver's trust score in the system is 4.72 (SD value is 1.01), when HMI 2 is on the screen, the driver's trust in the system score is 4.99 (SD value is 0.74), and when HMI 3 is on the screen, the driver's trust in the system score is 5.14 (SD value is 0.74). It can be seen that if the HMI does not display any transparency information, the driver has the lowest trust in the system, and when the HMI displays SAT 1, SAT 2, and SAT 3 related information at the same time, the driver has the highest trust in the system.

8.5.1.2 Workload

Since HMIs with different levels of transparency have different amounts of information, this implies that it is also necessary to explore whether the increase in the amount of information brings a greater workload to the driver. This is measured using the NASA-TLX (NASA Task Load Index) scale. As can be seen from Table 8.4, when the screen displays HMI 0, the driver's perceived workload is the largest. This is because HMI 0 does not visually distinguish information, and the driver cannot quickly obtain the information that they want in a short time. When HMI 1, HMI 2, and HMI 3 are displayed on the screen, the driver's perceived workload increases with the increase in information, but there is no significant difference in the results.

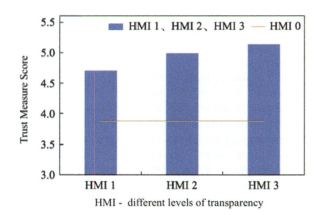

FIGURE 8.9

Trust scale score.

Table 8.4 NASA-TLX scale scores.

HMI	M	SD
HMI 0	3.67	1.19
HMI 1	3.32	1.15
HMI 2	3.35	1.29
HMI 3	3.54	1.35

8.5.2 Correlation analysis

8.5.2.1 Trust and situational awareness

The trust scale score above shows that HMI 3 has the highest trust level. Is this result related to the driver's situational awareness level? Since HMI 0 does not display information related to perception, understanding, and prediction, the analysis of the driver's situational awareness level is mainly based on HMI 1, HMI 2, and HMI 3, and the SAGAT scores of each SAT level under these three conditions are compared. The results show that when HMI 1 is displayed on the central control screen, the user's ability to perceive information is stronger than other solutions. When HMI 2 is displayed, the driver focuses on understanding the current driving behavior of the system, and his perception level is slightly reduced. When HMI 3 is displayed, the driver's perception level is slightly lower, the driver has the strongest ability to predict upcoming situations, and at the same time, the level of perception and understanding has improved. The correlation calculation was performed between trust and the three factors that affect situational awareness: perception, understanding, and prediction. The results are shown in Table 8.5.

It can be seen from the three levels of situational awareness that trust is significantly related to understanding information on the interface. It can be considered that in the current task, if HMI can help users obtain better situational awareness, especially when providing understanding information, users will have higher trust in the system.

8.5.2.2 Trust and workload

Similarly, to explore whether the workload will affect users' trust as transparency information increases, the correlation calculation between the total score of the trust scale, the total score of the NASA-TLX scale, and the scores of each option was calculated. The results are shown in Table 8.6.

Although there is no significant difference in workload between the three solutions, there is a correlation between workload and trust, indicating that when users face the information presented on the HMI, the greater the workload, the greater the trust in the system. Conversely, a lower workload corresponds to reduced trust. This correlation indicates a link between the user's attention occupied by the information displayed on the HMI and the degree of effort and trust the user puts into completing the task. Therefore when designing for transparency, to prevent users from reducing their trust in the system, design elements should aim to minimize the occupation of users' attention, such as the use of eye-catching colors, and so on, while reducing the user's effort required for task completion.

Table 8.5 Correlation analysis between trust and situational awareness.

Trust	Perception	Understanding	Prejudgment
r	0.004	0.371*	0.223
p	.98	.028	.128

*$p<.05$; **$p<.01$.

Table 8.6 Correlation analysis between trust and workload.

Trust	NASA-TLX	Attention	Effort level
r	−0.325*	−0.389**	−0.317*
p	.024	.006	.028

*$p<.05$; **$p<.01$.

8.5.3 Interview data

The main difference between these three design solutions lies in the different design elements. To explore which design elements affect trust, the participants were interviewed after the experiment. Interviews related to trust mainly start from three perspectives, namely, predictability, reliability, and loyalty. The interview questions are shown in Table 8.7, and the interview results are shown in Fig. 8.10. From the interview, it can be found that the reason why the trust score is low when the screen displays HMI 0 is because the driver cannot determine the distance between the rear vehicle and the autonomous vehicle and the speed of the rear vehicle when overtaking. There are two main reasons why participants have the highest trust in the information displayed by HMI 3. One is that the solution uses different colors to distinguish the dangerous levels of vehicle conditions, and the other is the use of yellow in the driver's blind spot. Blocks add predictive information that helps drivers determine whether the system has the ability to handle changes in vehicle conditions. HMI 2 only incorporates the second reason, which refers to the use of yellow blocks in the driver's blind spot to provide predictive information, while HMI 1 does not use colors to distinguish hazard levels. This is the reason why the trust scores of HMI 1 and HMI 2 are not high.

8.5.4 Experimental conclusions

This paper evaluates HMIs with different levels of transparency from three dimensions: situational awareness, trust, and workload. Experimental data shows that when the level of transparency is higher, it is predicted in advance that there is a car approaching from behind the driver, and different colors are used to distinguish the position of the approaching car relative to the autonomous vehicle, which greatly increases the driver's perception ability and the driver's awareness of automation. Trust in the system is also higher. In addition, the increased amount of information did not

8.5 Analysis of experimental results

Table 8.7 List of interview questions.

Interview angle	Interview questions
Predictability	In the test plan, when the vehicle condition changes, what kind of information displayed by the HMI can help you judge in advance that the autonomous driving system can handle the change?
Reliability	In the test scenario, what design elements make you believe that the autonomous driving system is reliable?
Loyalty	In the test scenario, what design elements can make you trust and motivate you to continue using the autonomous driving system?

Interview perspective	HMI Part 0 Interview Content	Interview content for HMI 1, HMI 2, and HMI 3 parts			
Predictability	"I don't really believe it, I feel it's not very useful. Even if the car's condition changes, there is no change on the screen, so I still need to observe it myself." ——Participant 1	"I see yellow blocks, and I can basically believe that this system can detect cars coming from behind." ——Participant 8	"The entire process, from the appearance of yellow blocks to the car overtaking me, was very clear and made me feel at ease." ——Participant 15	"From the color change of the car, it can be guessed that yellow is not dangerous, while red is a bit dangerous. When I saw the yellow color block, I thought there was no danger from cars coming from behind." ——Participant 16	
Reliability	"If you could tell me through the screen how fast the vehicle behind me is and how long it will take to overtake, I would have some confidence in my mind, and I could even decide whether to let it overtake or not." ——Participant 9	"I see that the distance between the car on the central control screen and me is the same as the real situation, so there is no doubt." ——Participant 4	"I think it's quite reliable. When the car next to me was closer, the small car on the central control screen turned red." ——Participant 10	"When the car next to me was overtaking, I was driving autonomous driving. When the car's condition was relatively normal, the car did not engage in autonomous driving or other behaviors, which made me feel more reliable." ——Participant 10	
Loyalty	"Actually, I wouldn't look at what's displayed on this screen when the car's condition is quite dangerous. When it's more dangerous, I don't dare to use autonomous driving, I'm afraid of accidents, and I don't know how intelligent this system can be." ——Participant 9	"The car on the central control uses colors to tell me if there is a dangerous situation at present. I think it is a good partner for me to drive, and I will make some judgments based on the information on the central control screen." ——Participant 1	"I saw the small car on the central control screen turn red, and I felt that the car next to me was closer, and then the color turned yellow again. I knew that this dangerous situation had disappeared, so I could safely continue to activate the automatic crossbow." ——Participant 2	"When the car coming from behind is quite far away from me, I can't see clearly, but the color block on the central control screen gives me a sense of prediction, which fills my blind spot." ——Participant 6	"This yellow color block makes me feel like when I start autonomous driving, I don't have to drive it myself because the car condition is under my control." ——Participant 12

FIGURE 8.10

Interview results.

have a significant impact on the driver's workload. Among the three HMIs used in this experiment, the following conclusions can also be drawn.

1. The level of situational awareness is positively related to the user's trust in the autonomous driving system, especially in understanding information. It can be considered that when users turn on autonomous driving, they pay more attention to the system's understanding of the current driving status displayed on the HMI, which shows that the system makes reasons for decisions.
2. There is a negative correlation between workload and user trust in the autonomous driving system. When the transparency level of HMI is increased, the occupation of user attention by interface transparency information and the degree of effort users put into completing tasks will reduce their trust in the system. When users turn on autonomous driving, they

focus on the road conditions and pay less attention to the information prompts on the interface. If the current decision-making of the autonomous driving system is consistent with the user's own understanding of the road conditions, the user will trust the system. Additionally, systems that users trust are intelligent and do not require users to expend significant effort to complete tasks.

8.6 Conclusion

This paper mainly studies the HMI design based on transparency in autonomous driving and introduces the SAT model developed by Chen into the autonomous driving scenario. It is found that the higher the transparency of the HMI, the stronger the user's situational awareness of the current environment, and the higher the trust in the system. After concluding that transparency will affect users' situational awareness and trust in the system, it goes beyond this and conducts a more detailed analysis of the correlation between situational awareness, workload, and trust, providing a basis for future research. Enhancing driver trust provides an idea for targeted transparency design.

References

[1] J.D. Lee, B.D. Seppelt, Human factors in automation design, Springer Handbook of Automation, Springer, Berlin Heidelberg, 2009, pp. 417–436.
[2] M.R. Endsley. Toward a theory of situation awareness in dynamic systems, Hum. Factors 37 (1995) 85–104.
[3] B.M. Muir, Trust in automation: Part I. Theoretical issues in the study of trust and human intervention in automated systems, Ergonomics 37 (11) (1994) 1905–1922.
[4] J.Y. Jian, A.M. Bisantz, C.G. Drury. Foundations for an empirically determined scale of trust in automated systems, Int. J. Cogn. Ergon. 4 (1) (2000) 53–71.
[5] J.Y.C. Chen, M.J. Barnes, Agent transparency for human-agent teaming effectiveness, in: 2015 IEEE International Conference on Systems, Man, and Cybernetics, IEEE, 2015, pp. 1381–1385.

Further reading

M.R. Endsley, From here to autonomy: lessons learned from human-automation research, J. Hum. Factors Soc. 59 (1) (2017) 5–27.
K. Hosono, A. Maki, Y. Watanabe, et al., Implementation and evaluation of load balancing mechanism with multiple edge server cooperation for dynamic map system, IEEE Trans. Intell. TransportatiSyst. (2021) 1–11. 99.
Internationals, Taxonomy and definitions for terms related to on-road motor vehicle automated driving systems [J3016_201401].
K.R. Johannsdottir, C.M. Herdman, The role of working memory in supporting drivers' situation awareness for surrounding traffic, Hum. Factors 52 (6) (2010) 663.
C. McKeown, Designing for situation awareness: an approach to user-centered design, Ergonomics Qual. Manag. J. 24 (2) (2013) 56. Available from: https://doi.org/10.1080/10686967.2017.11918512.

CHAPTER 9

Drivers' trust of the human-machine interface of an adaptive cruise control system

Jianmin Wang[1], Wenjuan Wang[1], Xiaomeng Li[2] and Fang You[1]

[1]Car Interaction Design Lab, College of Arts and Media, Tongji University, Shanghai, P.R. China [2]Centre for Accident Research and Road Safety-Queensland (CARRS-Q), Institute of Health and Biomedical Innovation (IHBI), Queensland University of Technology (QUT), Kelvin Grove, Brisbane, Australia

Chapter Outline

9.1 Introduction 115
9.2 Research question 116
9.3 Method 116
    9.3.1 Participants 116
    9.3.2 Experimental design 116
    9.3.3 Experimental procedure 117
9.4 Results 119
9.5 Limitations 120
9.6 Conclusion and discussion 120
Further reading 121

9.1 Introduction

To enhance driving comfort and improve safety, significant efforts have been made to develop advanced driver-assistance systems (ADAS), with adaptive cruise control (ACC) emerging as one of the most widely adopted in recent decades. As a semiautomatic system, ACC requires drivers to collaborate with vehicles. The confidence of the human driver on ACC plays a vital role in the cooperation process.

As a crucial part of the ACC system, the ergonomic design of human-machine interface (HMI) is a complex, interdisciplinary challenge. Besides the technical aspects, selecting interaction patterns that align with the mental model of the user poses a significant challenge. Trust is recognized as a primary concern in HMI design. Conducting real-world HMI tests usually demands substantial resources and may expose participants to unpredictable risks. One promising approach to address this issue is to provide users with a relatively realistic driving environment through driving simulators. Thus in this study, we used the videos recorded from real-world driving as driving scenarios in the driving simulator.

9.2 Research question

This paper presents an experimental study to evaluate drivers' trust in the ACC system's HMI under different cut-in conditions using a driving simulator. The study aims to investigate the factors that affect driver trust levels, providing suggestions for enhancing the design of ACC's HMI based on these findings.

9.3 Method

9.3.1 Participants

A total of 24 participants (12 male and 12 female) participated in the study. The participants' ages ranged from 21 to 30 years, with a mean age of 24 years (SD = 2.11). All participants possessed valid driver's licenses and had a basic understanding of how the ACC operates, as well as how to use it.

9.3.2 Experimental design

The cut-in event is a typical scenario of ACC and it is fairly common in day-to-day driving. Previous research has summarized the statistical characteristics of time headway (THW) (distance/velocity) in various cut-in situations, and it was found that the influence of THW on subjective driver states in ACC is significant. In addition to the THW, velocity is another important parameter in driving. Therefore, the real-world driving in the study was designed considering two independent variables, THW and velocity. From the real-world driving, six video scenarios were extracted that met the condition requirement of the simulator testing. The specific parameters of test conditions are shown in Table 9.1.

In the simulator test, a within-subjects experimental design method was adopted. The Latin square experimental design method was used to eliminate the influence of learning effects and experiment order effects. Participants were divided into six groups, each group consisting of four individuals. The experimental test sequence differed for each group, and each group completed six conditions in accordance with the designated order. The ACC function was turned on in all conditions, with the side vehicle cutting in from the left side at the same speed as the subject vehicle.

Table 9.1 Specific parameters of six conditions.

	Velocity (km/h)	THW (s)
1	60	1.2
2	60	0.7
3	50	1.2
4	50	0.7
5	30	1.2
6	30	0.7

9.3.3 Experimental procedure

Real-world driving was carried out using a professional automobile (Volvo S90). Videos capturing the driver's front view (Fig. 9.1) and the corresponding dashboard (Fig. 9.2) were recorded during real-world driving under various cut-in conditions.

Before the laboratory test, participants were instructed to read the ACC manual to enhance their familiarity with its operation and functionality. They were provided detailed explanations of the interface elements (Fig. 9.3) and interaction logic of the ACC system's HMI in the Volvo.

FIGURE 9.1

Video of driver's front view.

FIGURE 9.2

Corresponding dashboard video.

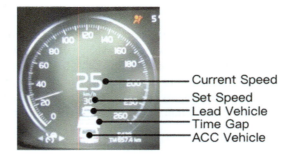

FIGURE 9.3

Interface elements of adaptive cruise control.

FIGURE 9.4

Testing scene.

Participants were informed that the ACC function would work normally in all scenarios and were tasked with making comprehensive driving judgments based on the environment and HMI display. When the participants were familiar with the ACC operation and HMI information, they were given a practice drive on the simulator for 2−3 minutes.

During simulator driving, videos of the front view and dashboard corresponding to different conditions were played in the driving simulator to immerse the participants in the driving experience as much as possible (Fig. 9.4). If participants perceived a scenario as risky, they had the option to press the foot brake pedal or turn the steering wheel to terminate the ACC automated driving. The entire driving process was recorded on video to capture drivers' behavior.

At the end of each test drive, the trustworthiness of HMI about ACC was assessed using a 12-question survey using a 5-point Likert-type scale (1 = totally disagree, 5 = totally agree). This questionnaire, comprising items adapted from previous studies and supplemented with new inquiries, aimed to gauge participants' perceptions. Sample items from the scale included: "I can assume that this HMI will display the correct information;" "I don't care about the display of the dashboard," etc. Responses were averaged to generate a reliable measure of trustworthiness, with higher scores indicating greater perceived trustworthiness. Subsequently a simple interview was conducted with the participants about their driving experience. Interview questions encompassed inquiries such as: "Did you notice the change of HMI when the side vehicle cut in?" and "Was the ACC system HMI helpful when you were driving?"

9.4 Results

Based on the analysis of recorded videos from the experiment, it was observed that all participants intervened by pressing the brake pedal, with none opting to steer the vehicle manually. The statistical result of drivers' braking behavior is shown in Fig. 9.5. As the THW decreased, the proportion of drivers who pressed the brake pedal to take over increased, indicating an increased distrust of the ACC. Besides, the proportion of brake generally increased with the decrease in velocity under both THW conditions. Since the lower velocity corresponds to a shorter distance between vehicles at a certain THW condition, the decrease in trust level along with the decrease in velocity implies that drivers might be more sensitive to the distance between vehicles instead of speed when they make a take-over decision.

The trust scores of HMI were compared under different conditions. Fig. 9.6 showed that regardless of velocity, the trust score when THW was 1.2 seonds was generally higher than that when THW was 0.7 seconds. This suggested again that drivers tended to be more confident of the HMI at large THW condition. For the same THW condition, the change of trust score did not follow a consistent pattern with the change of driving velocity.

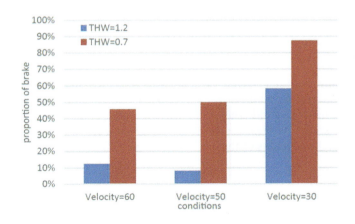

FIGURE 9.5

Proportion of people who press the brake in different conditions.

Chapter 9 Drivers' trust of the human-machine interface

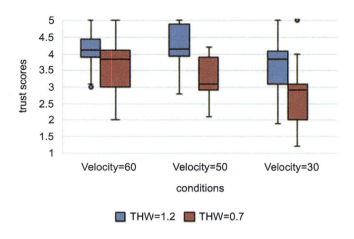

FIGURE 9.6

Boxplot of trust scores.

To identify the impact of THW and velocity on driver's confidence in the HMI, a repeated two-way ANOVA ($/ = 0.05$) was conducted, and the results show that both THW (F (2,138) = 3.06, $p < 0.05$) and velocity (F (1,138) = 3.91, $p < 0.05$) had a significant impact on the trust score, while the interaction effect was not significant (F (2,138) = 3.06, $p > 0.05$).

According to the interview results, it seemed that participants tended to check the dashboard when there was no perceived risk, and meanwhile, the velocity was stable. After the driving test, four participants suggested that it would be better if a voice prompt was provided so that they could quickly notice the change in the vehicle's state. Most of the participants indicated that they trusted their own judgment more than the HMI at the early stage of the test. However, as the test went on, trust in the ACC system HMI was enhanced.

9.5 Limitations

The limitations of using a driving simulator and video scenarios should be noted. Although the video was recorded from real-world driving, the perceived velocity and distance in the lab were still different from those in the real world. Therefore, future research in the field is necessary to generalize the findings of this study. Besides, the study only involved a cut-in situation as the test scenario, thus limiting the comprehensiveness of the evaluation of the HMI.

9.6 Conclusion and discussion

In this study, we introduced a methodology employing a driving simulator to assess the level of trust in ACC system HMIs in cut-in conditions. Our findings show that both velocity and THW have a significant influence on the driver's HMI trust level. Drawing from the experimental data, it is evident that designers should take into account THW and velocity values when developing ACC system HMIs. The values examined in the study provide references for HMI design to a certain extent.

In light of participant feedback, it is recommended that the HMI design not only incorporate the display of the current ACC state but also integrate multichannel cues. Interactive modes in multiple modalities, such as audio prompts, may outperform single-interface prompts. A comprehensive HMI design, combined with considerations of user experience, can enhance driver awareness of road conditions and improve trust in HMI.

Further reading

M. Beggiato, M. Pereira, T. Petzoldt, J. Krems, Learning and development of trust, acceptance and mental model of ACC. A longitudinal on-road study, Transp. Res. Part F Psychol. Behav 35 (2015) 75–84.

F. Ekman, M. Johansson, J. Sochor, Creating appropriate trust in automated vehicle systems: a framework for HMI design, IEEE Trans. Hum.-Mach. Syst. 48 (1) (2018) 95–101.

Z. Feng, X. Ma, L. Xia, X. Zhu, Z. Ma, Analysis of driver initial brake time under risk cut-in scenarios, in: The 14th International Forum of Automotive Traffic Safety, Changsha, China, 2018, pp. 113–122.

Y. Forster, F. Naujoks, A. Neukum, Increasing anthropomorphism and trust in automated driving functions by adding speech output, in: 2017 IEEE Intelligent Vehicles Symposium (IV), IEEE, Los Angeled, 2017.

R. Gu, X. Zhu, Summarization of typical scenarios of adaptive cruise control based on natural drive condition, in: Proceedings of the 11th International Forum of Automotive Traffic Safety, Chongqing, China, pp. 387–393 (2014).

J.-Y. Jian, A.M. Bisantz, C.G. Drury, Foundations for an empirically determined scale of trust in automated systems, Int. J. Cogn. Ergon. 4 (1) (2000) 53–71.

J.B. Lyons, C.K. Stokes, et al., Trustworthiness and IT suspicion: an evaluation of the nomological network, Hum. Factors J. Hum. Factors Ergon. Soc. 53 (3) (2011) 219–229.

F.W. Siebert, M. Oehl, et al., The influence of time headway on subjective driver states in adaptive cruise control, Transp. Res. Part F Traffic Psychol. Behav. 25 (Part A) (2014) 65–73.

F.M.F. Verberne, J. Ham, et al., Trust in smart systems: sharing driving goals and giving information to increase trustworthiness and acceptability of smart systems in cars, Hum. Factors 54 (5) (2012) 799–810.

G. Xiuyan, Experimental Psychology, 1st edn., People's Education Press, China, 2004.

CHAPTER 10

Take-over requests analysis in conditional automated driving

Fang You[1], Yujia Wang[1], Jianmin Wang[1], Xichan Zhu[2] and Preben Hansen[3]

[1]*Car Interaction Design Lab, College of Arts and Media, Tongji University, Shanghai, P.R. China* [2]*School of Automotive Studies, Tongji University, Shanghai, P.R. China* [3]*Department of Computer and Systems Sciences, Stockholm University, Stockholm, Sweden*

Chapter Outline

10.1 Introduction	123
10.2 Background	124
10.3 Highway hazard scenario analysis	125
10.3.1 Scenario analysis	125
10.3.2 Lane changing visual scanning analysis	126
10.3.3 Decision-making phase vision scanning analysis	127
10.3.4 Execution phase vision scanning analysis	128
10.3.5 Adjustment phase vision scanning analysis	128
10.4 Application and evaluation	129
10.4.1 Evaluation procedure	129
10.4.2 Results	130
10.5 Discussion and conclusion	131
References	131
Further reading	132

10.1 Introduction

With the continuous development of surrounding detection and control systems, self-driving cars have gradually developed from a concept to a reality. At the same time, the relative traffic safety problem has become more and more serious. Corresponding to automotive security, human driver in-vehicle behavior, and role definition, the car intelligence degree has attracted the attention of researchers.

Self-driving vehicles can be implemented on highways with full-speed range adaptive cruise control, enabling automatic speed control and vehicle-to-vehicle distance at full speed range. Lane-keeping assist ensures automatic control of the steering to keep the vehicle within the lane. In 2016, Tesla had several traffic accidents. In these cases, the activation of Tesla's autopilot mode, limited capacity of sensing the environment, and driving context resulted in driver casualties.

Further, these kinds of phenomena inspired researchers to cooperate on issues related to human drivers and vehicles in partial and conditional automation.

10.2 Background

The summary of levels of driving automation for on-road vehicles describes level 3 as "the driving mode-specific performance by an automated driving system of all aspects of the dynamic driving task with the expectation that the human driver will respond appropriately to a request to intervene." It does not define the situation or time in detail. Therefore the major problem of conditional automotive is to define when manual driving takes control over automotive driving.

The selection and handling of secondary task engagement, along with the needed reaction time to comply with a take-over request is one of the main research questions. This also affects the design of take-over requests concerning the human-machine interface (HMI). In similar studies of take-over requests, the issue starts in a highway scenario and is then researched actively in various other situations.

From a safety priority aspect, the early warning system should alarm at an early point, but the early warning system will cause the driver to think that no alarm or false alarm is required, thus reducing the effectiveness of the early warning system. This means that the problem of the take-over request time's upper limit is affected by several complex factors. Furthermore, accompanied by advances in technology, the upper limit of time required is changing. Some scholars have focused their research on take-over lower limit time. Some studies focus on the time required for manual driving to take over from automated driving at level 3 conditional automated driving. Using different driving conditions in an immersive driving simulator experimental setting, we studied and tested the time it takes for a complete take over. We studied the time used for the transition from automated driving to manual driving, before a simulated road hazard. These experiments were conducted in three steps. First, participants were given a practical exercise task (5–10 minutes) to get accustomed to the test setting and the simulated driving environment. Second, the participants performed the main experimental task involving the test car simulator in an automated driving mode. Finally, the participants were involved in a set of take-over required tasks in which the researchers implemented different time variables to be tested.

Damböck et al. (2012) and Ito et al. (2016) set the automotive driving speed to 100 km/hour, and the take-over request time was 4, 6, 8 seconds, and 5 and 7 seconds [1,2]. In the first study, 6-second conditions were used, and most of the drivers could take over the driving rights and complete all three scenarios under the driving task. In the 8-second condition case, all drivers completed the task, maintaining consistent driving performance levels throughout the process. In another study, when using the time conditions of 5 and 7 seconds, all six combinations (such as leaving the steering wheel or not, eyes monitoring the driving environment or not, etc.) of the drivers' state were able to successfully take over the driving rights and to avoid rear-end accidents. In the analysis of this limited study, it seems that the driver's safe completion of the driving right, which takes 6–8 and 8 seconds, is a good timeframe for an acceptable transition in a highway scenario.

van den Beukel et al. [3] simulated different jammed traffic situations by making the front car emergency braking. At the time conditions of 1.5, 2.2, and 2.8 seconds, the accident percentages were 47.5%, 20.8%, and 12.5%, respectively. Mok et al. (2015) set 2, 5, or 8 seconds before collision obstacles. Just a few drivers within the 2-second condition were able to safely navigate the road hazard situation, while the majority of drivers in the 5- or 8-second conditions were able to navigate the hazard safely.

In addition to the problem of take-over request elements, from the in-vehicle system giving drivers take-over requests to drivers passing the obstacles safely, the whole driving process needs to be considered. We tried to figure out how drivers acquire environment information in lane changing and different driving performances under different secondary task engagements. Studies based on these two aspects help to improve the safety of driving take-over requests This paper involves two interconnected studies: (1) monitoring and observing drivers' visual scan while driving to investigate drivers' needs when performing a lane-changing driving scenario, and (2) experimenting to test the success rate in a take-over request scenario under two different conditions: an electronic reading state and a voice chat state.

10.3 Highway hazard scenario analysis
10.3.1 Scenario analysis

Automation technology is still not mature enough and today a vehicle cannot entirely perform the driving task itself without human input. Drivers always need to keep the environment under observation, and be ready to take over if the driving situation requires it. Based on the in-vehicle interaction system, we tried to improve the safety of driving through the collaboration of humans and vehicles. When the autopilot is not able to complete the driving task, the driver needs to take over the driving control. For our analysis, we selected and designed a highway scenario wherein the driver encountered obstacles on the road (Fig. 10.1).

In the highway crash scene, the driver scenario is characterized as follows: when the driver observes the obstacles, the driver starts to slow down and sweeps the target lane direction on the external mirror to confirm that the lane is safe and secure. The driver turns the steering wheel and activates the lights to change the course of the way, and finally, moves the vehicle through the first traffic cone, and adjusts the vehicle to complete the lane.

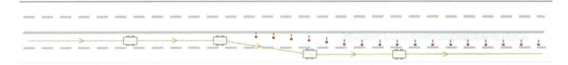

FIGURE 10.1

Highway hazard scenario schematic diagram.

10.3.2 Lane changing visual scanning analysis

A previous and related research study shows that about 90% of the information acquired during the driving process comes from the human visual system. After understanding the encountering road hazard highway scenario, we started to analyze the driver's visual characteristics in the person-car-environment frame. Through recording and gathering the driver's eye movement data, we summarized the driver's visual scanning mode during the lane-changing course. The objective was to extract information about the drivers' needs for in-vehicle HMI design.

We recruited 18 participants with the following requirements: a valid driving license and already having a private car. Gender distribution was as follows: males ($n = 9$) and females ($n = 9$). The participants reported that they owned their driving licenses from 1 to 13 years (M = 7.45, SD = 4.23). The participants belonged to diverse backgrounds, such as teachers, security guards, office staff, and so on. Our experiment utilized an eye movement device called Tobii Pro Glasses 2 and installed the software GoPro HERO 41 in the copilot to record the driver's head and hand movements. We recorded both the internal and external environment when the driver performed the task by using the camera inside the driver's eye movement instrument. At the end, the participants were invited to score the degree of interest in the area of interest (AOI).

As for the data collection of the eye movements, we measured and analyzed the eye fixation points. Table 10.1 the eye movement area is divided into six AOI, as shown in Fig. 10.2.

The classification of the lane is divided based on Worrall and Bullen's [4] stages of the lane change process, and the analysis of lane changing decision-making stages. The preparation phase of

Table 10.1 Right-to-left lane changing eye-tracking data.

AOI area	Total visit count	Percentage fixated (%)
A	521	100.00
B	226	100.00
C	96	38.89
D	57	38.89
E	46	27.78
f	41	33.33

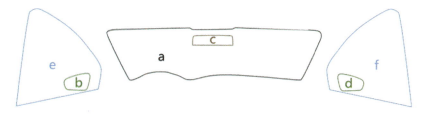

FIGURE 10.2

Area of interest. (A) Front view. (B) Left outer rearview mirror. (C) Rearview mirror. (D) Right outer rearview mirror. (E) Left side window. (F) Right side window.

the lane change is subjected to two elements, which are time and distance. Where time is the duration the vehicle needs to approach the target lane, and distance is the location of the vehicle from the demarcation zone of the two lanes.

The execution phase of the lane-changing activity starts when the driver decides to turn to the right, and normally, this is done when most of the subjects start to turn the steering wheel. In the adjustment stage, the lane-changing task is finished, and the vehicle returns to its normal running state in the target lane.

In lane-changing situations (Fig. 10.2), the front view and left rear view mirrors get the most attention. The driver's observation and participants' vision scanning were categorized into three groups based on whether they utilized the front view or left outer rear: (a) observing the front view and left outer rear (three variations), (b) observation was more from the front view and left outer rear; and (c) observation was less than the front view and left outer rear, that is, observation was only from the front view (Fig. 10.3).

10.3.3 Decision-making phase vision scanning analysis

In the decision-making phase, when the external environment is relatively simple, participants looked at the front view area and left the outer rear area to confirm the opportunity to change lanes. Vision scanning alternated between the front view and the left outer rear.

When the external environment is more complex and the obstacles are closer to the vehicle, the driver may be in a "tense mode" and need to be alert to confirm safe passing in multiple areas. Participants are not only concerned with the front view, and left outer rear but also pay attention to other areas, such as the right outer rear and rearview mirror.

When the external environment is more complex and obstacles are far from the vehicle, the driver observes the road far ahead, monitoring changes in road conditions. In this situation, the participants fix their eyes only on the front view.

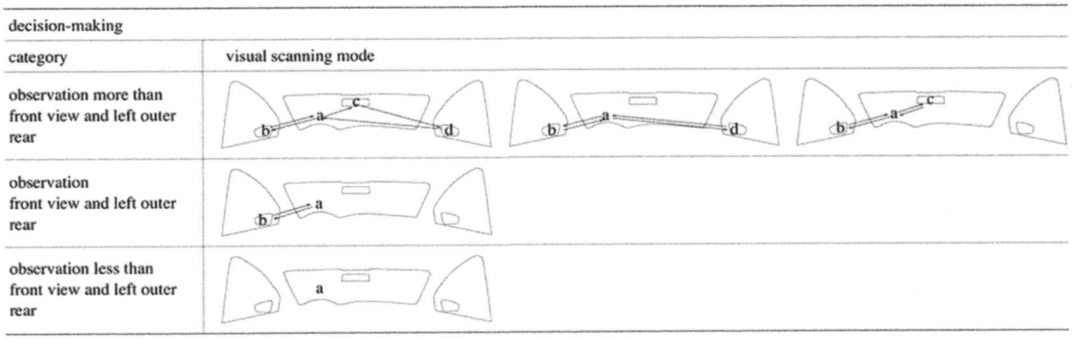

FIGURE 10.3

Visual scanning mode summary of the decision-making phase. (A) Front view. (B) Left outer rearview mirror. (C) Rearview mirror. (D) Right outer rearview mirror.

10.3.4 Execution phase vision scanning analysis

When the original lane appears closer to the interference factor, it will make participants vigilant, watching the right rearview mirror more often. When vehicles, pedestrians, or nonmotor vehicles appear in close distance in front of the road of the experimental car, it results in persistent observation, and then the participants fix their eyes only on the front view. Most of the participants in the execution phase, observed the left rearview mirror several times to look for cars behind. Overall, in the lane-changing scenario, Fig. 10.3. depicts the visual scanning mode summary of the decision-making phase.

In the take-over request analysis in the conditional automated driving 235 execution phase, the road is relatively calm (no or few vehicles). Furthermore, the environment outside the car did not contain any particular events that may attract the drivers' attention. This resulted in participants only scanning the front view and two areas of the left side of the rearview mirror to confirm pathways and driving safety (Fig. 10.4).

10.3.5 Adjustment phase vision scanning analysis

This case involves more complicated traffic conditions, and the interference factor around and close to the experimental vehicle caused participants to scan multiple areas to confirm driving safety. The driving speed of this group of participants was significantly higher or lower than the other two groups of participants (Fig. 10.5).

When, in the road ahead, interference factors, such as vehicles, pedestrians, or nonmotor vehicles, appear close to the experimental car, the test participants fix their eyes only on the front view. After changing lines, a vehicle appearing on the original road caused participants to make a security observation of the environment by scanning the right rear mirror and inside the rearview mirror.

execution category	visual scanning mode
observation more than front view and left outer rear	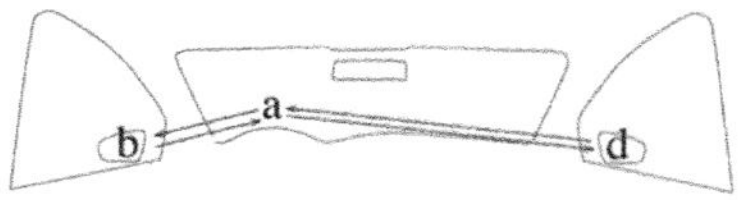
observation front view and left outer rear	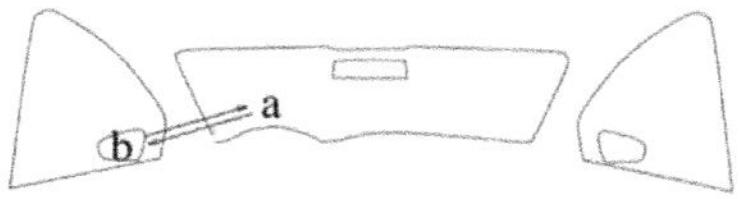
observation less than front view and left outer rear	

FIGURE 10.4

Visual scanning mode summary of the execution phase. (A) Front view. (B) Left outer rearview mirror. (C) Rearview mirror. (D) right outer rearview mirror.

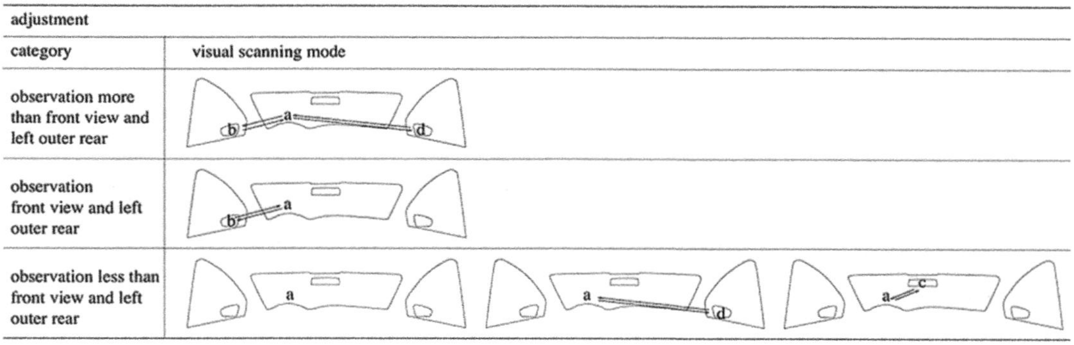

adjustment category	visual scanning mode
observation more than front view and left outer rear	
observation front view and left outer rear	
observation less than front view and left outer rear	

FIGURE 10.5

Visual scanning mode summary of the adjustment phase. (A) Front view. (B) Left outer rearview mirror. (C) Rearview mirror. (D) Right outer rearview mirror.

10.4 Application and evaluation

After the data analysis of the three different scenarios of phases in lane changing as described above, we found that distant obstacles attract the participants' attention and generate interference that will cause problems for the driver to complete the driving task. When encountering hazardous scenarios on the highway, they will reverse and bypass the obstacles. When the driver cannot monitor the surrounding driving environment properly, traffic accidents may increase.

When applying our findings to a take-over request problem, we suggest that to overcome the problem, the obstacle needs to be marked in advance. If drivers could understand lane conditions faster, it would be helpful in a take-over safety situation.

This will draw the attention of the driver and result in a more unproblematic traffic condition. Head-up displays have already been proven to improve information access and reading speeds compared with classic head-down displays. While integrating lane-changing processes and highway hazard scenario analysis, it is necessary to mark the lane line that contains obstacles to guide drivers to quicker and more clear driving. We designed the heads-up display in the windscreen for the highway hazard scenario, including lane-safety states, traffic cone azimuth, and sound for notice regain driving control (Fig. 10.6).

10.4.1 Evaluation procedure

The simulated driving course contains three different parts and two alternative control transitions. In the first part, participants were introduced to the virtual simulation driving environment and hardware facilities (5 minutes). In the second experimental task, they switched to an automotive mode, which involved the car set to 80 km/hour speed. The task duration was 10 minutes. In the third experimental task, the participants encountered a take-over request within three time intervals (4, 6, or 8 seconds), where the participants needed to regain driving control.

130   Chapter 10 Take-over requests analysis in conditional automated driving

FIGURE 10.6

The human-machine interface prototype (*left*). The participant is shown in two situations: reading condition (*middle*) and take-over requests to avoid obstacles in lane changing (*right*).

Table 10.2 Take-over requests success rate under electronic reading and voice chat conditions.

Time required (s)	Electronic reading condition take-over request success rate (%)	Voice chat condition take-over request success rate (%)
4	33.33	66.66
6	83.33	100.00
8	83.33	100.00

Distraction events were based on common driving distractions in Chinese traffic situations. When the vehicle was in automotive mode, the participants were asked to perform different types of nondriving tasks, so the participants were in a distractive state. Participants' distraction task was introduced under two different conditions: (a) an electronic (mobile) reading condition in which the participants were free to browse in an SNS application, and (b) a voice chat condition, in which participants used a phone app to engage in a voice chat. In a pilot study, researchers studied six participants, mainly university teachers and students, three females and three males. The participants were between 19 and 28 years old, with driving experience ranging from 1 to 12 years.

10.4.2 Results

The driver's performances under the two distraction conditions of the electronic reading state and voice chat are shown in Table 10.2 below. Regarding the electronic reading condition and voice chat condition rates, the driving performances are the same at take-over time in both the 6-second and 8-second conditions. However, in the 4-second condition, the success rate for participants in the voice chat state take-over requests was better compared with the success rate for electronic reading state take-over requests, which were 33.33% and 66.66%, respectively.

Furthermore, in the electronic reading condition, participants were engaged in a relatively strong secondary task (driver take-over). When they needed to regain driving control, they would put down the phone and then put their hands on the wheel to take over. In the voice chat condition, some participants were in different body postures: one hand on the steering wheel, and with a mobile phone on the other hand, which made it easier to regain manual driving. During the experiment, when the participants received the take-over request, the behavior of the participants to manage the operation was ordered in the following categories: (a) the phone placed in the legs; (b) hands to take the steering wheel; (c) while others took the phone in their hand, and (d) while yet others were holding the phone and steering wheel.

10.5 Discussion and conclusion

This paper investigated the visual driver behavior within the complex human-vehicle-environment system, considering various environmental interference factors. The study shows that different driving behaviors are influenced by drivers' attention and decision-making processes.

In particular, when the interference factors, such as vehicles, pedestrians, or nonmotor vehicles, are far from the experiment car and there are few obstacles on the road, the driver will manage the interference factor through the front view. When the interference factors are near the experimental car and numerous obstacles of various types are present, the driver will engage in saccadic eye movements between multiple areas.

In our take-over request experiment, though the sampling for the study was low, it indicated different performance behaviors within the secondary task engagement. During the electronic reading condition, not all participants managed to finish regaining driving control, not even within the 8-second interval. For the electronic reading condition, all participants completed the state transition from manual driving to automated driving in 6 seconds. Finally, different body gestures were encountered and discovered, which affected the take-over request success rate. In future research, more subjects will be needed, and additional secondary task engagements will need to be tested to discover and explain more effective take-over request procedures.

References

[1] D. Damböck, K. Bengler, M. Farid, L. Tönert, Übernahmezeiten beim hochautomatisierten Fahren, Tagung Fahrerassistenz, München, 2012.
[2] T. Ito, A. Takata, K. Oosawa, Time required for take-over from automated to manual driving, in: SAE 2016 World Congress and Exhibition, SAE International, New York, 2016.
[3] A.P. van den Beukel, M.C. van der Voort, The influence of time-criticality on situation awareness when retrieving human control after automated driving, in: 16th International IEEE Annual Conference on Intelligent Transportation Systems, Hague, 2013.
[4] R.D. Worrall, A.G. R. Bullen, An empirical analysis of lane changing on multilane highways. Highway Research Record 303, 30–43. National Research Council, Washington, DC (1970).

Further reading

V. Melcher, S. Rauh, F. Diederichs, H. Widlroither, W. Bauer, Take-over requests for automated driving, in: 6th International Conference on Applied Human Factors and Ergonomics, Elsevier B.V., Las Vegas, 2015.

B. Mok, M. Johns, K.J. Lee, D. Miller, D. Sirkin, P. Ive et al., Emergency, automation off: unstructured transition timing for distracted drivers of automated vehicles, in: IEEE 18th International Conference on Intelligent Transportation Systems, IEEE, New York, 2015.

SAE levels of driving automation. <http://www.sae.org/servlets/pressRoom? OBJECT_ TYPE = PressReleases& PAGE = showRelease&RELEASE_ID = 3544>.

R. Singh Tomar, S. Verma, Safety of lane change maneuver through a priori prediction of trajectory using neural networks, J. Netw. Protoc. Algorithms 4 (1) (2012) 4−21.

H. Summala, D. Lamble, M. Laakso, Driving experience and perception of the lead car's breaking when looking at in-car targets, J. Accid. Anal. Prev. 30 (4) (1998) 401−407.

X. Wang, L.I. Yan, Characteristics analysis of lane changing behavior based on the naturalistic driving data, J. Traffic Inf. Saf. 1 (2016) 17−22.

PART 3

Research on intelligent cockpit HMI design under automated driving scenarios

CHAPTER 11

Interactive framework of a cooperative interface for collaborative driving

Jun Zhang[1,2], Yujia Liu[1,2], Preben Hansen[3], Jianmin Wang[2] and Fang You[2]

[1]*College of Design and Innovation, Tongji University, Shanghai, P.R. China* [2]*Car Interaction Design Lab, College of Arts and Media, Tongji University, Shanghai, P.R. China* [3]*Department of Computer and Systems Sciences, Stockholm University, Stockholm, Sweden*

Chapter Outline

11.1 Introduction 135
11.2 Framework of cooperative interface 137
    11.2.1 Human autonomy team 138
    11.2.2 Level of interaction 143
    11.2.3 Dynamic external factors 144
11.3 Conclusion 144
References 145
Further reading 146

11.1 Introduction

The development of automobile intelligence makes advanced driver assistance systems (ADASs) shift toward fully automated vehicles. ADAS improves human perception and control ability, shifting the driver's role from physically controlling the vehicle to supervising driving tasks and automation systems. Supervising automation implies a shift from a behavioral task to a cognitive task. However, this shift in tasks also brings new problems. What kind of mechanism can keep individuals in the loop when they step away from the control interface and simultaneously collaborate with the intelligent system to control the car?

According to SAE's Level 0−Level 5 autonomous driving classification, starting from Level 3, the guidance and control tasks are performed by the system. However, at this stage, drivers have to supervise the driving tasks of the system and be ready to take over control at any time. In other words, before real Level 5 autonomous driving is realized, humans and machines share control of the vehicle. Shared control means that in certain critical states, driving rights will be switched between man and machine. In this critical state of handover, there are currently two main research directions: (1) The first is a takeover, which is the black-and-white switching mode. Although takeover is one method, even if there is enough time for the driver to take over the car, it may fail because of the inability to quickly restore situation awareness. The second refers to cooperative driving, which is a model of human-machine integration. There are various perspectives on cooperative driving, which include vehicle-to-vehicle cooperation, vehicle-to-infrastructure cooperation, and cooperation between the driver and the intelligent system within the car to form a joint team to perform driving tasks together.

Walch et al. [1] proposed the use of a cooperative interface to collaborate at the abstract level (plan, decision-making, etc.) to overcome the limitations and uncertainties of the system. For example, the car cannot accurately identify whether a person is blocking the front or it is a flying plastic bag. In a scenario like this, only the driver needs to approve the system to execute the next strategy. Subsequently, the system controls the car in accordance with the established strategy. It appears that the cooperative interface does not concentrate on specific car control rights, but on the cooperation of human and machine cognitive dimensions, involving the exchange of opinions within their respective realms of thought.

Essentially, the cooperation between humans and intelligent systems is based on cooperation in the cognitive dimension of the two worlds (the world felt by the machine and the world felt by the human). To build an interactive framework for the cooperative interface, we may first highlight two human abilities: (1) cognitive ability, which refers to cognitive tasks (perception, consciousness, reasoning, and judgment), and (2) behavioral ability, which refers to the activities that a person's body functions can complete. When a machine is an extension of the human body's abilities, we can call it a tool. If it is an extension of cognitive ability, we would prefer to call it a partner, as it appears to have observed what we have seen, contemplated what we have thought, and executed what we have wanted to do.

As shown in Fig. 11.1, in manual driving, inspired by Guy A. Boy's [2] framework of car driving activity as a perception-cognition-action dynamic regulation loop, the driver first performs

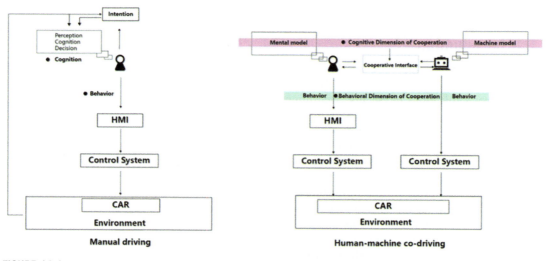

FIGURE 11.1

Structure comparison between manual driving and human-machine codriving.

cognitive tasks, such as perception, cognition, decision-making, etc., based on his own intentions, and then performs behavioral tasks and interacts with the control system through a human-machine interface and controls the movement of the car.

It can be seen from Fig. 11.1 that this structure changes the interactive objects that the driver faces, and the human interactive objects change from a control system to a virtual driver. Obviously, they need an interface for communication. However, this interface is not for controlling the vehicle, but for two cognitive systems to communicate cognitive information (such as intention, perception, and decision-making) during the execution of tasks. The medium that carries this kind of communication is regarded as a cooperative interface.

This shift means that people and machines need to be considered together, rather than separate entities linked together through human-machine interface. At this time, the view of human-machine interaction changes from the human-machine system view to the joint cognitive system view. The joint cognitive system considers cognition as a goal-oriented interaction between people and artifacts to generate work in a specific context and at the level of ongoing work. We claim that in human-machine cooperative driving, people and intelligent systems form a joint cognitive and decision-making team system. Automation and people in the team must be coordinated as a joint cognitive body. The result of cognition directly outputs the strategy of decision-making and affects behavior. Therefore the cognitive dimension of cooperation is an important dimension of team cooperation between humans and agents. The cooperative interface will be the carrier of this coordination. The present study mainly aims to propose a framework for cooperative interface interaction in collaborative driving based on cognitive information exchange. The study further aims to clarify the cognitive interaction logic between humans and agents, and attempts to answer the following research questions:

1. What kind of cognitive information do humans and machines need to communicate?
2. Does the information from the agency need to be designed for transparency?
3. What is the interaction mode of the cooperative interface?

11.2 Framework of cooperative interface

Inspired by Flemisch's hierarchical framework of human-machine codriving cooperation, which divides driving tasks into four levels: navigation level, maneuver level, trajectory level, and control level, and Wang's [5] noncritical spontaneous situations (NCSSs) cooperation framework, which is intended to provide an overview over the most relevant elements that influence the NCSSs. Combining these two frameworks, we propose an interactive framework of the human-computer cooperative driving interface.

As shown in Fig. 11.2, the framework consists of three parts: the human autonomy team, the four levels of interaction in the driving process, and the dynamic external factors. We believe that the study of the overall framework elements and the influence factors of interface information can provide some inspiration to the interaction design. The following contents will explain the elements of the framework in detail.

Chapter 11 Interactive framework of a cooperative interface for collaborative driving

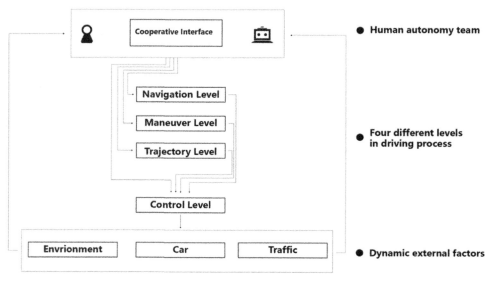

FIGURE 11.2

Interaction framework of cooperative driving interface.

11.2.1 Human autonomy team

11.2.1.1 Team

According to Salas et al. [6], a team is "a distinguishable set of two or more people who interact dynamically, interdependently, and adaptively toward a common and valued goal, who have each been assigned specific roles or functions to perform and who have a limited life span of membership." Therefore a team is a goal-oriented combination, where members can be human or nonhuman. In smart cars, humans and smart systems can form a human-autonomy team. Hocthen has put forward the minimum requirements for teamwork:

1. Each one strives toward goals and can interfere with others on goals, resources, procedures, etc.
2. Each one tries to manage the interference to facilitate the individual activities and/or the common task when it exists.

His definition implies that everyone has a goal. To achieve the goal, two or more agents interfere with each other and adjust their actions.

This interference can be cognitive or behavioral. Fiebich et al. [9] summarized the cooperation between the cognitive dimension and the behavioral dimension, pointing out that the minimum premise of the cognitive dimension cooperation is: (1) two (or more) agents perform actions to pursue the same goal, and (2) the agents know that they have the same goal. The premises of cooperation in the behavioral dimension are: (1) Two (or more) agents coordinate their behavior in space and time. (2) It can be observable from the outside. (3) Bringing about a change in the environment.

Synthesizing their viewpoints, "team" can be regarded as an adaptive dynamic system, and the members adjust their state through interaction to achieve a common goal. In the driving task, the human and machine share control of the vehicle, and the team's goal is constantly changing at different levels, such as at the navigation level, maneuver level, trajectory level, and control level, of the driving task based on the corresponding intention. Zimmermann et al. [10] proposed five layers of cooperation elements for human-machine cooperation in highly complex and unpredictable dynamic scenarios during driving including the user and machine intention, the mode of cooperation, the dynamic task and action allocation, the human-machine interface and the contact between human and machine.

To sum up, in the team tasks, both parties of the human-machine team are goal-oriented. Both of them contribute their ownabilities, communicate perceptions, plans and decisions with each other, and adjust their own status to achieve the goal of successfully completing the tasks.

11.2.1.2 Cooperative interface

The cooperative interface is the carrier of human-machine cognitive information communication. Team cognition is a cognitive activity at the team level, formed by team members through clear communication activities (such as talking and face-to-face communication.). In other words, team cognition is the product of interaction between team members. Research shows that effective communication and coordination are positively related to team performance. From the human-machine team, members exchange the cognitive resources of both parties through team cognitive activities with the aim reach a common understanding.

Walch et al. [3] believe that the purpose of the cooperative interface is to keep people in the loop and realize real-time communication between team members. The purpose is to improve situation awareness and the understanding of intentions and actions by both parties. In addition, they further put forward four basic requirements for designing a cooperative interface, respectively, mutual predictability, directability, shared situations representation, and calibrated trust in automation. Wood et al. [11] consider that in order to promote teamwork, the cooperative interface needs to provide a shared status representation, including information about the partner's status, plans, goals, and activities, as well as information about the current task status, environment, and situation. Zimmermann et al. [10] concluded that the human-machine cooperative interface should (1) display the intention of the user and the machine and (2) convey the reconfiguration of the dynamic adaptive system (Fig. 11.3).

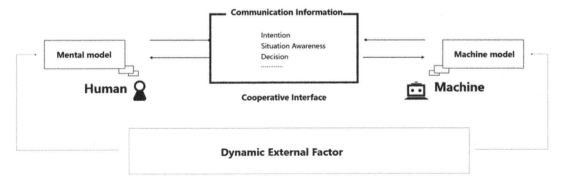

FIGURE 11.3

Human-machine cognitive information exchange.

Based on the above analysis of the literature, the main cognitive information elements that need to be communicated in the human-machine cooperative interface include intention, situation awareness, planning, decision-making, task assignment, and cooperation mode. Furthermore, the transparency of information can increase people's understanding of machines and increase trust in machine decisions. Finally, the kind of interaction mode used to facilitate the communication of this information is also an important aspect. Therefore the following section discusses the cooperative interface from the three aspects of cognitive information: element composition, transparency, and interaction mode.

11.2.1.3 Cognitive information
11.2.1.3.1 Intention

The individual's intentions determine the action goals, plans, and cognitive tasks. Intentions are determined by external attributes, such as environment, self-vehicle dynamics, and surrounding traffic, and are affected by internal factors. For drivers, internal factors are their motivation, short-term goals, mood, experience, etc. For the system, internal factors may be preset codes of conduct.

Understanding the intentions of others in the team and the intentions that can be characterized as understood are considered key factors for cooperation. This understanding is a two-way process. Both parties need to establish common beliefs and joint intentions through interactive activities. Hoc et al. [12] divide cooperative activities into action level, planning level, and metalevel by hierarchy. Therefore intentions can occur at any level. The action level is a short-term task. Intentions change rapidly, which can be communicated using verbal commands, such as rapid lane changes and overtaking. The planning level is a medium- and long-term task, and the two parties can exchange intentions and coordinate plans, such as navigation and route planning. As a long-term team model, metalevel is based on training and experience.

However, intention is a mental activity and cannot be directly observed. To recognize the intention of the other party, it is necessary to use an interface, which could be voice, text, etc., or measurement and reasoning based on behavior indicators. Bengler et al. [13] proposed that the machine must infer intentions in the following three ways: (1) The machine must measure and interpret human intentions according to its tasks and goals; (2) The machine needs to communicate its intentions in a reliable way, allowing users to obtain Information; and (3) The machine needs to obtain information, regardless of whether the information has been noticed and processed by the human-computer interaction partner.

In summary, understanding the intentions of cooperating members exerts an important impact on team joint activities. Consequently, the cooperative interface needs to accurately convey and represent intentions in a suitable way, such as through visual, auditory, and tactile cues.

11.2.1.3.2 Situation awareness

In general, situation awareness is a person's perception, understanding, and prediction of the surrounding changing environment. It determines the person's next decision and action. Thus, it is an extremely important psychological activity. Endsley defines three levels of situation awareness. On the first level, the driver perceives each element in the environment. On the second level, the driver has a comprehensive understanding of the current situation. On the third level, the driver predicts the possible future development of these elements. For nonhuman actors, the ability to provide mechanisms for predicting future actions based on prior knowledge, perception, and prediction is called autonomy situation awareness.

From the perspective of a team, situation awareness currently does not have a unified definition. The most influential related theory is the situation awareness research based on the expansion of individuals to teams, such as Endsley's shared situation awareness, Salas et al.'s [7] team situation awareness model as well as Shu and Furuta's mutual awareness model. The other one is based on system situation awareness research, for instance, Stanton's distribution situation awareness theory.

However, team situation awareness is more than just combining situation awareness of individual team members. Team situation awareness consists of collective awareness of the team. Salas et al. [8] believe that the key to team situation awareness lies in the information interaction process of team members. The situation awareness obtained by an individual forms a team situation awareness through team interaction, and the individual situation awareness can be modified in real time. Consequently, team members must have situation awareness related to their roles and goals, as well as situation awareness associated with other team members. Anticipation of the behavior and information needs of other team members in the team is of much importance for effective team situation awareness. The situation awareness among members is distributed and exchanged in the team through team processes, which include communication, coordination, etc. Therefore team situation awareness includes three aspects, respectively, situation awareness of a single member, situation awareness shared between team members, and situation awareness of the entire team combined.

Since the results of human-machine situation awareness may be different, the two parties can cooperate to make up for each other's possible shortcomings. For example, Walch et al. [1] studied how to make the driver's perception cooperate with the automation system through a collaborative interface, aiming to overcome the limitation of automatic perception in the overtaking scene. Chao et al. [4] studied people's predictions of the intentions of other traffic participants and integrated the prediction judgment into the operation of the automation plan level to make up for the lack of judgment ability of the automation system.

Nevertheless, in the human-autonomy team, the driver's overreliance on ADAS will lead to a mental model of overdependence, which will make the driver spend less time focusing on the road and more time engaged in nondriving tasks. Once the machine has insufficient capacity to correctly understand the current situation, and the driver lacks situation awareness, being unable to quickly take over the vehicle, it will cause a serious accident. Kridalukmana et al. [14] put forward a supporting situation awareness model of multiagent cooperation to provide support for man-machine team cooperation in driving.

Briefly, situation awareness is an important cognitive activity of information acquisition and analysis. For the elements of the current changing scenario, the two parties should fully exchange information and reach a consistent understanding with the aim of taking the correct action. The cooperative interface plays a role in full communication, allowing both parties to complement each other and promote the formation of shared situation awareness.

11.2.1.3.3 Decision

The intelligent system undertakes part of the cognitive work and can make decisions. Adaptive automation can replace the operator's perception as well as assist or replace decision-making processes and actions. However, the effectiveness of this automated decision-making will have an impact on the human mental model, resulting in excessive trust or distrust. In cooperation between humans and agents, they need to exchange plans and decisions for the following steps. Walch et al. [1] investigated the interaction of the cooperative interface to achieve the next step plan when the system remains uncertain or limited. First, the system will alert the driver and inform him of restrictions or uncertainties. Second, it will provide the driver with solutions, such as overtaking, route selection, takeover, and other decisions, and finally wait for the driver to approve one of them. If the driver chooses one of the recommendations, the automation will be executed automatically, thus avoiding manual takeover. If the driver does not select any item, the system will automatically downgrade and allow the driver to resume manual driving.

11.2.1.4 Information transparency

As the degree of intelligence increases, agents can make complex decisions, and such decisions and actions may be challenging to understand. Therefore, proper characterization of the reasons for the overthrowing process and the behavior behind the decision contributes to people's understanding. One method is to design the machine transparency module so that the behavior, reasoning, and expected results of the intelligent robot are transparent to the operator. As a result, people's understanding of machines can be enhanced. Therefore transparency can provide a mechanism for understanding and monitoring agent behavior, including the predictability of recent actions.

Chen et al. [15] took inspiration from the three-level situation awareness model and developed an agent transparency model based on situation awareness called the system action transparency model. It supports human understanding of machines by emphasizing the intent, reasoning, future plans, and uncertainties associated with intelligent machines. The SAT framework is divided into three levels:

Level 1: Intelligent machines provide people with basic information about their current status, goals, intentions, and plans. Besides, they can help people perceive the current actions and plans of intelligent machines.

Level 2: Intelligent machines provide people with their reasoning process and the environmental constraints considered when planning their behavior. It benefits people by helping them understand the current behavior of the machine.

Level 3: Intelligent machines provide people with predictions for the future, their consequences, and the possibility of success and failure, including uncertainties. It is helpful for people to predict future results.

Based on the requirements for the duration of the driving task, the time is often extremely short, and the driver may not have enough time to absorb the transparency information provided by the machine, especially the first level of SAT information. Kridalukmana et al. [14] proposed a time-constraint-driven transparency framework, dividing the theme of transparency into four categories: interpretation and performance, planning, decision-making and coordination, as well as intention and outcome prediction. According to time constraints, the transparency theme is set on priority and visibility.

In general, transparency exerts an important role in explaining the behavior and decision-making of agents, which can improve the trust of the driver in the machine and the performance of the team. Therefore, in the cooperative interface, appropriate transparency design for information is of great necessity.

11.2.1.5 Interactive mode

The interaction between humans and smart products is widely used in multimodalities, such as gestures, voice, and expressions. Multimodal interaction integrates human vision, hearing, touch, and other channels, which can completely simulate the natural way of interacting between people. As said by Norman [16]: "The machine should not allow the operator to leave the loop, but should continue to interact with it in a normal and natural way."

Bengler et al. [13] proposed that multimodal interaction methods should be established, including auditory, tactile, and visual channels, instead of reducing the information between humans and machines to visual interfaces. Numerous models of cars are also equipped with gesture recognition technology, which operates the overall display screen located in front of the driver through gesture recognition. Voice is the mode of expressing the user's intention. In human-computer interaction, the user's voice information is translated into a certain semantic after being accepted by the agent to help the agent understand the user's intention. Flemisch et al. [17] proposed the Horse-mode model, drawing inspiration from the collaboration between humans and horses aided by the use of reins. People can use the tactile interface to convey the driving intention through force to change the level of automation. Additionally, vehicle-mounted robots or virtual robots can also be used as an emerging interactive method that can communicate their intentions with the driver in an anthropomorphic way. For example, the positive or negative expressions of virtual characters can improve the driving behavior of the driver. Research by Johannes-Maria Kraus et al. [19] proves that the introduction of anthropomorphic design functions may lead to an increase in people's trust in autonomous driving systems.

In addition, from the perspective of the interaction medium between the driver and the car, in future smart car applications, interaction is no longer limited to media that require physical control, such as touch screens, knobs, and levers. In fully automatic driving mode, traditional control equipment may be hidden or disappear and replaced by other methods. Any physical equipment in the car may become an interactive medium. For example, in terms of media materials, the material itself, flexibility, and curvature of the inner surface of the car realize diversified entities and build new forms of interaction.

Overall, the future smart car is not only a means of transportation, but also a mobile place for emotional interaction that can satisfy people's multilevel needs. The design of multimode cross-sensory systems should fully consider the design of the cooperative interface.

11.2.2 Level of interaction

In a driving task, long-term tasks and short-term tasks are based on the length of time, which usually refers to the three-layer hierarchical control structure of "strategy, tactics, and control." Flemisch et al. [17] proposed four levels of classification on the three-level model, including navigation level, maneuver level, trajectory level, and control level.

The navigation level refers to the long-term plan of the driver to choose the destination. The interaction frequency is not high and the interaction time is not long. The driver can interact with the intelligent system through multimodal interaction, with the aim of selecting the destination.

The maneuver level represents a medium-term plan. It refers to the driver's intermittent handling of the detection and response tasks of incidents that occur in road conditions such as turning left and right, overtaking, and merging.

The trajectory level is a short-term route plan that contains space and time dimensions. For the same scene, humans and intelligent systems have the same intentions, but may adopt different path motion strategies. For example, in the overtaking scenario, the system provides a variety of path planning options, allowing the driver to select one of them through the touch interface or take over driving the vehicle in accordance with his own intentions.

The control level is the level with the most frequent interaction. The driver directly controls the lateral and longitudinal movement of the vehicle through the interface.

The cooperation at these four levels does not remain static, but can include dynamic changes at each level of cooperation, and the four loops are related to each other.

11.2.3 Dynamic external factors

According to Guy A. Boy's description, car driving activity is defined as a continuous dynamic loop of regulation between (1) inputs, coming from the road environment, and (2) outputs, corresponding to the driver's behaviors implemented into the real world via the car, which generate (3) feedback in the form of new inputs, requiring new adaptation (that is, new outputs) from the driver, and so on. Compared with the human-machine team, we can consider the input part as dynamic external factors, environment, vehicle, and traffic conditions.

11.2.3.1 Environment

Environment refers to the external conditions of the vehicle, including weather and roads. Among them, weather conditions affect the perception of humans and intelligent systems, such as rain, snow, and heavy fog. The road conditions affect the control strategies of people and intelligent systems, such as the width, curve, and type of the road.

11.2.3.2 Vehicle

Vehicles are dynamically changing and are composed of internal and external factors. Internal factors include the state of the vehicle and its own performance, and external factors include changes in position and speed caused by humans as well as intelligent systems controlling it. In addition, these changes also change the traffic environment and affect the intentions and decisions of other traffic participants.

11.2.3.3 Traffic conditions

Traffic conditions refer to all participants in the traffic system, including various other vehicles and pedestrians on the road. These elements share road resources with the vehicle, and their location changes, speed, and intentions affect the vehicle's action decisions.

The above dynamic influencing factors affect the situation awareness of people and intelligent systems, consequently affecting their intentions and control plans.

11.3 Conclusion

The present study distinguishes human-computer cooperation from the cognitive dimension and behavioral dimension, and combines it with driving tasks to propose a general cooperation framework, concentrating on the human-computer cognitive information exchange before and during the cooperation process. The human-machine team is goal-oriented, and joint intentions can point to any level of driving task cooperation. We believe that cognitive cooperation occurs earlier than behavioral cooperation. Therefore, we did not consider cooperative driving from the perspective of driving task cooperation level, but from the perspective of cognitive information exchange. Moreover, we considered the information elements needed for this kind of cognitive information exchange and its influential factors.

References

[1] M. Walch, T. Sieber, P. Hock, M. Baumann, M. Weber, Towards cooperative driving: involving the driver in an autonomous vehicle's decision making 2016. Available from: https://doi.org/10.1145/3003715.3005458.

[2] G.A. Boy (Ed.), The Handbook of Human-Machine Interaction: A Human-Centered Design Approach, CRC Press, 2017.

[3] C. Wang, A framework of the non-critical spontaneous intervention in highly automated driving scenarios, in: Proceedings of the 11th International ACM Conference on Automotive User Interfaces and Interactive Vehicular Applications, ACM, 2019, pp. 421–426.

[4] E. Salas, T.L. Dickinson, S.A. Converse, S.I. Tannenbaum, Toward an understanding of team performance and training, in: R.W. Swezey, E. Salas (Eds.), Teams: Their Training and Performance, Ablex, Norwood, 1992, pp. 3–29.

[5] A. Fiebich, N. Nguyen, S. Schwarzkopf, Cooperation with robots? A two-dimensional approach, in: C. Misselhorn (Ed.), Collective Agency and Cooperation in Natural and Artificial Systems. PSS, 122, Springer, Cham, 2015. Available from: https://doi.org/10.1007/978-3-319-15515-9_2.

[6] M. Zimmermann, K. Bengler, A multimodal interaction concept for cooperative driving, 2013 IEEE Intelligent Vehicles Symposium (IV), 2013, pp. 1285–1290. Available from: https://doi.org/10.1109/IVS0.2013.6629643.

[7] M. Walch, K. Mühl, J. Kraus, T. Stoll, M. Baumann, M. Weber, From car-driver-handovers to cooperative interfaces: visions for driver–vehicle interaction in automated driving, in: G. Meixner, C. Müller (Eds.), AutomotiveUser Interfaces. HIS, Springer, Cham, 2017, pp. 273–294. Available from: https://doi.org/10.1007/978-3-319-49448-7_10.

[8] D.D. Woods, E. Hollnagel, Joint Cognitive Systems: Patterns in Cognitive Systems Engineering, CRC Press, Boca Raton, 2006. ISBN 9780849339332.

[9] J.M. Hoc, Towards a cognitive approach to human-machine cooperation in dynamic situations, Int. J. Hum. Comput. Stud. 54 (2001) 509–540.

[10] K. Bengler, M. Zimmermann, D. Bortot, M. Kienle, D. Damböck, Interaction principles for cooperative human-machine systems, IT—Inf. Technol. 54 (4) (2012) 157–164. Available from: https://doi.org/10.1524/itit.2012.0680.

[11] E. Salas, C. Prince, D.P. Baker, L. Shrestha, Situation awareness in team performance: implications for measurement and training, Hum. Factors 37 (1995) 1123–1136.

[12] E. Salas, D.E. Sims, C.S. Burke, Is there a "big five" in teams work? Small Group. Res. 36 (5) (2005) 555–599.

[13] M. Walch, M. Woide, K. Muehl, M. Baumann, M. Weber, Cooperative Overtaking: Overcoming Automated Vehicles' Obstructed Sensor Range Via Driver Help, 2019, pp. 144–155. <https://doi.org/10.1145/3342197.3344531>.

[14] C. Wang, M. Krüger, C.B. Wiebel-Herboth, "Watch out!": prediction-level intervention for automated driving, in: 12th International Conference on Automotive User Interfaces and Interactive Vehicular Applications, 2020.

[15] R. Kridalukmana, L. Hai Yan, M. Naderpour, A supportive situation awareness model for human-autonomy teaming in collaborative driving, Theor. Issues Ergon. Sci. (2020). Available from: https://doi.org/10.1080/1463922X.2020.1729443.

[16] J.Y.C. Chen, K. Procci, M. Boyce, J. Wright, A. Garcia, M. Barnes, Report No. ARL-TR-6905 Situation awareness based agent transparency, U.S. Army Research Laboratory, Aberdeen Proving Ground, MD, 2014.

[17] D.A. Norman, The 'problem' with automation: inappropriate feedback and interaction, not 'over-automation', Philos. Trans. R. Soc. B Biol. Sci. 327 (1241) (1990) 585–593. Available from: https://doi.org/10.1098/rstb.1990.0101.

[18] F.O. Flemisch, K. Bengler, H. Bubb, H. Winner, R. Bruder, Towards cooperative guidance and control of highly automated vehicles: H-mode and conduct-by-wire, Ergonomics 57 (3) (2014) 343–360.

[19] J.M. Kraus, et al., Human after all: effects of mere presence and social interaction of a humanoid robot as a co-driver in automated driving, in: Adjunct Proceedings of the 8th International Conference on Automotive User Interfaces and Interactive Vehicular Applications, 2016.

Further reading

T.M. Allen, H. Lunenfeld, G.J. Alexander, Driver information needs, Highw. Res. Rec. 366 (1971) 102–115.

M. Aramrattana, T. Larsson, J. Jansson, C. Englund, Dimensions of cooperative driving, ITS and automation, 2015 IEEE IntelligentVehicles Symposium (IV), 2015, pp. 144–149.

C.E. Billings, Aviation Automation: The Search for a Human-Centered Approach, Erlbaum, Hillsdale, 1996.

Y. Bitan, J. Meyer, Self-initiated and respondent actions in a simulated control task, Ergonomics 50 (5) (2007) 763–788. Available from: https://doi.org/10.1080/00140130701217149.

K. Christoffersen, D.D. Woods, 1. How to make automated systems team players, Adv. Hum. Perform. Cognit. Eng. Res. 2 (2002) 1–13. Available from: https://doi.org/10.1016/S1479-3601(02)02003-9.

N.J. Cooke, J.C. Gorman, C.W. Myers, J.L. Duran, Interactive team cognition, Cogn. Sci. 37 (2) (2013) 255–285. Available from: https://doi.org/10.1111/cogs.12009.

M.R. Endsley, E.O. Kiris, The out-of-the-loop performance problem and level of control in automation, Hum. Factors J. Hum. Factors Ergon. Soc. 37 (2) (1995) 381–394. Available from: https://doi.org/10.1518/001872095779064555.

M.R. Endsley, M.M. Robertson, Situation awareness in aircraft maintenance teams, Int. J. Ind. Ergon. 26 (2) (2000) 301–325. Available from: https://doi.org/10.1016/S0169-8141(99)00073-6.

M.R. Endsley, From here to autonomy: lessons learned from human-automation research, Hum. Factors J. Hum. Factors Soc. 59 (1) (2017) 5–27.

E. Hollnagel, D.D. Woods, Cognitive systems engineering: new wine in new bottles, Int. J. Man-Mach. 18 (6) (1983) 583–600.

E. Hollnagel, D.D. Woods, Joint Cognitive Systems: Foundations of Cognitive Systems Engineering, Taylor and Francis, Boca Raton, 1995.

E. Hutchins, How a cockpit remembers its speeds, Cogn. Sci. 19 (3) (1995) 265–288. Available from: https://doi.org/10.1016/0364-0213(95)90020-9.

A.T. Jones, D. Romero, T. Wuest, Modeling agents as joint cognitive systems in smart manufacturing systems, Manuf. Lett. (2018). Available from: https://doi.org/10.1016/j.mfglet.2018.06.002.

Further reading

M. Kyriakidis, C.V.D. Weijer, B.V. Arem, R. Happee, The deployment of advanced driver assistance systems in Europe, ITS World Congress 2015 Proceedings, SSRN, Bordeaux, France, 2015. Available from: https://doi.org/10.2139/ssrn.2559034.

X. Li, A. Vaezipour, A. Rakotonirainy, S. Demmel, O. Oviedo-Trespalacios, Exploring drivers' mental workload and visual demand while using an in-vehicle hmi for eco-safe driving, Accid. Anal. Prev. 146 (2020) 105756.

O. McAree, W.H. Chen, Artificial situation awareness for increased autonomy of unmanned aerial systems in the terminal area, J. Intell. Rob. Syst. 70 (1–4) (2013) 545–555. Available from: https://doi.org/10.1007/s10846-012-9738-x.

R. Parasuraman, T. Bahri, J. Deaton, J.G. Morrison, M. Barnes, Theory and design of adaptive automation in aviation systems. Naval AirWarfare Center,Warminster, PA,Technical report NAWCADWAR-92 033-60 (1992).

SAE, Taxonomy and definitions for terms related to driving automation systems for on-road motor vehicles, Standard J3016_201609, SAE International, 2016.

M. Saffarian, J.C. de Winter, R. Happee, Automated driving: human-factors issues and design solutions, Proc. Hum. Factors Ergonomics Soc. Annu. Meet. 56 (1) (2012) 2296–2300. Sage, Los Angeles.

M. Walch, M. Colley, M. Weber, Driving-task-related human-machine interaction in automated driving: towards a bigger picture, in: The 11th International Conference, 2019.

Y. Shu, K. Furuta, An inference method of teams situation awareness based on mutual awareness, Cogn. Technol. Work. 7 (4) (2005) 272–287.

N.A. Stanton, M.S. Young, G.H. Walker, The psychology of driving automation: a discussion with professor Don Norman, Int. J. Veh. Des. 45 (3) (2007) 289–306. Available from: https://doi.org/10.1504/IJVD0.2007.014906.

N.A. Stanton, Distributed situation awareness, Theor. Issues Ergon. Sci. 17 (1) (2016) 1–7.

N. Strand, J. Nilsson, I.M. Karlsson, L. Nilsson, Semi-automated versus highly automated driving in critical situations caused by automation failures, Transp.Res. Part. F. Traffic Psychol. Behav. 27 (2014) 218–228.

C.D. Wickens, S.E. Gordon, Y. Liu, An Introduction to Human Factors Engineering, Addison-Wesley Educational Publishers Inc, New York, 1997.

CHAPTER 12

Design methodologies for human-artificial systems: an automotive augmented reality headup display design case study

Cuiqiong Cheng[1], Fang You[1], Preben Hansen[2] and Jianmin Wang[1]

[1]Car Interaction Design Lab, College of Arts and Media, Tongji University, Shanghai, P.R. China [2]Department of Computer and Systems Sciences, Stockholm University, Stockholm, Sweden

Chapter Outline

- 12.1 Introduction 149
- 12.2 Related works 150
- 12.3 Interaction design method framework 150
  - 12.3.1 Six-phase model of interaction design method framework 150
  - 12.3.2 Three perspectives of interaction design method framework 151
- 12.4 Automotive augmented reality head-up display design case study 152
  - 12.4.1 Project description 152
  - 12.4.2 Project implementation process 152
- 12.5 Conclusion 154
- References 154

12.1 Introduction

In the transition stage of intelligent products, technology is constantly evolving and intelligent products are becoming more complex. Interaction design plays an important role in reducing the complexity of intelligent products and improving user acceptance of them. With an emphasis on interaction design, academia has continuously developed interaction design methods and approaches since 2010s.

In this paper, we presented a systematic and comprehensive framework for interaction design method framework (IDMF), intended to be utilized as a practical design tool for collaborative team use. The IDMF design process comprises six distinct phases and involves interdisciplinary expertise.

12.2 Related works

In the software engineering discipline, researchers have proposed different life-cycle models, such as the waterfall model, spiral model, etc. In the design discipline, Zimmerman et al. [1] proposed six components of the design process, while Norman and Draper [2] introduced user-centered design methods, and Cooper et al. [3] developed goal-directed design methods. Additionally, numerous companies have begun establishing their own design process.

However, most design processes are just part of the whole product development process. Missing designers in the early process will have a degree of further negative influence on the whole process. Therefore this paper attempts to put forward the design process from a macro point of view, to elaborate product design systematically. In the IDMF, we propose that design runs through the entire product development process, with the aim of generating new creative designs based on emerging technologies to address user needs and solve problems effectively.

12.3 Interaction design method framework

The IDMF is composed of 102 different design methods distributed into six phases, involving interdisciplinary knowledge from design, information, and business domains. Serving as a comprehensive and systematic design method model, the IDMF is intended to be utilized as a practical design tool for collaborative teams, facilitating effective design processes:

1. Enhance the understanding of the methods used in human-artificial systems design projects.
2. Develop and expand single units of methodologies or processes within the IDMF for future use.
3. Utilize the methods established through experience and from real commercial projects to develop better design methodologies for teaching purposes.

12.3.1 Six-phase model of interaction design method framework

The six phases of the IDMF are as follows (Fig. 12.1):

1. Market research and design research. Analyze commercial markets, existing designs, and emerging technologies to gain insights into market needs and industrial situations.
2. User research. Conduct surveys and analysis to understand users' needs, motivations, and scenarios, identifying behavior patterns to inform intelligent product design.
3. Business model and concept design. Build a business model to effectively deliver value to users. Define with the stakeholders what service should be provided to users. Create innovative concept designs based on the business model, user scenarios, and emerging technologies.
4. Information architecture and design implementation. Organize information to build an information architecture that can be understood clearly by the user. Design task flows based on the understanding of user behavior and create prototypes for information layout and feedback mechanisms.

12.3 Interaction design method framework

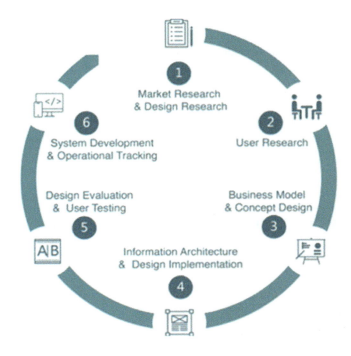

FIGURE 12.1

Six-phase model of interaction design method framework.

5. Design evaluation and user testing. Gather feedback from users and stakeholders to refine prototypes. Use evaluation frameworks and testing methods to find usability issues and iterate designs.
6. System development and operation tracking. Actively communicate design instructions to programmers, track operational data to analyze user behavior, and make changes for the next iteration.

The six phases of the IDMF are designed to help design practitioners learn design thinking, and understand the design process. The six phases of the IDMF are neither isolated nor linear, but interrelated, forming a spiral and iterative process of development. In practice, practitioners can adapt the framework to their needs and do not have to follow the IDMF phase by phase as a strictly linear process, but can view it as a flexible loop in the iterative process.

12.3.2 Three perspectives of interaction design method framework

A successful product is desirable, viable, and buildable. It must balance business and technology concerns with user concerns. Design, business, and information thinking are both important perspectives in successful product development. These three perspectives are represented as follows:

- **Design:** Focuses on user models and product designs. Identifying user goals, needs, and how to meet them.
- **Business:** Focuses on the business model and plan. Determining how to sell our service to target users, achieve profit, and sustain a business.
- **Information:** Focuses on content, data, and information technology. Identifying core technologies to be used and estimating technical feasibility.

It provides a structure for designers to acquire basic interdisciplinary knowledge and explore human-artificial systems design challenges. In each phase of the IDMF, different phases from design, business, and information perspectives are included.

12.4 Automotive augmented reality head-up display design case study

This section describes how to use the IDMF in the human-artificial systems design project undertaken at the anonymous automotive company, G.

12.4.1 Project description

This project aims to design an augmented reality head-up display (AR-HUD) for connected and automated vehicles (CAVs) with an advanced driver assistance system (ADAS) to improve driving safety and driving experience. This was a brand-new design project. We cooperated with a business manager, a project manager, and several developers from company G. Thanks to the IDMF, we designed the project process together and proposed emerging technology-based designs (Fig. 12.2).

12.4.2 Project implementation process

To make our study more manageable, we focus on Loop 1 of the project process.

FIGURE 12.2

Company G's augmented reality head-up display design process according to the six-phase interaction design method framework model.

12.4.2.1 Phase 1: market research and design research
In this phase, we employed design methods tailored to different perspectives. From an information perspective, we conducted thorough literature reviews to explore emerging technologies related to CAVs. Using competitor analysis from a business perspective, we assessed the strengths and weaknesses of competitors and identified potential opportunities for our enterprise. Additionally, utilizing the critical incident technique from a design perspective, we engaged drivers to recall and share significant problems they encountered while driving. Furthermore, we preliminary determined business objectives, product functions, and target users.

12.4.2.2 Phase 2: user research
The design methods selected for this phase include questionnaire surveys, in-depth interviews, and naturalistic observation from an information perspective, with storyboard and scenario design employed from a design perspective. Through questionnaire surveys on safe driving, we obtained initial insights into factors that affect safe driving, users' demand, and driving behavior. Our findings revealed that the driving environment in China is complicated and drivers are easily distracted. To better understand drivers' need for safe driving, we conducted in-depth interviews with nine drivers and sorted out typical driving scenarios and difficulties. We learned drivers' real behavior and reactions through naturalistic observation. Subsequently, utilizing storyboard techniques we visualized user expectations regarding information and functionalities, providing valuable insights during discussions with company G. By using scenario design, we meticulously created possible scenarios to enhance the narrative and bring the design concepts to life.

12.4.2.3 Phase 3: business model and concept design
The design methods selected in this phase are brand positioning from a business perspective, affinity diagram from an information perspective, and blueprints from a design perspective. We recognized that safety is the most crucial aspect for users so we positioned our product under the brand attribute of "safety." We converted drivers' needs into notes and organized them into affinity diagrams, revealing four primary information needs: traffic information, environmental information, vehicle status, and driver status. Together with company G, we crafted a blueprint to map the entire driving experience to identify potential opportunity points.

12.4.2.4 Phase 4: information architecture and design implementation
The design methods selected in this phase are organization system design from an information perspective, layout design, paper prototyping, interface style design, and interface component design from a design perspective. A clear organization system could help drivers find needed information immediately. We divided information into three categories: (1) core information, which is about safety, such as ADAS; (2) secondary accessibility information, which is personalized, such as navigation; and (3) random information, which is real-time information about the road and surroundings based on networking and big data. To reduce drivers' workload, we divided the layout of HUD into three fixed blocks: warning area, normal area, and auxiliary area. Paper prototyping allowed us to simulate the flow of information switching quickly. For testing design in a real situation, we designed several sets of interface styles and interface components.

12.4.2.5 Phase 5: design evaluation and user testing

The design methods selected in this phase are cognitive walkthrough and usability testing from a design perspective. Since, drivers' attention is valuable during driving, especially for novice drivers, we used cognitive walkthroughs to identify how easy it is for novice drivers to obtain information with the AR-HUD. In addition, we conducted usability testing on a driving simulator to figure out usability issues. Although most of the participants completed the tasks successfully, there were several things we needed to improve in the next loop.

12.4.2.6 Phase 6: system development and operation tracking

Since the design of Loop 1 needs improvement, we did not enter Phase 6 and instead transitioned directly into Phase 4 of Loop 2.

From the automotive AR-HUD case study, the usability and feasibility of applying IDMF to a human-artificial systems design have been verified.

12.5 Conclusion

This paper presents a systematic and comprehensive IDMF for human-artificial systems. To facilitate designers' work in close collaboration with other stakeholders and propose emerging technology-based designs, we deconstructed the IDMF according to the three dimensions of design, information, and business, which offer basic interdisciplinary knowledge. We offer a practical guide to using the IDMF by providing a high-level summary of the automotive AR-HUD design.

References

[1] J. Zimmerman, J. Forlizzi, S. Evenson, Taxonomy for extracting design knowledge from research conducted during design cases, in: Proceedings of the Futureground; 2004.
[2] D.A. Norman, S.W. Draper, User Centered System Design: New Perspectives on Human-Computer Interaction, L. Erlbaum Associates, Hillsdale, 1986.
[3] A. Cooper, The Inmates are Running the Asylum, Indianapolis, IN, Sams, 1999.

CHAPTER 13

Design factors of shared situation awareness interface in human-machine co-driving

Fang You[1], Xu Yan[1], Jun Zhang[1,2] and Wei Cui[1,2]

[1]Car Interaction Design Lab, College of Arts and Media, Tongji University, Shanghai, P.R. China [2]College of Design and Innovation, Tongji University, Shanghai, P.R. China

Chapter Outline

13.1 Introduction 156
13.2 Related works 156
    13.2.1 Situation awareness 156
    13.2.2 Shared situation awareness 157
    13.2.3 Interface designs of human-machine co-driving 158
13.3 Design methods 158
    13.3.1 Abstraction hierarchy analysis 158
    13.3.2 Abstraction hierarchy analysis of shared situation awareness human-machine interaction interfaces 159
    13.3.3 Four factors of shared situation awareness interface 160
13.4 Simulation experiment 160
    13.4.1 Experimental subjects 160
    13.4.2 Experimental design 161
    13.4.3 Selection of experimental scenarios and requirement analysis of shared situation awareness information 161
    13.4.4 Experimental environment 163
    13.4.5 Experimental process 164
    13.4.6 Experimental evaluation method 164
13.5 Results 165
    13.5.1 Eye movement 166
    13.5.2 Situation awareness 167
    13.5.3 Usability 167
    13.5.4 Task response time and task accuracy 168
    13.5.5 Interview 169

13.6 Discussion	170
13.7 Conclusions	172
Further reading	172

13.1 Introduction

With the development of intelligent driving technologies, advanced automated systems are becoming equipped with driving abilities that can coexist with those of human drivers, both of which are capable of independently controlling the vehicles to various extents [1]. The American Society of Automotive Engineers (SAE) defines six levels (L0−L5) of driving automation. L2 automated driving systems can control the vehicle horizontally or vertically, but the whole process still requires close supervision and observation of the vehicle's behavior and surrounding environment by human drivers. When the control of vehicles is gradually transferred to intelligent systems, the cooperation between human drivers and intelligent systems has higher requirements. Walch et al. [2] argue that there are four basic requirements for human-machine collaborative interfaces: mutual predictability, directability, shared situation representation, and trust and calibrated reliance on the system. Among the four requirements, achieving a shared situation representation stands out as the primary challenge in human-machine cooperation in intelligent driver's compartments. Here intelligent systems can obtain information about the driver by recognizing the driver's intentions and predicting their driving behavior, but the information acquired by the driver from the automated systems presents a paradox of heavy burden and little information.

Therefore research on human-machine shared situation representation in intelligent driving, especially research on how automated systems can effectively share their situation awareness with human drivers, is of vital significance for the safety, credibility, and cooperation efficiency of human-machine co-driving in the future, but there is a lack of research in the field to guide the design of shared situation awareness (SSA) interfaces in human-machine co-driving. This paper aims to enhance the SSA of drivers and intelligent systems in driving scenarios. Using abstraction hierarchy analysis to break down the design goal layer by layer, generic element types of shared situation representation interfaces are proposed to guide the design of human-machine collaborative interfaces and apply them to design cases.

13.2 Related works

13.2.1 Situation awareness

In its early stages, situation awareness (SA) primarily centered around individuals. It is an essential concept in the ergonomics study of individual tasks and is defined as the ability to perceive, understand, and predict elements in the environment in certain temporal and spatial contexts.

During automated driving, drivers are likely to take their eyes off the road and are often unable to monitor the driving situation. They need to actively scan relevant areas of the driving scenarios to maintain SA acquisition and an adequate SA level.

As research on SA goes deeper, it has been extended from the individual level to the team level, indicating the shared understanding of a situation by team members at some point (Fig. 13.1). The SA of a team is deemed as an important factor that influences team output, evaluates team performance, and improves team performance.

13.2.2 Shared situation awareness

SSA is critical for teams to cooperate efficiently and adapt to dynamic challenges when team members are working together to complete a task, and this is true in various fields that require cooperation to complete activities, such as railway operations, nuclear power plant management, military teamwork, and crisis management.

SSA reflects the extent to which all team members accurately understand the information required to achieve the goals and subgoals related to their joint mission. Studies have shown that for drivers engaged in nondriving-related tasks during automated driving, it takes at least 7 seconds to acquire sufficient SA. Once the system sounds an alarm, the driver needs to take a series of actions, including shifting their attention from nondriving-related tasks to the traffic-related environment, acquiring SA, making decisions, and making a move. Some of the actions are dealt with simultaneously, which significantly increases the driver's burden. With the enhancement of automation capabilities, the system can perform a wider range of driving tasks, with drivers naturally reducing their attention to driving tasks. The obtained SA information is insufficient, resulting in the driver being unable to monitor the vehicle and traffic environment and deal with emergencies properly. The ability of drivers to understand the driving situation is weakened, thereby affecting driving safety. Therefore the sharing of SA between the driver and the intelligent system in the driving environment, especially the SA sharing of the system with the human, is of great importance and will help the driver to understand the decision of the vehicle, predict the behavior of the vehicle, and make an early judgment.

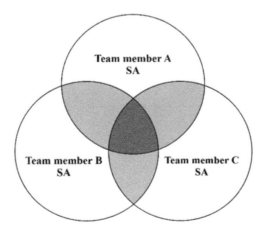

FIGURE 13.1

Team situation awareness model.

From Endsley M R. A model of inter-and intrateam situation awareness: Implications for design, training, and measurement[J]. New trends in cooperative activities: Understanding system dynamics in complex environments, 2001: 46–67.

13.2.3 Interface designs of human-machine co-driving

It has been established that situation awareness shared between humans and computers is crucial for safe driving, especially in potentially dangerous emergency scenarios. However, there is much research on how the vehicle or intelligent driving system shares the SA information with human drivers and the methods or channels through which this information is transmitted. Kirti Mahajan [3] believed that the warning sound issued by the system could help drivers realize the danger and respond to the takeover request more quickly. Cristy Ho et al. [4] demonstrated that the prompt of tactile vibration has a positive effect on the driver's attention returning to the right direction. It is believed that each stage of automated driving should have an appropriate interface to ensure that the driver can establish accurate SA. Therefore the multichannel design of vision, hearing, and touch can help the driver obtain the SA of the intelligent driving system. However, the visual channel is still the main source for drivers to obtain information in driving scenarios. Therefore it is crucial to study how the human-machine interface (HMI) helps human drivers to obtain the SA of an intelligent vehicle.

Wang's research showed that human-system cooperation in interface design and implementation of predictive information could improve the experience and comfort of automated driving effectively. Guo et al. formulated three principles for HMI design to improve efficient cooperation between drivers and the system. They are "displaying the driving environment," "displaying the intention and available alternatives," and "providing users with a way to select alternatives." Zimmermann et al. proposed that in human-machine cooperation, the goal of the HMI should be to display and infer the intention of users and machines and to convey dynamic system task allocation. Kraft et al. proposed four principles and applied them to the design of the HMI to achieve collaborative driving. They are "displaying the driving environment," "explaining the intentions of the partners," "suggesting actions," and "successful feedback." There have been many studies on how to promote human-machine cooperation, but there is still a lack of relevant studies on how to improve the HMI design of SSA between human and intelligent systems during driving.

Based on the above SSA and HMI design theories, we put forward some design factors of SSA HMI in automated driving of SAE L2–L3 and provide direct suggestions for rapid HMI design practice in related scenarios.

13.3 Design methods
13.3.1 Abstraction hierarchy analysis

The abstraction hierarchy (AH) is an analysis tool used to disassemble work domains and unravel hierarchical structures. It is the main method for work domain analysis and is widely applied in the design of complex interfaces in the field of human-machine interaction. The AH describes the work domain according to five levels of abstraction (Table 13.1), functional purposes, values and priority measures, purpose-related functions, object-related processes, and physical objects. Each layer is connected to the next by a structural framework. The purpose of the decomposition hierarchy AH is to decompose a system into subsystems, each subsystem into functional units, each unit into subsets, and each subset into components, to analyze the specific

Table 13.1 Descriptions of abstraction hierarchy.

Old labels	New labels	Description
Functional purposes	Functional purposes	The purposes of the work system and the external constraints on its operation
Abstract functions	Values and priority measures	The criteria that the work system uses for measuring its progress toward functional purposes
Generalized functions	Purpose-related functions	General functions of the work system that are necessary for achieving the functional purposes
Physical functions	Object-related processes	Functional capabilities and limitations of physical objects in the work system that enable the purpose-related functions
Physical form	Physical objects	Physical objects in the work system that afford the object-related processes

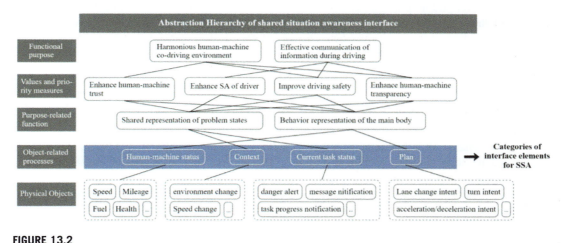

FIGURE 13.2

Abstraction hierarchy of shared situation awareness interface.

functional architecture and morphological output of the system. The analysis process of the AH is based on aspects such as the environment, technology, and human capabilities and relies on experienced domain experts.

13.3.2 Abstraction hierarchy analysis of shared situation awareness human-machine interaction interfaces

The AH approach has been widely used in the design of complex systems in fields such as aviation and shipbuilding but is less commonly applied in the design of interfaces for increasingly complex intelligent cars. This paper proposes the four types of information that should be included in an SSA interface (Fig. 13.2) through the AH method to facilitate more efficient communication and cooperation between human drivers and intelligent vehicles. The whole analysis process of AH was jointly discussed by five domain experts and scholars. They extensively discussed how to organize the five levels of AH when the design goal is to enhance SSA. After reasoning, analysis, and induction, the AH results are summarized, as shown in Fig. 13.2. The first hierarchy of the AH system aims to achieve an effective communication mechanism and a harmonious human-machine co-driving environment. The second hierarchy values enhancing the driver's SA of the environment and enhancing the transparency of the information passed to the driver by the system, thus boosting trust in human-machine cooperation and improving driving safety. The third hierarchy describes functions related to the purposes (i.e., the interpretation of general functions) and realizes the value dimension mentioned in the previous hierarchy through the shared representation of the problem or task state and that of the activities of the actors in the cooperative team. The fourth hierarchy describes functions related to the objects (i.e., further reasoning, analysis, and generalization of the general functions mentioned in the previous hierarchy), thus obtaining the following four types of information: human-machine state, context, current task status, and plan, which help the system to deliver its shared scenarios to the driver and help the driver understand the connotations of the system decision. The final hierarchy of the model contains specific representation elements, such as speed, fuel level, environmental changes, hazard warnings, and lane change intentions. The elements are to be presented on the interface of which scenarios need to be analyzed and selected according to the specific needs of the scenario.

13.3.3 Four factors of shared situation awareness interface

This paper, through the above analysis, reasoned and proposed four generic types of interface elements for the shared situation representation of human-machine collaborative interfaces, namely, human-machine state, context, current task status, and plan or motive. The human-machine state explains the current level, attribute, or normal state of both or one of the parties of SSA, such as the display of automated and manual driving, fuel level, speed, and other driving states of the car. The context contains both temporal and spatial dimensions. The temporal context describes the changes taking place in the intelligent vehicle-mounted system over a period of time, and the spatial context explains the relationship between the car and the surrounding vehicles. The current task status, related to specific tasks, explains the progress of the tasks, such as displaying the status of "searching" when using the search function to explain to the user the transparency of the system's operation. The plan shows the reasons why the system is about to manipulate the vehicle to perform a specific behavior, enhancing the transparency of information in human-machine communication.

A driving scenario contains four modules: human, intelligent system, environment, and task. The human and the intelligent system communicate through the HMI, which contains the four factors of SSA when completing a task (Fig. 13.3).

13.4 Simulation experiment

13.4.1 Experimental subjects

A total of 16 subjects (8 males and 8 females) aged from 25 to 40 were selected (mean = 32.1, SD = 4.61), all of whom had 3–13 years of driving experience (mean = 7.3, SD = 3.11), with normal visual acuity.

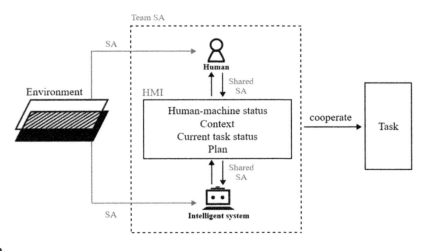

FIGURE 13.3

Abstraction hierarchy of shared situation awareness interface.

Table 13.2 Comparison of experimental schemes.

	Human-machine status	Context	Plan	Current Task status
Scheme 1	√	√	√	√
Scheme 2	√	√	√	/
Scheme 3	√	√	/	/

Notes: *The symbol "√" means the scheme contains this element. The symbol "/" means the scheme does not contain this element.*

13.4.2 Experimental design

A within-subjects experiment was adopted in this study. The experimental variable for the three schemes was the number of types of SSA human-machine collaborative interface elements included in the interface (Table 13.2). Scheme 1 contained all four interface elements, Scheme 2 contained three types of elements (human-machine state, context, and current task status), and Scheme 3 contained only two types of elements (human-machine state and context).

13.4.3 Selection of experimental scenarios and requirement analysis of shared situation awareness information

The design and experiment were carried out by selecting typical driving scenarios to verify the effectiveness of the HMI design using the four elements of an SSA interface proposed above.

A potentially dangerous scenario of a vehicle coming from the rear was selected in this experiment. In this scenario, the system needs to share the dangerous intentions of the vehicle from the

162 Chapter 13 Design factor of shared situation awareness interface

rear that it senses and the safety status of the vehicle with the driver in time to help the driver react appropriately to avoid risks and improve driving safety.

According to the aforementioned four types of interface elements for shared situation representation and taking into account the driving environment, road characteristics, and driving habits of the scenario with a vehicle approaching from the rear, the information requirements for the scenario were obtained, as shown in Table 13.3.

The overtaking process includes the following five stages in chronological sequence (Fig. 13.4). The vehicle coming from the rear in the left lane is about to enter the field of vision, the vehicle with dangerous intention is approaching from the rear in the left lane, the vehicle in the left lane is abreast of the present vehicle, the vehicle in the left lane overtakes the present vehicle and gradually drives away, and the vehicle in the left lane is about to leave the field of vision. The design of

Table 13.3 Descriptions of levels of abstraction.

Factors	Descriptions
Human-machine status	Display the current automated driving status, current vehicle speed, etc.
Plan	Indicative message of the presence of vehicles behind
Current task status	Prompt for the level of vehicle hazard intent
Context	Position change of the oncoming vehicle from behind, the relative positional relationship between the oncoming vehicle and own vehicle

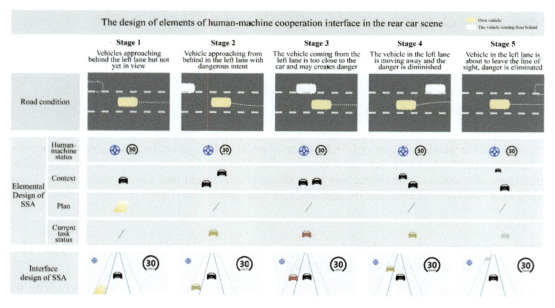

FIGURE 13.4

Design of elements of human-machine cooperation interface in the rear car scene.

interface elements for the five stages of the scenario was visualized (not all four types of information are included in each stage) based on information requirements.

- **Human-machine status:** The automated driving status—the icon of using the steering wheel (blue color means the automated driving function is on); the current speed—a more common circular icon containing a number that varies according to the actual speed.
- **Context:** A car approaching from the rear—the icon of two cars, with their relative positions in sync with the experiment process.
- **Plan (only in the first stage):** A potentially dangerous vehicle is about to appear in the view from the left rear—a gradient color block displayed in the left rear lane on the interface.
- **Current task status:** The level of danger the driver is in—color added to the vehicle on the interface, with gray, yellow, and red indicating no danger, mildly dangerous, and highly dangerous, respectively.

A prototype interface for each stage can be obtained by combining the aforementioned design elements through a reasonable layout that can be used in the subsequent evaluation experiment. A prototype interface was designed for each scheme (Fig. 13.5) for subsequent experiments.

13.4.4 Experimental environment

The experiment was based on a self-developed driving simulation system that includes three modules: vehicle, scenario, and driving monitoring and analysis (Fig. 13.6). The vehicle module is the test bench simulating driving, composed of parts such as acceleration and deceleration pedals, seats, steering wheel, and rearview mirror. The scene module builds simulation scenarios based on Unity software to simulate factors such as the weather, road conditions, and surrounding vehicles in the

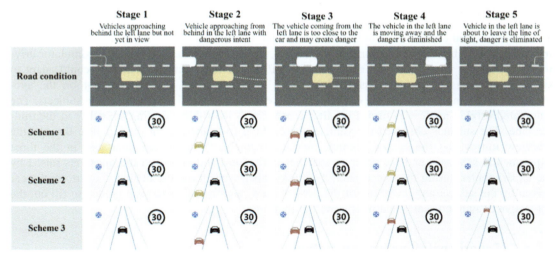

FIGURE 13.5

Low-fidelity prototype of the testing interface.

FIGURE 13.6

Experimental site.

real driving environment. The module of driving monitoring and analysis monitors the behaviors of the user's head and feet and records driving data. Eye-tracking devices were used in the experiment to capture dynamic eye movement data.

13.4.5 Experimental process

The subjects filled out the basic information form and the informed consent form before the experiment. They were briefed on the purpose and task of the experiment, as well as about simulator driving. The subjects completed safe driving at 30 km/h for 2 min to ensure that they could adapt to simulator driving and maintain safe driving in a task-free state. The subjects put on the eye-tracking device and completed the calibration. The experimenter assigned tasks to the subjects. The automated driving mode was turned on after a beep, with the subject driving smoothly in the middle lane. The subjects observed the vehicle approaching from the rear in the left lane through the rearview mirror and the central control interface (namely, the designed interfaces in the three experimental schemes, and the order of schemes is random) and answered the questions raised by the experimenter during the experiment. The task finished when the car behind it overtook the car driven by the subject and drove out of the field of view, and the subjects were guided to fill out the subjective evaluation scales and interviewed at that time.

13.4.6 Experimental evaluation method

The subjective evaluation scales of the subjects and objective data analysis were adopted in the experiment, coupled with postexperiment interviews. The subjective scales included the situation awareness global assessment technique (SAGAT) scale used to test the level of SA and the after-scenario questionnaire (ASQ) used to assess the usability of a scheme. Objective data included task response time and task accuracy, which were used to test task performance.

13.4.6.1 Situation awareness global assessment technique
The SAGAT scale evaluated the level of SA by examining three dimensions: perception, understanding, and prediction. Subjects needed to answer SA questions that had correct answers, such as how many times the color changed of the car in another lane on the screen during the experiment just now. One point was awarded if the subject answered each question correctly; otherwise, there were no points.

13.4.6.2 After-scenario questionnaire
The ASQ is available after each task and can be performed quickly. The three questions in ASQ refer to three dimensions of the usability of the central control interface, namely, ease of task completion, time required to complete tasks, and satisfaction with support information, which were scored using 7-point graphic scales. The options were set from 1 to 7, with higher scores referring to a higher level of performance.

13.4.6.3 Task response time and task accuracy
The task response time refers to the time from the start of the task to the time when the subject first notices the approaching car behind them, which was collected by the subject pressing a button on the steering wheel. Task accuracy refers to the subjects' correct judgment of whether the vehicle is in potential danger and the danger level (low, medium, and high) in stage 2. The accuracy data were collected by thinking aloud, in which the experimenter asked the subjects questions when the task reached stage 2, and the subjects responded promptly. In fact, in stage 2, the vehicle is in potential danger, but the level is not the highest. If the subject correctly judges the dangerous situation of stage 2, the accuracy of this task is considered 100%; otherwise, the accuracy is 0.

13.4.6.4 Quick interview
After all tasks were completed, the subjects were interviewed briefly and quickly. The questions included whether they noticed the difference between the different schemes, how they viewed the three schemes, and which scheme they thought was the best and why. Furthermore, the subjects were asked how much they trusted the information prompts on the interface subjectively. The purpose of the interview is to gain insight into more potential subjective feelings beyond the scale assessment as a supplement and possible explanation of the assessment.

13.5 Results
One-way analysis of variance (ANOVA) with post hoc pairwise comprehension (Tukey's Honestly Significant Difference) was conducted for the experiment (alpha = 0.05, effect size $f = 0.46$) with the scheme as the within-subject factor. The experimental results are divided into the following parts.

13.5.1 Eye movement

The eye movement of all subjects in the process of completing the task was superimposed. According to the thermodynamic diagram of the eye movement of the three schemes (Fig. 13.7), the central control area of Scheme 3 had more concentrated sight lines than that of Scheme 1, on both of which sight lines stayed longer than that of Scheme 2. Specifically, there were more scattered weak hot zones near the central control area of Scheme 1, and the hot zones near the central

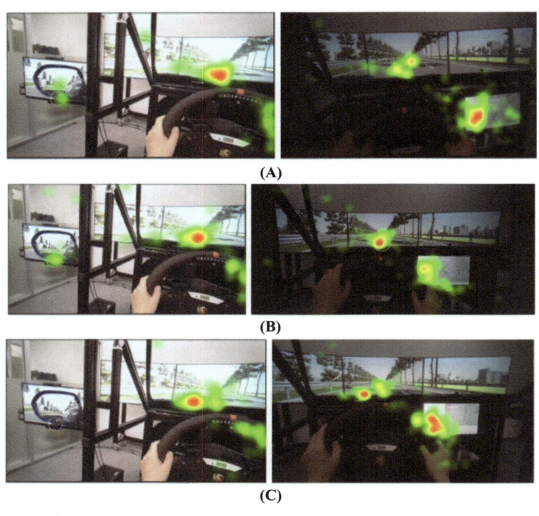

FIGURE 13.7

Superimposed schematic diagram of the subject's eye movement heat zone: (A) Scheme 1; (B) Scheme 2; and (C) Scheme 3.

control area of Scheme 2 were too weak to form a concentrated red area, while the hot zones in Scheme 3 were concentrated in the location where the car appeared on the screen, indicating that the red car on the interface of Scheme 3 is more attractive to the subjects. In terms of the rearview mirror area, Schemes 1 and 2 presented hot zones, while Scheme 3 did not, suggesting that the subjects in Scheme 1 and Scheme 2 had a more comprehensive observation of vehicles coming from the rear in the real environment and a relatively better perception of the environment.

13.5.2 Situation awareness

The scores of the SAGAT scales after normalization processing (Fig. 13.8) showed that the overall SA scores of Scheme 1 and Scheme 2 were the same, slightly lower than Scheme 3 (0.74 ± 0.06) by 0.03 points. In terms of prediction, however, Scheme 1 scored higher than Scheme 3 by 0.19 points, and Scheme 2 scored higher than Scheme 3 by 0.16 points, indicating that the subjects had a stronger ability to predict the danger behind them when using Scheme 1, which is also one of the factors that have a great impact on driving safety in dangerous driving situations. Nonetheless, more effort is required in perception and understanding, maybe due to the large number of elements and changing colors on the interface.

13.5.3 Usability

Scores of the ASQ scale reflect the usability of the design schemes. As shown in Fig. 13.9, Scheme 1 scored the highest in terms of the overall ASQ score, followed by Scheme 2 and then Scheme 3. The scores of satisfaction with support information of the three schemes varied drastically, with the largest gap of 1.13 points. Overall, the scores of various indicators in Scheme 1 were balanced, with an overall score of 6.15 ± 0.31 points (out of 7 points), reaching a relatively high level, indicating that Scheme 1 outperforms the other two schemes in terms of usability. Through one-way

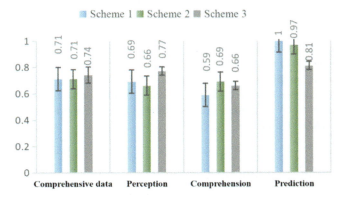

FIGURE 13.8

Situation awareness assessment results.

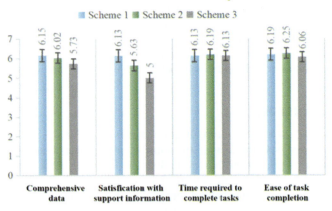

FIGURE 13.9
Usability assessment results.

Table 13.4 Results of the analysis of correlation.

		Satisfaction with support information
Prediction	r	0.488**
	p	.000 < .001

**p < .01.

ANOVA, only the dimension "satisfaction with support information" showed a significant difference among the three schemes (F (2.45) = 4.442, p = .017 < .05, effect size f = 0.46). Furthermore, the Post hoc tests (Tukey's Honestly Significant Difference) revealed that there was a significant difference between Scheme 1 and Scheme 3 (p = .013 < .05), with no significant difference between other schemes.

Correlation analysis was conducted between SAGAT and ASQ data (including each subdimension). The results (Table 13.4) showed that there was a moderate positive correlation (r = 0.488**, p = .000 < .001) between "prediction" of SA and "satisfaction with support information" of usability, which suggested that the effective design of help information could improve the situation awareness and prediction level of human drivers. Other than that, there was no correlation among the subdimensions of SAGAT and ASQ.

13.5.4 Task response time and task accuracy

Task response time and the accuracy of identifying dangerous conditions are important factors reflecting task performance, while task performance, to some extent, reflects the effects of the HMI design. According to the statistics (Fig. 13.10), the values of task accuracy of the three schemes

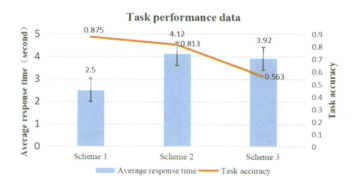

FIGURE 13.10

Average reaction time and correct rate results.

Table 13.5 Results of the analysis of variance.

	Scheme (mean ± standard deviation)			F	p
	1	2	3		
Reaction time	2.50 ± 0.75	4.12 ± 0.75	3.92 ± 0.64	24.217	.000**

**p < .01*.

were 87.5% (Scheme 1), 81.3% (Scheme 2), and 56.3% (Scheme 3), indicating that the subjects in Scheme 1 had the highest accuracy in identifying potentially dangerous conditions. In Scheme 3, most of the subjects thought that the danger happened earlier, and it was found, from the follow-up interview, that the subjects could not accurately perceive the change in hazard because the color of the icon of the car representing the current task status on the interface was always red.

The average task response time of Scheme 1 was 2.5 s, that of Scheme 2 was 4.12 s, and that of Scheme 3 was 3.91 s. The subjects in Scheme 1 performed better in terms of response time compared with those in Schemes 2 and 3. In other words, subjects in Scheme 1 made more rapid and effective judgments about approaching vehicles based on their observations. According to the one-way ANOVA of the data on the subjects' task response time, the results (Table 13.5) showed a significant difference among the three schemes (F (2.45) = 24.217, p = .000 < .001, effect size f = 0.46). Post hoc tests (Tukey's Honestly Significant Difference) revealed that the response time of Scheme 1 was significantly lower than that of Schemes 2 and 3 (p = .000 < .001), with no significant difference between Schemes 2 and 3 (p = .694 > .05).

13.5.5 Interview

The interview responses were coded, analyzed, and summarized by two researchers and verified to be 82.5% interrated congruency, meeting the congruency requirement. The main conclusions are shown in Table 13.6. We also annotated the proportion of subjects who mentioned this content in

Table 13.6 The results of interviews.

	Question	Answer 1 (rate)	Interview results Answer 2 (rate)	Answer 3 (rate)
1	Notice the differences?	Yes (16/16)		
2	Prefer which scheme and why?	Scheme 1 (12/16) Prompt for the coming car and help to estimate riskColor change demonstrates state of dangerFeeling more reliable	Scheme 2 (3/16) Color change demonstrates state of dangerColor block is not visible enough to be observed for a short time	Scheme 3 (1/16) Not used to too many color changes in the interface personally
3	Attitude of subjective trust	Trust Scheme 1 the most (12/16)	Trust Scheme 1 and Scheme 2, but not Scheme 3 (3/16)	Trust actual observation more than interface display (1/16)
4	Other subjective feelings	Driving habits may affect experimental observations (3/16)	Red is more likely to attract attention (15/16)	Being red all time will bring a sense of tension (15/16)

all subjects. In general, all subjects indicated that they noticed the differences between the schemes, and most (12/16) of the subjects in the interview had a stronger subjective preference for Scheme 1. They believed the yellow block element (plan) helped them predict cars coming from behind. The color of the other car (current task status) on the interface during the process made them more aware of the danger level they were facing at that time. Additionally, 12/16 of the subjects believed that the system reliability was higher when Scheme 1 was presented, resulting in higher subjective trust. Meanwhile, 3/16 of the subjects said that they observed the car conditions through the rear-view mirror instead of paying attention to the HMI during the automated driving process, so they may have missed some information during the experiment. However, this long-term inherent driving habit is not easy to change.

13.6 Discussion

The main purpose of this paper was to study how the interface design of vehicles with automated driving functions can share SA with human drivers better. We proposed four factors of the SSA interface by applying the analytical approach of AH and relying on expert experience and evaluated the experimental design in terms of situation awareness, usability, and task performance. Overall, Scheme 1, containing the complete four SSA interface factors (human-machine state, context, plan, and current task state), had better comprehensive performance, followed by Scheme 2, containing three factors, and Scheme 3, containing only two kinds of factors, had the worst performance.

In terms of situation awareness, there was little difference in the total score among the three schemes, with Scheme 3 scoring slightly higher than the other two schemes. However, by comparing the scores of the three subdimensions, it could be seen that Scheme 3 mainly got higher scores in the "perception" dimension than the other two schemes, which led to a high total score, but there was no better performance in the "comprehension" and "prediction" dimensions. What we found in our interviews was that the subjects thought that the colors on the screen would attract their attention. Red would make them more likely to notice changes, particularly. Scheme 3 happened to be the scheme where the red element appeared for the longest time, which may impact perception. In addition, Scheme 1 got full marks in the "prediction" dimension, which meant that all subjects obtained the meaning of the yellow color block representing "plan" accurately. This result suggests that in future interface design, a transparent presentation of the "plan" element could help drivers obtain a better level of prediction of driving activity.

For usability, Scheme 1 is generally higher than Schemes 2 and 3, especially in the dimension of "satisfaction with support information." In the "time required to complete tasks" and "ease of task completion" dimensions, the scores of the three schemes are very close; although Scheme 2 is slightly higher than Scheme 1, there is no significant difference. From the perspective of experimental design, it may be because the task is too simple to cause time urgency and difference in task difficulty.

In terms of task performance, the task reaction time and accuracy mainly measure whether the subjects can capture the important prompt information quickly and accurately to illustrate the helping effect of the design scheme. The task reaction time of Scheme 1 was significantly shorter than that of the other two schemes, which may be because the subjects observed the interface, understood the meaning of the yellow block element (plan), and predicted the incoming car correctly. This suggests that an early presentation of the "plan" factor for potential hazards on the interface may help drivers to react quickly to driving situations. Task accuracy is higher in Scheme 1 compared with Scheme 2, and Scheme 2 is higher compared with Scheme 3. Scheme 1 showed two more of the four factors and achieved the highest data in terms of task accuracy. This suggests that we could consider adding "current task status" and "plan" factors on the interfaces to help drivers make more accurate judgments about current driving activities, such as the safety status of the vehicle in the future.

Interviews and eye-tracking heat maps were used to assist in the interpretation of the overall experimental results. Through the interview, the subjects' evaluation can explain some of their subjective feelings. Most of the participants agreed that Scheme 1 gave them a better prompt for the car coming from behind. However, some people mentioned that due to their long-term driving habits, they may have habitually watched the rearview mirror and missed the prompt information from the center control interface. Some subjects mentioned that the yellow block is not obvious enough, which reminds us that when designing, we should consider the user's past experience, and when displaying important information, we should consider the proper use of color, shape, motion effect, and so on to make it more easily perceived. For example, using more eye-catching colors or adding an animated display for icons on the interface, but paying attention to the duration and avoiding meaningless visual occupancy. The eye movement heat map showed the main areas where the subjects' eyes focus. Subjects in Scheme 1 not only observed the central control interface fully but also took the road conditions in the rearview mirror into account, which was closer to the expected observation state in the experiment.

This study completed the whole experimental research based on a driving simulator, which has the advantages of being easy to operate, high safety, and saving time. However, the three screens in front that simulate the external road environment are not large enough to fully cover the visual field of the subject's eyes, which weakens the realism of driving. In the future, we will consider increasing the screen size or using a wraparound screen to create immersive driving. In addition, the "plan" factor only appears in Stage 1 of the experimental design, and it is not visually obvious, which may affect the evaluation results of Scheme 1. In the future, we will consider adding the design of the "plan" factor reasonably in more stages, such as adding the forecast route of the vehicle on the interface from Stage 2 to Stage 4. In addition, we will consider expanding driving scenarios, such as lane changing and overtaking, to evaluate the design value of the four factors of the SSA interface in more typical driving scenarios.

13.7 Conclusions

Based on the SSA theoretical background and AH analysis method, we proposed four factors of interface design to enhance human-machine SSA. Through experiments, we verified that the factors "plan" and "current task status" contribute to the level of SA prediction, usability, and task performance in the human-machine co-driving scenario of automated driving. In addition, statements of the interview implied that the interface containing the four factors had a positive impact on the subjective trust attitude of humans toward the automated system.

References

[1] C. Zong, C. Dai, D. Zhang, Human-machine interaction technology of intelligent vehicles: current development trends and future directions, China J. Highw. Transp 34 (6) (2021) 214.
[2] M. Walch, K. Mühl, J. Kraus, et al., From car-driver-handovers to cooperative interfaces: Visions for driver–vehicle interaction in automated driving, Automotive User Interfaces: Creating Interactive Experiences in the Car (2017) 273–294.
[3] K. Mahajan, D.R. Large, G. Burnett, et al., Exploring the benefits of conversing with a digital voice assistant during automated driving: a parametric duration model of takeover time, Transp. Res. Part. F. Traffic Psychol. Behav. 80 (2021) 104–126.
[4] C. Ho, C. Spence, The Multisensory Driver: Implications for Ergonomic Car Interface Design, 1st ed., CRC Press, Boca Raton, FL, 2017.

Further reading

M. Baumann, J.F. Krems, Situation awareness and driving: a cognitive model, in: P.C. Cacciabue (Ed.), In Modelling Driver Behaviour in Automotive Environments, Springer, London, UK, 2007.
J. Cohen, A coefficient of agreement for nominal scales, Educ. Psychol. Meas. 20 (1960) 37–46.
S. Debernard, C. Chauvin, R. Pokam, S. Langlois, Designing human-machine interface for autonomous vehicles, IFACPapersOnLine 49 (2016) 609–614.
L. Dixon, Autonowashing: the greenwashing of vehicle automation, Transp. Res. Interdisc. Perspect. 5 (2020) 100113.

M.R. Endsley, Direct measurement of situation awareness: validity and use of SAGAT, Situation Awareness Analysis and Measurement, Lawrence Erlbaum Associates Publishers, Mahwah, NJ, USA, 2000, pp. 147–173.

M.R. Endsley, Situation awareness global assessment technique (SAGAT), in: Proceedings of the National Aerospace and Electronics Conference (NAECON), Dayton, OH, USA, 23–27 May 1988, pp. 789–795.

M.R. Endsley, Toward a theory of situation awareness in dynamic systems, Hum. Fact. J. Hum. Fact. Ergon. Soc. 37 (1995) 32–64.

M.R. Endsley, W.M. Jones, A model of inter and intra team situation awareness: implications for design, training and measurement. New trends in cooperative activities: understanding system dynamics in complex environments, Hum. Fact. Ergono. Soc. 7 (2001) 46–47.

D.L. Fisher, W.J. Horrey, J.D. Lee, M.A. Regan (Eds.), Handbook of Human Factors for Automated, Connected, and Intelligent Vehicles, 1st ed., CRC Press, Boca Raton, FL, USA, 2020. Chapter 7.

C. Guo, C. Sentouh, J.-C. Popieul, J.-B. Haué, S. Langlois, J.-J. Loeillet, et al., Cooperation between driver and automated driving system: implementation and evaluation, Transp. Res. Part. F. Traffic Psychol. Behav. 61 (2019) 314–325.

J3016_202104, Taxonomy and Definitions for Terms Related to Driving Automation Systems for On-Road Motor Vehicles, SAE International, Warrendale, PA, USA, 2016.

A. Khurana, P. Alamzadeh, P.K. Chilana, ChatrEx: designing explainable chatbot interfaces for enhancing usefulness, transparency, and trust, in: Proceedings of the 2021 IEEE Symposium on Visual Languages and Human-Centric Computing (VL/HCC), St. Louis, MO, USA, 10–13 October 2021, pp. 1–11.

A.K. Kraft, C. Maag, M. Baumann, Comparing dynamic and static illustration of an HMI for cooperative driving, Accid. Anal. Prev. 144 (2020) 105682.

J.R. Lewis, Psychometric evaluation of an after-scenario questionnaire for computer usability studies, ACM Sigchi Bull. 23 (1991) 78–81.

T. Louw, R. Madigan, O.M.J. Carsten, N. Merat, Were they in the loop during automated driving? Links between visual attention and crash potential Injury Prevention, Inj. Prev. 23 (2017) 281–286.

T. Louw, N. Merat, Are you in the loop? Using gaze dispersion to understand driver visual attention during vehicle automation, Transp. Res. Part. C. Emerg. Technol. 76 (2017) 35–50.

Z. Lu, X. Coster, J. de Winter, How much time do drivers need to obtain situation awareness? A laboratory-based study of automated driving, Appl. Ergon. 60 (2017) 293–304.

A.D. McDonald, H. Alambeigi, J. Engström, G. Markkula, T. Vogelpohl, J. Dunne, et al., Toward computational simulations of behavior during automated driving takeovers: a review of the empirical and modeling literatures, Hum. Fact. 61 (2019) 642–688.

N. Merat, B. Seppelt, T. Louw, J. Engström, J.D. Lee, E. Johansson, et al., The "out-of-the-loop" concept in automated driving: proposed definition, measures and implications, Cognit. Technol. Work. 21 (2019) 87–98.

D. Park, W.C. Yoon, U. Lee, Cognitive states matter: design guidelines for driving situation awareness in smart vehicles, Sensors 20 (2020) 2978.

E. Salas, C. Prince, D.P. Baker, L. Shrestha, Situation awareness in team performance: implications for measurement and training, Hum. Fact. 37 (1995) 1123–1136.

J. Sauro, J.R. Lewis, Chapter 8—Standardized usability questionnaires, Quantifying the User Experience, 2nd ed., Morgan Kaufmann, Burlington, MA, USA, 2016, pp. 185–248.

T. Schneider, J. Hois, A. Rosenstein, S. Ghellal, D. Theofanou-Fülbier, A.R.S. Gerlicher, Explain yourself! Transparency for positive UX in autonomous driving, in: Proceedings of the 2021 CHI Conference on Human Factors in Computing Systems (CHI '21), Yokohama Japan, 8–13 May 2021, Association for Computing Machinery: New York, NY, USA, 2021. Article 161, pp. 1–12.

N.A. Stanton, P.M. Salmon, G.H. Walker, D.P. Jenkins, Cognitive Work Analysis: Applications, Extensions and Future Directions, CRC Press, Boca Raton, FL, USA, 2017, pp. 8–19.

S. Valaker, T. Hærem, B.T. Bakken, Connecting the dots in counterterrorism: the consequences of communication setting for shared situation awareness and team performance, Conting. Crisis Manag. 26 (2018) 425–439.

C. Wang, T.H. Weisswange, M. Krüger, Designing for prediction-level collaboration between a human driver and an automated driving system, in: AutomotiveUI '21: Proceedings of the 13th International Conference on Automotive User Interfaces and Interactive Vehicular Applications, Leeds, UK, 9–14 September 2021; Association for Computing Machinery: New York, NY, USA, 2021; pp. 213–216.

J. Xue, F.Y. Wang, Research on the relationship between drivers' visual characteristics and driving safety, in: Proceedings of the 4th China Intelligent Transportation Conference, Beijing, China, 12–15 October 2008, pp. 488–493.

M. Zimmermann, K. Bengler, A multimodal interaction concept for cooperative driving. In Proceedings of the 2013 IEEE Intelligent Vehicles Symposium (IV), Gold Coast, QLD, Australia, 23–26 June 2013; pp. 1285–1290.

CHAPTER 14

Automotive head-up display interaction design based on a lane-changing scenario

Chenxi Jin, Fang You and Jianmin Wang
Car Interaction Design Lab, College of Arts and Media, Tongji University, Shanghai, P.R. China

Chapter Outline

- 14.1 Introduction 175
- 14.2 Background 176
- 14.3 Research on lane-changing scenario 176
  - 14.3.1 Preliminary investigations 176
  - 14.3.2 Study on environment, behavior, and psychology 177
  - 14.3.3 Observation 178
- 14.4 Automotive head-up display interaction design 178
  - 14.4.1 Head-up display information filtration and organization 179
  - 14.4.2 Head-up display interface design 179
- 14.5 Conclusion 181
- Further reading 181

14.1 Introduction

With the wide use of computer and network technology in the field of vehicle transport and the development of car technology, the internal space, human-machine interaction (HMI) design, function operation, and interactive process of the vehicle are undergoing significant revolutionary changes. Presently, both industrial and academic researchers are paying more attention to artificial intelligence studies aimed at enhancing the interaction between humans and automobiles via smart cars.

The primary objective of vehicle safety and driving-assistance technology is to enhance driving safety. This technology relies on sensors installed in vehicles and on roads to collect information about the vehicles, road conditions, and environmental factors. Consequently the system can provide the drivers with advice and warnings based on this data.

This paper will mainly focus on the application of head-up display (HUD) technology for vehicle safety and driving assistance in lane-changing scenes. The advantage of HUD is that the driver can access necessary information without having to lower or turn their heads. By looking forward through the HUD, drivers can seamlessly integrate exterior views with the information displayed on the HUD screen. This decreases the frequency of head drooping and turning and increases driving safety.

14.2 Background

HUD technology can be traced back to the 1960s, when it was initially used for optical and radar aiming purposes. Over time it found widespread application in aircraft, particularly towards the end of the 1960s. By the 1970s, HUD technology had expanded its use in conveyors, civil aircraft, helicopters, and the space shuttle. The display system enables drivers to view critical information on the windshield without lowering their heads, facilitating the timely observation of emergency situations and prompt action-taking while driving.

HUD technology was later ported to vehicles. The technology was first used in cars in 1988. GM launched the first HUD for the car, which displays speed and other useful data on the standard production windshield. In 2012, Pioneer Corporation introduced a navigation system that projects a HUD in place of the driver's visor that presents animations of conditions ahead, a form of augmented reality (AR). Subsequently, many car manufacturers, such as BMW and Volkswagen, designed their own head systems. BMW has developed vision cameras that can even read temporary or permanent road signs as well as overhead signs. They can project the temporary speed limit or other hazard information onto the HUD.

Internally, the use of HUD systems in aerospace has undergone a more in-depth study, but the research for vehicular HUD systems has just begun. In 2011, the domestic Dongfeng Peugeot 508 brought the HUD into the ordinary family. It can help the driver grasp condition and navigation information without diverting their gaze from the road, aligning better with the driver's visual tendencies and significantly enhancing driving safety.

14.3 Research on lane-changing scenario

14.3.1 Preliminary investigations

According to the literature, the annual number of deaths in road traffic accidents in China remains alarmingly high. In 2014, China ranked seventh in the total number of deaths, following cerebrovascular, respiratory system, and other diseases. According to the Ministry of Public Security Traffic Management Bureau news, in 2014 China had 264 million motor vehicles, and the number of traffic accident deaths was 58,080. Although the number is decreasing every year, it still cannot be underestimated.

Based on this data, we conducted an investigation into the causes of traffic accidents. According to the annual road traffic accident reports released by the Shanghai Municipal Public Security Bureau Traffic Police Corps, it can be observed that in motor vehicle accidents, the majority of drivers involved have between 6 to 10 years of driving experience, followed by those less than 5 years of experience. Additionally, violations by motor vehicle drivers in traffic accidents are predominantly related to not giving right of way by rules and violating traffic signals.

Not giving right of way by rules mainly includes causing accidents with normal lane vehicles when changing lanes, causing accidents with the opposite lane or normal lane vehicle when overtaking, causing accidents with the traffic going straight when turning a corner through a no-traffic light cross, and so on.

In addition, we also had a questionnaire survey and in-depth interviews for some novice drivers. The problems and difficulties encountered in actual driving include the inability to predict traffic light changes at traffic light intersections, difficulty in changing lanes, sudden appearance of motorized and nonmotorized vehicles, reduced visibility during hazardous weather conditions, and difficulties with reversing maneuvers.

According to the above analysis, we have identified five key scenarios: reversing, lane changing, crossroads, roadside trails, and crowded roads. In this paper, we focus on HUD design for the lane-changing scenario.

14.3.2 Study on environment, behavior, and psychology

The problem of road traffic safety has always been a focus of concern. The road traffic system consists of humans, vehicles, and roads. In the past, people focused mainly on the research of vehicle and road characteristics. Until the 1950s, people began to concentrate their attention on the characteristics of traffic safety effects. During the 1960s, the United States, Japan, and other countries launched extensive research on various characteristics of drivers, including their psychological activities while driving. In recent years, with the breakthrough in human physiological and psychological data collection technology, the application of psychology in the field of transportation has become more widespread.

In the three elements, the driver is the core of the system. In the process of driving, the driver has to complete three driving actions, namely, the perception, the judgment, and the driving operation. First, drivers have to obtain the information through visual, auditory, and tactile perception. Then they have to think and judge the traffic situation. Finally, they have to drive the car on their hands and feet. In this process, the environment encompasses factors such as the vehicle, road, weather conditions, and some other emergencies. The driver's psychology and behavior are greatly influenced by the quality of the environment. Negative emotions can lead to psychological fluctuations, resulting in slowed or erroneous driving behavior, with potentially dangerous consequences. To conclude, the coordination of the environment, people, and vehicles is the key to achieving the safety requirements of road traffic systems.

The environment plays a significant role in influencing people's mood and performance, while emotion involves complex psychological changes and subjective experiences. When drivers encounter unfriendly environments, they often experience stress responses. Stress, in this context, refers to the unpleasant actions triggered by stress. The response to stress represents an attempt to restore balance when coping with environmental challenges. Insufficient coping abilities can lead individuals to experience stress responses. At the beginning of a stressful moment, due to the survival instinct, the person will mobilize all the physical responses to cope with the immediate threat and take the appropriate measures and actions. At this stage, if the person doesn't effectively respond to the environmental stimulation, the subject may regard the stimulation as a threat, resulting in anger, fear, anxiety, and other emotional responses. These negative emotions will have constraints on people's behavior.

For instance, in the lane-changing scenario, congestion and limited space can induce anxiety in drivers if they struggle to execute the maneuver swiftly. anxiety perception, muscle strength, and control. Prolonged feelings of anxiety can impair perception, decrease muscle strength, and lead to errors in vehicle control. Consider another scenario: novice drivers may experience psychological pressure when changing lanes near large trucks, leading to heightened tension. In addition, overload behavior can also cause stimulus constraints. In the road or expressway ramp before changing lanes, the drivers have to observe not only the front and rear of the vehicle but also the signs and instructions and control the speed of the vehicle, causing slow reaction force. There is a psychological phenomenon known as "psychological confrontation," wherein individuals experience a loss of control when certain factors interfere with or hinder their actions, leading to a sense of restriction and unpleasant emotion. At such times, individuals may initially react by attempting to regain control of the situation and restore their sense of freedom. In the lane-changing scenario, the behavior of other vehicles in other lanes can obstruct the normal driving of drivers in their lane, leading to a sense of psychological confrontation. Instinctively drivers may resist allowing other vehicles to change lanes, refraining them from slowing down or avoiding such behavior, which can result in vehicle collision. Weather conditions can also influence the degree of stress reactions. For instance, hazy or rainy weather with low visibility may make the drivers psychologically uncomfortable, depressed, and other negative emotional reactions of discontent.

The above discussion highlights how an unfriendly environment can produce a psychological impact on individuals and constrain their behavior. However, a favorable environment for safe driving may also harbor hidden dangers. When the traffic environment is exceptionally good, external environmental stimulation is minimal. At this time, individuals may experience physiological and mental relaxation. Once stimulation occurs, due to the paralytic psychological condition, the low sensory ability and the slow reaction rate may cause accidents.

14.3.3 Observation

To gain a deeper insight into drivers' behavior during lane changes, we conducted several observations and used v-box technology to record driving processes under different weather and road conditions to do the research.

During driving, lane-changing behavior involves three processes: perceiving information, judging information, and performing operations. To perceive information drivers use both visual and auditory cues to obtain driving information, such as the current car speed, the vehicle distance from a nearby car, the speed of the following car, etc. Judging information involves using the current data to decide whether to change lanes, while performing an operation involves the physical action of changing lanes. We observed that new drivers are relatively nervous and not skilled, especially in a lane-changing situation, since they are not familiar with the driving speed and the safe distance between two cars. We also found that experienced drivers exhibited a greater frequency of lane-changing behavior, and they tended to change lanes in more complicated situations. Drivers' perception of car speed on the highway is weaker than on an ordinary city road because of the excessive speed on the highway. Moreover, driving blind is another problem during the process of changing lanes. Drivers usually need to turn around to see the blind spot, causing accidents. According to our observations of driving in different weather conditions and brightness, we found that low visibility at night on rainy and foggy days increases the difficulty of changing lanes.

14.4 Automotive head-up display interaction design

When driving a car, 80% of traffic information is achieved by vision. Usually, a driver needs 4−7 seconds to look down at the dashboard and then look forward again. During this process, the time the driver spends on the board is 3−5 seconds. This gap of time is called blind-vision time, which causes driving distraction and is very dangerous. According to statistics, about 30% of traffic accidents are related to this gap in blind-vision time. If there is a car running on the expressway at 100 km/hour, it will run 100 m forward during the visual and action distraction time, which is caused by looking down, reading, and looking up. Therefore, HUDs prove to be very helpful in reducing visual distractions.

14.4.1 Head-up display information filtration and organization

14.4.1.1 Information filtration

Due to advancements in technology, the information displayed on the HUD interface has become more extensive and diverse. However, an excess of information may aggravate the difficulty of recognition and response, making information filtration more and more important.

Information mainly includes status information and function information. Status information describes the car status and transportation circumstances, such as speed, indicator light, temperature, oil consumption, traffic conditions, location, and so on. Function information is based on the analysis of status information that can instruct drivers to complete the driving mission. The purpose of filtering is to select useful information for the driver.

In lane-changing scenarios, the purpose of the safety and driving-assistance system is to help drivers avoid unsafe factors and perform lane-changing safely. According to previous observations and interviews, information regarding the distance between cars, current speed, target speed, lane-changing path, and blind area is considered important.

14.4.1.2 Information organization

Information organization involves structuring and defining the interface layout, workflow, and interaction behaviors based on an interaction framework. It facilitates the efficient organization of information, enabling the driver to find the needed information quickly and accurately. Excellent information organization is essential for maintaining the balance between interaction and safety.

In lane-changing scenarios, information can be divided into several modules and then displayed on the interface in a tiled way. Based on the filtered information, we categorize them into the speed module, danger remaindering module, and blind-area module. Besides, we further classify these modules into constant shown modules and random shown modules. The speed module is a constant shown module. The danger reminder module and blind-area module are only shown when dangers and blind areas appear.

14.4.2 Head-up display interface design

Overall, the HUD interface should observe the general design rules such as clarity, readability, unification, beauty appreciation, and so on. Here we discuss visual symbols, visual color, and visual motion in design.

14.4.2.1 Visual symbols

Visual symbols include character symbols and graphic symbols. Character symbols include numbers and text. Numbers are used for reminding information such as speed and distance, while text is used for reminding information such as destination and street names. Since text needs more time for drivers to read, it is not so much recommended in HUD design. The merit of graphic symbols is their readability and strong instruction. For example, in the BMW-HUD interface design (Fig. 14.1), they use a fork-road symbol to convey dangerous information. Besides, the white arrow shows the direction that the car should keep to the right. In addition, we should also pay attention to the size of the symbols. In the BMW-HUD interface design (Fig. 14.1), they make "62" much bigger than "km/hour" because "62" is the primary information. Finally, it is essential to maintain visual consistency across all symbols, including typefaces, to ensure a cohesive user interface.

In lane-changing scenarios, acceleration serves as an example. As shown in Fig. 14.2, four examples are presented. A questionnaire was constructed to determine which option is more effective. The results showed that arrow symbols are more readable and intelligible than character symbols.

14.4.2.2 Visual color

In the 1980s, scientists demonstrated that color not only affects physiological reactions but also triggers emotional reactions. Blindfolded testers were instructed to enter three differently colored rooms: when they entered the red room, their pulse pressure increased by 12%, while in the blue room, their pulse pressure decreased by 10%. However, in the yellow room, their pulse pressure remained consistent. Hence, careful consideration should be given to color selection during the design process.

FIGURE 14.1

BMW-head-up display interface design.

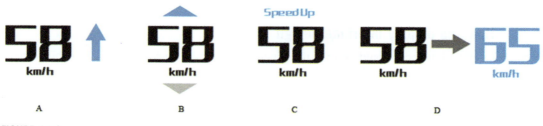

FIGURE 14.2

Acceleration examples.

In lane-changing scenarios, we chose red and yellow to indicate danger and blue and green to indicate speed and instruction. Because in daily transportation, red means stop, yellow means caution, and green means go, these colors accord with people's general cognition. The design in BMW-HUD (Fig. 14.1) uses red to indicate danger.

Besides, the saturation of color should be reduced a little to merge with reality. And the color lightness can automatically change with the environment.

14.4.2.3 Visual motion

Visual motion in HUD interface design holds significant importance. For example, using glints to signal dangers in lane-changing scenarios can be more effective than color changes in capturing the driver's attention. However, the frequency of glints can lead to different results. A high frequency might induce nervousness in the driver, affecting the driver's behavior. Therefore, a low frequency of glint is preferable. When designing visual motion, it is essential to integrate psychological insights and prioritize simplicity and readability.

14.5 Conclusion

This paper analyzes the relationship between environment, mindset, and behaviors during lane-changing scenarios and finds that unfriendly environmental stimulation may lead to drivers' unstable emotions. Consequently, this can prolong response time and increase driving error rates. According to the analysis, we put forward information organization and design rules for HUD interface design. Critical information elements in lane-changing scenarios include distance between cars, current speed, target speed, lane-changing path, and blind spot information.

Further reading

N. Yang, H.T. Dong, S. Yang, Study on installation of head-up display on aircraft. Electron, Opt. Control. 04 (2007) 117–118.

Shanghai Municipal Public Security Bureau Traffic Police Corps, 2014 Shanghai traffic accidents description, Traffic Transp. (03)(2015) 75–77.

Tao Pengfei, Modeling of driving behavior based on the psychology field theory, Jilin Univ. (2014) 2–3.

Z.F. Hu, Y.L. Lin, Environment Psychology, China Architecture & Building Press, Beijing, 2012, pp. 117–119.

Z.F. Hu, Y.L. Lin, Environment Psychology, China Architecture & Building Press, Beijing, 2012, pp. 127–129.

A. Schmidt, W. Spiessl, D. Kern, Driving automotive user interface research, IEEE Pervasive Comput (1) (2010) 85–88.

Q.S. Zeng, H. Tan, Research on head-up display navigation system, in: Proceedings of the User Friendly 2014 UXPA, 2014.

A. Min, L. Yuhong, Q. Xiaohong, et al., The impact of color on human physiology and psychology, China J. Health Psychol (2015) 317–319.

CHAPTER 15

Human-computer collaborative interaction design of intelligent vehicles—a case study of HMI of adaptive cruise control

Yujia Liu[1,2], Jun Zhang[1,2], Yang Li[3], Preben Hansen[4] and Jianmin Wang[2]

[1]College of Design and Innovation, Tongji University, Shanghai, P.R. China [2]Car Interaction Design Lab, College of Arts and Media, Tongji University, Shanghai, P.R. China [3]Karlsruher Institute of Technology, Baden-Württemberg, Germany [4]Department of Computer and Systems Sciences, Stockholm University, Stockholm, Sweden

Chapter Outline

15.1 Introduction 184
15.2 Related work 184
    15.2.1 Taxonomy of automated driving systems 184
    15.2.2 Human-computer collaborative interaction in automated driving 185
15.3 Human-engaged automated driving framework 186
    15.3.1 Full human 186
    15.3.2 Full automation 187
    15.3.3 Driver assistance 187
    15.3.4 Human supervision 187
    15.3.5 Collaboration driving 188
15.4 Case 188
    15.4.1 Application scenario design 188
    15.4.2 Information architecture design 190
    15.4.3 Interface design 193
15.5 Interface experiment 194
    15.5.1 Experimental design 194
    15.5.2 Participants 195
    15.5.3 Apparatus 195
    15.5.4 Task and procedure 195
    15.5.5 Measures 196
    15.5.6 Results 196
    15.5.7 Discussion 198

15.6 Conclusions	199
References	199
Further reading	199

15.1 Introduction

The advancement of technologies, such as deep learning, machine learning, and cloud computing based on big data and artificial intelligence (AI), has propelled automated driving as an essential area of research, technology, and economic development. As computers become more adept at predicting human thoughts and behaviors, it is crucial for them to provide feedback to users, enabling them to make correct predictions about the next intentions and behaviors of the intelligent vehicle. This shift in the human-vehicle relationship from supply and demand to collaboration underscores the pressing need for better collaboration and interaction between humans and vehicles.

In the contemporary era of the rapid development of intelligent vehicle technology, this article focuses on how to better handle the interaction between humans and intelligent vehicles. The primary objective includes improving the safety and work efficiency of intelligent driving, while also improving the user experience. We proposed a human-computer interaction (HCI) framework for intelligent vehicles to guide the design of a human-machine interface (HMI). Second, it explores the engagement of human and intelligent vehicles at different stages. Using the adaptive cruise control (ACC) function as an example, we analyze the cut-in scenario design and establish an information architecture. In addition, based on the analysis of information exchange, we redesigned the HMI. Finally, the study conducts usability tests and administers questionnaires (Likert scales from 1 to 5) to evaluate interface elements in an experimental setting.

15.2 Related work

15.2.1 Taxonomy of automated driving systems

Intelligent vehicles are equipped with advanced onboard sensors, controllers, actuators, and other devices, as well as modern communication and network technology, to realize the exchange and sharing of intelligent information between a car and X (cars, roads, people, clouds, etc.). Its functionalities in environmental perception, intelligent decision-making, collaborative control, and other areas fulfill people's desire for safe, efficient, comfortable, and energy-saving driving. In the SAE J3016 standard, the progression of vehicle autonomy from full manual operation to full automation is delineated through six stages: no automation, driver assistance, partial automation, conditional automation, high automation, and full automation.

The development of domestic intelligent vehicle technology takes into account two pathways: intelligence and networking. The integrated development of intelligence + networking aims to eventually replace human drivers with autonomous systems capable of achieving all driving tasks. The Chinese Society of Automotive Engineering has formulated five levels of intelligence in intelligent vehicles based on the current generally accepted definition of the SAE classification in the

United States, and considering the complexity of China's road traffic conditions: driving assistance (DA), partially automated (PA) driving, conditional automated (CA) driving, highly automated (HA) driving, and fully automated (FA) driving. The specific content corresponding to each level has been outlined in Table 15.1. This intelligence classification provides guidance for the specific technological path in the development of intelligent vehicles.

15.2.2 Human-computer collaborative interaction in automated driving

In the environment of intelligent vehicles, more and more scholars and research institutions in the field of HCI have become involved in the study of intelligent collaborative interaction between humans and machines.

Ranney [1] proposed a human-machine cooperation strategy model for intelligent vehicle systems in 1994, which was decomposed into three levels of operations, tactics, and strategy. Salas believes that the key to situational awareness of teamwork lies in the information interaction process of team members. Team members must have situational awareness related to their goals and surroundings, as well as that of other team members. Parasuraman [2] described the four stages of human information processing in the interaction between humans and automated driving: information acquisition, information analysis, decision-making, and action implementation. In the scope of automated driving research, engagement is a communication process between humans and vehicles. They are formed in the process of achieving (shared) goals. The relationship of mutual cooperation and engagement requires interaction between humans and intelligent systems.

Table 15.1 Taxonomy of automated driving in China.

Intelligent level	Definition	Control	Supervision	Responding to failure
Driving assistance (DA)	Provide support for one operation of direction and acceleration and deceleration through environmental information; all other driving operations are carried out by humans.	Human and System	Human	Human
Partially automated driving (PA)	Provide support for one or more operations in direction and acceleration and deceleration through environmental information; all other operations are carried out by humans.	Human and System	Human	Human
Conditional automated driving (CA)	All driving operations are completed by the unmanned driving system, and the driver needs to provide appropriate intervention according to the system request.	System	System	Human
Highly automated driving (HA)	All driving operations are completed by the unmanned driving system. Under certain circumstances, the system will make a response request to the driver, and the driver cannot respond to the system request.	System	System	System
Fully automated driving (FA)	The unmanned driving system can complete all the operations in the road environment that the driver can complete without the driver's intervention.	System	System	System

15.3 Human-engaged automated driving framework

Many types of research on the HCI of intelligent vehicles are based on the interface between humans and machines (Fig. 15.1A). The conventional perspective treats the human driver and the intelligent vehicle as separate entities, each with its own set of interests and considerations. This way of thinking is considered from the perspectives (interests) of both human and intelligent vehicles, and lacks a collaborative development mindset, failing to recognize the potential synergy between humans and intelligent vehicles. The concept of engagement can be substituted into the role of the user, so that the user will not be overly dependent on the system in the process of interaction, ensuring continued perception of the surrounding environment. The human-engaged automated driving (HEAD) framework (Fig. 15.1B) is based on the intelligence levels of both human and vehicle, hierarchically integrated into the engagement design concept. This framework reveals new interactive methods that are changing, and finally shaping a collaborative interaction design framework.

Among them, different levels of automated driving intelligence are matched with human drivers who have different levels of engagement in AI, resulting in five stages: full human, full automation, driver assistance, human supervision, and collaboration driving (Fig. 15.1B).

15.3.1 Full human

Full human focuses on driving issues entirely from the human—driver perspective. The human driver treats the car as just a mechanical product and deals with driving issues based on their cognition in the current environment. Humans are still advanced complex creatures. When dealing with driving issues, the impact of various factors, such as safety, cognition, society, culture, and environment, are taken into consideration. Human intelligence is not the product of informatization and

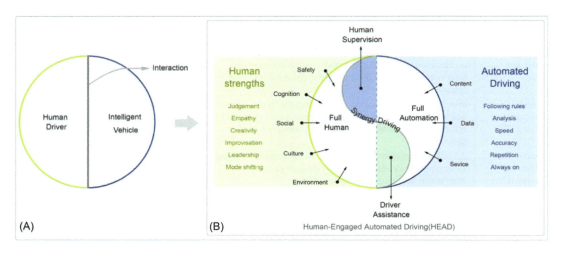

FIGURE 15.1

Theoretical inference and framework of human-engaged automated driving.

automation; rather humans are inherently wise. This wisdom stems from the integration of the aforementioned elements, serving as the core and the driving force behind intelligence. For instance, within society, there exists a normative relationship of thoughts and behaviors that are commonly recognized between individuals and among members of society. This dynamism involves the classification, analysis, and deconstruction of facts in the group, addressing value-based issues such as security, privacy, prejudice, substitution, and inequality.

The reason why human wisdom is smarter than AI is that humans possess the capacity to foresee events, reversals, and unforeseen circumstances that the intelligences cannot predict. Humans are endowed with their own unique abilities of judgment, empathy, creativity, improvisation, leadership, and adaptability, which collectively embody the essence of human spirit and ingenuity.

15.3.2 Full automation

Full automation focuses on addressing driving challenges through the intelligence of automated vehicles. Machines possess outstanding capabilities in data processing, processing speed, repeatability, and analysis. Achieving intelligence requires the integration of three key aspects: content, data, and service. Content produced by individuals caters well to personalized needs, but may not adequately address the broader needs of most people. Therefore it is crucial to collect data from groups of people. Furthermore, the accumulation of data results in patterns and trends within a group, which determines the generation of content and services in the intelligent future.

The inherent capabilities of automated driving are compliance, analysis, speed, accuracy, repetition, and consistency. Because it uses mathematical logic as the basic thinking logic, it can be used as an extension of some human abilities, such as accuracy. Automated driving can make analysis and behavioral responses in an extremely short time, avoiding the risks associated with human errors or lapses in judgment.

15.3.3 Driver assistance

Driver assistance focuses on integrating intelligent systems into conventional vehicles. In this setup, automated driving systems serve as assistants to human drivers, reducing information-intensive, repetitive, or tedious tasks. By leveraging sensors and cameras installed on vehicles, these systems can compensate for human perceptual limitations and extend human physiological capabilities. The intelligent system of vehicles can detect the surrounding environment, communicate risks, and alert the drivers to unknown factors. It helps humans handle the conversion of relevant information, but at the same time, the ultimate decision-making authority remains with humans. In addition, in scenarios where the machine cannot be discriminated against or when there is a lack of trust in the machine, it can be taken over by the driver, who retains absolute control over the operation of the vehicle.

15.3.4 Human supervision

Human supervision focuses on engaging human oversight with intelligent vehicles to accomplish driving tasks. In this setup, AI-driven intelligent vehicles replace human drivers, while information systems within the network constantly perform tasks and provide feedback for review and supervision. This arrangement liberates humans from manual labor, allowing them to oversee and manage driving tasks remotely.

However, the judgment of AI sometimes conflicts with human judgment because AI tends to prioritize achieving the optimal solution or finding the best path to complete the tasks, even attempting to anticipate human intentions and making decisions on behalf of humans. Moreover, machines do not have the same deep moral and ethical considerations as humans, so when AI and human opinions are not unified, humans retain the highest decision-making power.

15.3.5 Collaboration driving

Collaboration driving represents an optimal balance of the desired human-computer (vehicle) interaction. AI, at its core, aims to emulate human cognition, and Synergy driving seeks to infuse human wisdom and human thinking into AI algorithms. This is not to recreate the intelligence of a machine but to construct the intelligence of autonomous driving in accordance with human logic. In the future, intelligent communication needs to construct data that serves human values and contents in the human-machine medium and the environment. These data can meet the ethics and judgment of the intelligent medium in society.

The collaboration between the wisdom of humans and the intelligence of automated driving enriches driving activities, which is mutually beneficial. In this balanced state, the intelligent system of autonomous driving not only fulfills predetermined tasks but also improves the boundary of human collective cognition and humanity by improving human capabilities and potential.

15.4 Case

How to make the interaction between humans and vehicles more efficient and harmonious with the help of intelligent driving assistance systems has always been an issue being discussed in the industry. This article proposes a framework for human-machine collaborative driving (HMCD), discusses the interaction of human and intelligent vehicles in different driving statuses, and analyzes the design requirements of HMI. In this case, the driving scenario of the ACC function (there is another vehicle cutting into the main lane from other lanes and driving in front of it) is an example to analyze the interaction flowchart under human-machine collaboration, establish an information architecture, and finally carry out the prototype and interface design, expounding how to practice engagement in human-machine collaborative interactive interface design.

15.4.1 Application scenario design

ACC is a driver assistance system that controls the longitudinal movement of the vehicle. It is one of the main functions required for automated driving vehicles and can reduce the workload for the driver. However, many interface designs for the ACC function confuse people about the automated driving vehicle when they are used, increasing the workload and not being able to achieve cooperative driving. Good interface design can improve driving efficiency and, at the same time, promote the mutual symbiosis of both human drivers and automated driving. This case hopes to use the driving scenario under the ACC function as an example to establish an information architecture with human-computer engagement to help the designer prototype, interface design, and iteration. Its objective is to verify the role of the HEAD system in guiding the interaction of human and automated driving.

In ACC function research, vehicle cut-in represents a typical application scenario. In this case, the driving scene is divided into two stages, with the front wheel on the right side of the auxiliary test vehicle (F) passing the lane line as the dividing point, the first stage before passing, and the second stage after passing. The first stage includes two statuses: (1) cruise original setup, and (2) cut-in A. The second stage also includes two statuses: (1) cut-in B and (2) following. In cruise status, the driver sets the ACC to set the vehicle speed; in cut-in A status, the F car starts to overtake and starts preparing to cut into the main test lane; in cut-in B status, the F car starts to cut into the current lane; in following status, the S car follows the F car at the speed of the F car. The status description of two vehicles in two phases is shown in Fig. 15.2.

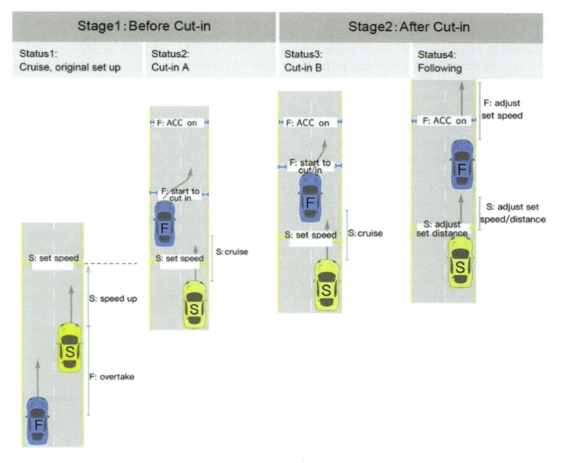

FIGURE 15.2

Cut-in scenario design of adaptive cruise control function.

15.4.2 Information architecture design

In the cut-in scenario of ACC function, a flowchart of the human driver and the ACC system engagement parts are established, as shown in Fig. 15.3. It analyzes the driver's use of ACC in different road scenarios involving physical operations and cognitive states to clarify the specific areas where human drivers and automated driving are engaged. This may reduce the potential dangers caused by the driver's ignorance of the driving environment and the capabilities of the vehicle's ACC function. It also has the potential to increase the trust of humans for automated driving and even improve the drivers' cognitive abilities.

The purpose of designing the in-vehicle information architecture is to find more deep and accurate ACC function design points based on driver cognition in typical scenarios. Combined with driver behavior prediction, the driving scenario is analyzed in stages and role dimensions to predict under which driving tasks the driver needs the help of interfaces (when the information should appear). Based on the augmented reality head-up display (AR-HUD) and windshield head-up display (W-HUD) areas, we can analyze the time and form of the information and the interaction between the two screens. When the external driving environment changes (such as when the vehicle cuts in), the driver obtains information from the external environment and can adjust the internal function (ACC) according to the current driving scene. This causes a series of changes in the information inside and outside the vehicle. This case aims to analyze the information architecture design displayed by the time sequence and the dimensions of different engagement roles. In terms of design, it is presented as the relationship between the AR-HUD and W-HUD information elements.

Combining the driver's visual glance analysis with the vehicle's driving trajectory, the driving trajectory of the vehicle under monitoring is obtained (Fig. 15.4). The approximate order of this trajectory is as follows: left rear mirror/interior mirror—left window—front left windshield lane—lane lane—the current lane, as shown in Fig. 15.7. In the figure The origin indicates the node of the vehicle's driving trajectory. In this context, 1−4 denotes the driver's identification range, while 4−6 represents the common identification range between the driver and the vehicle. When the vehicle is in the 4−5 area, the driver and the vehicle need to make a judgment on whether the state of the vehicle will affect the use of the ACC function of the vehicle, and whether the driver needs to adjust the ACC set speed and set distance. Therefore in this case, we used 4−5 as an area for design assistance to help the driver determine the track movement of the front car, enhance situational awareness, reduce driver workload, and make effective judgments and actions in a timely manner.

To provide a more intuitive representation of the changes in information architecture, icons and symbols are used to express various types of information, as shown in Table 15.2.

Based on the four statuses of this scenario, the flowchart of the in-car information architecture design is shown in Fig. 15.5:

1. Driving scene analysis of the trajectory of the auxiliary vehicle (representing the timeline of the information structure in the vehicle).
2. Four stages analysis diagram in the cut-in scenario.
3. Driver's vision diagram.
4. HUD information layout and linkage information design prototype on W-HUD and AR-HUD.

15.4 Case 191

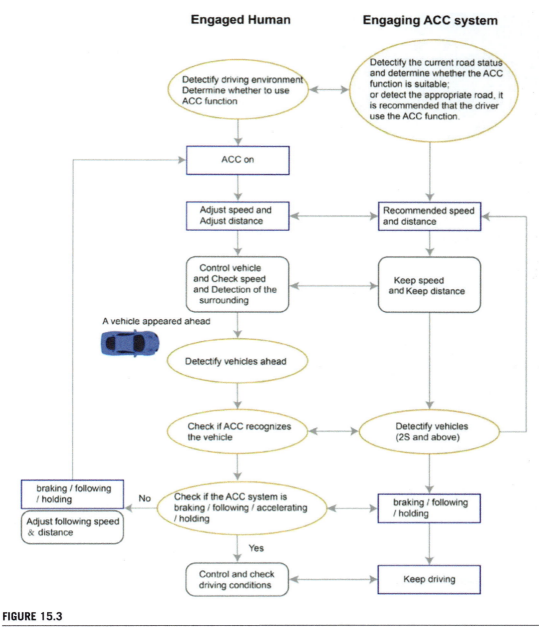

FIGURE 15.3

Engagement flowchart of human and adaptive cruise control system.

192 Chapter 15 Human-computer collaborative interaction design

FIGURE 15.4

Driving trajectory of synergy driving for engagement.

Table 15.2 Correspondence table of information and graphic symbols.

Information	External information icon	Internal information icon
Set speed	□	□
Set distance	○	○
Detection Information	△	△

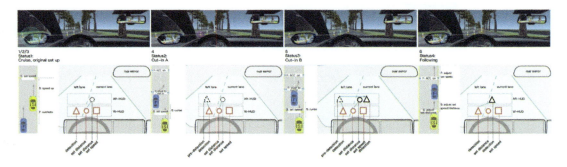

FIGURE 15.5

Flowchart of information architecture design.

15.4.2.1 Status 1: cruise original setup

The first stage of the cut-in driving scenario is the status when the F car is driving behind the S car. After the test vehicle reached a certain speed, the ACC function was turned on. At this time, the S vehicle was in cruise mode. The driver made the initial settings of the ACC function based on the current driving environment and previous experience using ACC, including setting speed and setting time.

The driver monitored the driving trajectory of the F vehicle, as shown in the vehicle driving trajectory points 1—3. Dotted lines indicate related changes in the same information. In status one,

AR-HUD displays only the time interval information, which indicates the actual distance of the time interval bar on the road; W-HUD displays the set speed, time interval, and identification information at this time.

15.4.2.2 Status 2: cut-in A

At this status, the front car is driving in the left lane. The driver needs to determine whether the left-side vehicle is overtaking or changing lanes, whether it will affect the use of the ACC function, and whether it is necessary to adjust the set speed and set time interval. At this stage, the driver monitors the driving trajectory of the F vehicle, as shown in the vehicle driving trajectory point 4. At this status, the S car is still in cruise mode. The vehicle needs to assist the driver in predicting the driving trajectory of the vehicle in front, informing the driver of the vehicle's working status, whether the front car is recognized, and so on. The recognition at this status is a predetection state, which only informs the driver that the vehicle has recognized the F vehicle, but has not switched to following mode.

15.4.2.3 Status 3: cut-in B

At this status, the F car travels across the lane line to the front of the S car, and the driver monitors the driving trajectory of the F car as shown by the vehicle trajectory point 5. At this time, the S car has switched from cruise status to following status, and the AR-HUD will continue to display information on identification, and the identification information changes from predetection to detection, enhancing the prompt of ACC status change.

15.4.2.4 Status 4: following

At this status, the F car is driving in front of the S car. The driver monitors the driving trajectory of the F car, as shown in vehicle trajectory node 6. The driver can adjust the set speed and set time interval according to the driving scene at this time, and the detection information is continuously displayed to indicate that the ACC function is in a continuous working state.

15.4.3 Interface design

Many vehicles with automated driving functions are equipped with the ACC function and have a corresponding design interface. The basic information includes the ACC set speed, set time interval, and vehicle identification information. The ACC information is mainly displayed on the instrument panel. Although the design performance of ACC varies among different car manufacturers, a relatively stable design plan has been established. Based on the previous work of the CarLxD laboratory at Tongji University, China, we integrated the same ACC information on the driving simulator, which was displayed on the instrument and the HUD, as shown in Fig. 15.6A.

Based on the information architecture outlined in the previous introduction, the original interface design is iterated (Fig. 15.6B). In the iterative design scheme, the real-time speed information is outlined by hollow circles, which makes the speed information more visible, distinguishing the real-time speed from the ACC set speed. For the "spot the front vehicle" element, when the system has recognized the front vehicle, the vehicle icon changes from white to blue. In the AR-HUD, a "predetection" state is introduced for the front vehicle, and the "blue semicurved base" is iterated into a more obvious "marquee" form. When the vehicle recognizes a vehicle in the front or next

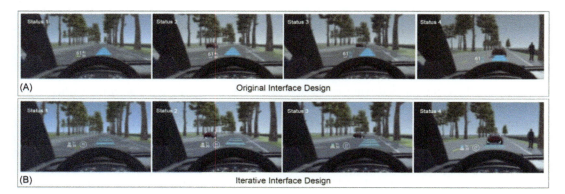

FIGURE 15.6

Original interface design and iterative interface design.

lane, the detection information is displayed in white, and a white checkbox appears. When the vehicle changes from cruise mode to following mode and the detection information is blue, a blue checkbox appears. For the "set distance" element, reducing the five-time intervals to three-time intervals can more clearly visually determine the distance change caused by the time interval increase and decrease. In the AR-HUD, the judgment of the ACC function mode is added. When the vehicle is in cruise mode, the time interval rectangle is hollow; when the vehicle is in follow mode, the time interval rectangle is solid. The rectangular display changes the distance following the car in front of it in real time. Visually, compared with the full solid design of the original interface, the iterative design weakens the rectangular display, enhances transparency, and increases the gradient effect. At the same time, the judgment of the ACC function mode has been added to the "detection information" and "time interval" design elements. A comparison of the two schemes of the design elements is shown in Fig. 15.7.

15.5 Interface experiment

In this experiment, the HUD design interface of the ACC function was used as the original interface design group, which was compared with the iterative interface design group. The experiment was performed on a driving simulator to analyze the results. Design element questionnaires were used to analyze whether the information architecture and interface design effectively guide design decisions and improve driver safety when using ACC functions.

15.5.1 Experimental design

The experiment was conducted in a within-subjects design with one independent variable comparing the original and iterative interface design. Participants underwent training in both conditions. Conditions were counterbalanced to minimize the sequence effect.

15.5 Interface experiment

Information	Original Interface Design			Iterative Interface Design				
	AR-HUD	W-HUD	Changes	Engaged Human	Engaging computing			
					AR-HUD	W-HUD	Changes
Vehicle Speed	None	 61 km/h	Change in real time	1) Control vehicle 2) Check speed 3) Detection of the surrounding 4) Control and check driving conditions	None	 61 km/h	Change in real time
Set Speed	None	 30	Increase or decrease at 5km / h	Adjusting Speed	None	 30	Increase or decrease at 5km / h
Set Distance		None	Increase or decrease at 5 levels of distance	Adjusting Distance	 Cruise Mode	 Following Mode	 Increase or decrease at 3 levels of distance
Identification Information	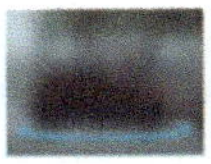	 30	Unidentified / identified	1) Detectify vehicle 2) Check if ACC recognizes the vehicle 3) Check if the ACC system is braking / following / accelerating / holding	 Pre-detection	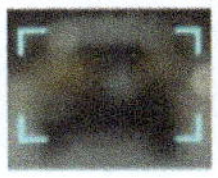 Detection	 Undetected / pre-detected / detected

FIGURE 15.7

Comparison of original and iterative design elements.

15.5.2 Participants

Ten university students and staff members (six females) aged 22–30 (M = 25.9, SD = 2.47) were recruited. Most of the participants have automotive user interface design experience and can be considered to have a certain ability to judge the interface in the experiment.

15.5.3 Apparatus

The experiment was completed on a driving simulator. The driving simulator provides a very realistic driving environment. The driver can fully realize vehicle driving control through the steering wheel and pedals and interact with other vehicles on the road. The hardware part of the experimental device includes two sets of Logitech G29 (including a steering wheel and three-foot pedals), two computers and six monitors. The software component of the experimental device includes Unity 5.0 and Unity-related plug-ins that can simulate the real driving environment. The test measurement tool contains the participant information sheet, the introduction of the purpose and process of the experiment, AR-HUD and W-HUD introductions and example pictures, ACC function introduction videos and user manuals, and questionnaires about relevant interface design elements.

15.5.4 Task and procedure

Participants were asked to sign a letter of consent and an information sheet. Later they were introduced to the procedure of the study. Participants were randomly assigned to either the original or iterative group. Conditions were counterbalanced to minimize the sequence effect. Participants were instructed in the use of ACC and then allowed to perform auto-driving and manual driving exercises on the driving simulator (to avoid learning effects, the driving scenario during the practice was different from the driving scenario used in the actual experiments), and each mode was tested for at least 15 minutes. Once participants became familiar with the use of the simulator and the use of the ACC function settings, the formal experiment began. The test scenario was not revealed for the participants prior to the experimental tasks. Once the subjects were prepared for testing, the tester issued mission instructions to the subjects. At the end of the experiment, the tester dictated the system usability scale (SUS), design element scoring, and interview scales. After completing the questionnaire, they proceeded to the next experiment. The test drive time was about 2–5 minutes, the time to perform each driving task we about 15–20 seconds, and the entire experimental process lasted about 50–60 minutes.

15.5.5 Measures

Questionnaires are considered a simple and economical way of obtaining and analyzing data. In this experiment, the questionnaire combined closed questions (rated on a 5-point Likert scale from 1 to 5) and open questions to obtain both quantitative data and qualitative data. The questionnaire is based on the design elements of: set speed, detection information, set distance, and mode identification. It lists a total of 16 questions to test drivers' evaluation of original and iterative interface design.

15.5.6 Results

Data were checked using the Kolmogorov–Smirnov test and analyzed using the T-test. The comparison of SUS between the original group and the iterative group is shown in Fig. 15.8.

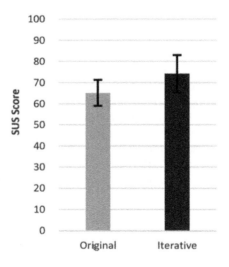

FIGURE 15.8

Comparison chart of system usability scale system availability scores under different groups.

The average score of the original group was 62.3 points (SD = 6.1), which was lower than the score of the iterative group of 74.3 points (SD = 8.8), and the difference was significant ($P = .016 < .05$). Fig. 15.9 summarizes quantitative data results.

15.5.6.1 Set speed

In the original group, two participants indicated that they did not see the symbol of set speed. In the acquisition of the set speed element information, compared with the original group (M = 1.90, SD = 0.57), the mean value of the set speed in the iterative design scheme is significantly higher (M = 4.11, SD = 0.20). Simple main effects analysis showed a significant difference ($P = .012 < .05$).

15.5.6.2 Detection information

The mean values of the iterative group (M = 3.20, SD = 0.49; M = 2.80, SD = 0.47; M = 2.60, SD = 0.31) for the three closed questions: (1) whether the front car was recognized? (2) the influence of the front car on the use of the ACC, and (3) the degree of attracting attention, are higher than the original group (M = 2.7, SD = 0.76; M = 2.20, SD = 0.57; M = 1.90, SD = 0.55), but it is not very significant ($P = .557$; $P = .279$; $P = .354$). In the original group, the scores of the participants were much more discrete than iterative, indicating that the subjects had a large difference in their cognition of original. Regarding the question on the impact on the attention of driving information, the iterative group provided participants with more information and thus increased their impact on subjects' attention.

The results of the open questions indicate that in the original group, 3 of the 10 participants did not perceive the identification information. In the iterative group, all the participants felt the detection symbol [P3]: "In the process of judging the track of the preceding car, the decision time is not enough. The white box icon can increase the driver's decision time." [P8]: "The white box indicates that the front car and the own car are in different lanes; the blue box indicates that the front car and the own car are in the same lane and it can lock with the car target." However, two participants tended to the original. [P5]: "The blue color is integrated with the car, and will not cause too much distraction for the driver, but feel that the difference of detection status is very necessary."

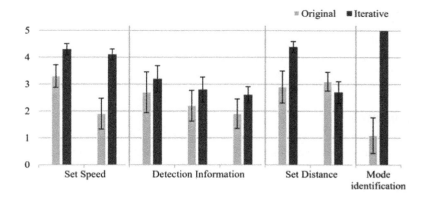

FIGURE 15.9

Quantitative score results in questionnaire.

15.5.6.3 Set distance

Regarding the question on judging the distance from the front car according to the blue rectangle, the mean value of iterative (M = 4.40, SD = 0.2) is much higher than original (M = 2.90, SD = 0.6). But the difference is not significant ($P = .153 > .05$). Regarding the question on the influence of the blue rectangle on attention, The average score of original (M = 3.10, SD = 0.35) and iteration (M = 2.70, SD = 0.4) is basically the same ($P = .42$), which shows that although the iterative design scheme increases the recognition status of the time rectangle, it does not cause much attention of the driver.

In the open question about hollow rectangles and solid rectangles, one participant did not notice the change from hollow rectangles to solid rectangles. Regarding the question on design schemes that tend to display synchronously with time−distance rectangles or different displays, four participants agreed design should have a differentiated display. [P1]: "Different from AR-HUD's dynamic time−distance display method, W-HUD gave the driver an iconic information prompt to know their original settings." [P9]: "W-HUD and AR-HUD are better expressed in a consistent manner, and it is not easy to cause misunderstanding and confusion"; the remaining participants indicated that they have no special influence.

15.5.6.4 Mode identification

The results of the Iterative (M = 5.00, SD = 0.00) are significantly better than the Original (M = 1.10, SD = 0.66), and the difference is significant ($P = .0002 < .05$), all participants can judge the mode of ACC function at that time by Iterative interface design.

The results of the open question indicate that in the original design scheme, the participants' perception of the mode of the ACC function in the driving scene at that time was quite confusing. During the car, following target was lost due to lane departure, and it returned to cruise mode.

15.5.7 Discussion

The SUS results show that the participants in the experiment believe that the interface design of the iterative group (74.3 points > 62.3 points) is more usable. From the questionnaire of design elements, the average score of set speed in the iterative group is higher than 3 points, which illustrates that this design can provide the driver the information more effectively than in the original group. The detection information in the iterative group promotes the driver by evaluating the forward vehicle state more effectively. Besides that, it has a positive impact on using the ACC function. However, except for the question of whether the front car was recognized, the average scores of the rest of the questions are lower than 3 points, indicating that this design element needs to be further improved.

15.5.7.1 Set distance

The reduction of five-time intervals to three-time intervals increases the difference in each time interval, which is more conducive to the driver's judgment of the time interval. The use of solid rectangles and hollow rectangles enhances the current mode of ACC, helping the driver estimate the ACC status. On the other hand, this design does not affect the driver's attention greatly. The iterative interface design is superior to the original design scheme in terms of ACC function time display, but the design method of AR-HUD and W-HUD asynchronous display still needs further design and testing.

15.5.7.2 Mode identification

Regarding the judgment of the ACC functional mode, the test results of the iterative design scheme are better than the original interface design. Compared with the original design, the results show that the internal design improves the driver's usability and helps the driver clarify the status of ACC.

It can be concluded that in the design element scoring and driver interview questionnaires, the results indicate that the iterative interface design is superior to the original interface design in four aspects: set speed, detection information, set distance, and mode identification. This experiment can prove that the interaction concept in the HEAD framework is used as a guide to build an information architecture that can be jointly engaged by humans and machines, helping designers to perform better prototypes of interface designs and iterations.

15.6 Conclusions

Based on the research of intelligent vehicles from a human-computer collaborative interaction design perspective, this paper proposes a theoretical framework based on engaged human-computer collaborative driving. The theoretical framework, combined with the analysis of the ACC, elucidate the interaction flowchart under human-machine collaboration. The information architecture discussed the interface design practices, and usability tests and interface elements-related questionnaires (Likert scale from 1 to 5) were established in the experiment. Experiment results show that the interactive interface design guided by the HMCD framework can improve the usability of the system, allowing users to understand the system more clearly through more appropriate information on the interface, improving the collaboration efficiency of human-vehicle interaction.

References

[1] T.A. Ranney, Models of driving behavior: a review of their evolution, Accident; Anal. Prev. 26 (6) (1994) 733−750.
[2] R. Parasuraman, T.B. Sheridan, C.D. Wickens, A model for types and levels of human interaction with automation, IEEE Trans. Syst., Man, Cybern. - Part. A: Syst. Hum. 30 (3) (2000) 286−297.

Further reading

F. Biondi, I. Alvarez, K.A.J.I.Jo.H.-C.I. Jeong, Human−vehicle cooperation in automated driving: a multidisciplinary review and appraisal. 2019. 35(11−15): p. 1−15.
J. Brooke, SUS - a quick and dirty usability scale, Usability Evaluation Ind. (1996) 189.
C. Cheng, F. You, P. Hansen, J. Wang, Design methodologies for human-artificial systems design: an automotive AR-HUD design case study, in: Proceedings of the 2nd International Conference on Intelligent Human Systems Integration (IHSI 2019): Integrating People and Intelligent Systems, San Diego, California, USA, February 7−10, 2019, pp. 570−575.

F. You, J. Zhang, J. Wang, M. Fu, Z. Lin, The research on basic visual design of head- up display of automobile based on driving cognition, in: HCI International 2019—Posters, 2019, pp. 412-−420.

S. Eduardo, E. Dana, C. Sims, Shawn Burke, Is there a "big five" in teamwork? Small Group. Res. 36 (5) (2005) 555−599.

J. Liao, P. Hansen, C. Chai, A framework of artificial intelligence augmented design support, Human−Comput. Interact. 35 (5-6) (2020) 511−544.

Frank J. Massey, The Kolmogorov−Smirnov test for goodness of fit, Publ. Am. Stat. Assoc. 46 (253) (1951) 68−78.

M. Matell, J. Jacoby, Is there an optimal number of alternatives for Likert-scale items? Effects of testing time and scale properties, J. Appl. Psychol. 56 (1972) 506−509.

F. Naujoks, Y. Forster, K. Wiedemann, A.Neukum, A human-machine interface for cooperative highly automated driving, in: N.A. Stanton et al. (Eds.), Advances in Human Aspects of Transportation, 2017, pp. 585−595.

E. Ohn-Bar, M.M.J.I.T.o.I.V. Trivedi, Looking at humans in the age of self-driving and highly automated vehicles, 2016, pp. 90−104.

Redesigning work in an age of automation. https://www.slideshare.net/planstrategic/redesigning-work-in-an-age-of-automation, 125030010.

X. Ren, C. Silpasuwanchai, J. Cahill, Human-engaged computing: the future of human−computer interaction, CCF Trans. Pervasive Comput. Interact. 1 (1) (2019) 47−68.

Towards a new era of big data content services. <http://www.cbbr.com.cn/article/117835.html>.

J. Wang, Z. Cai, P. Hansen, Z. Lin, Design exploration for driver in traffic conflicts between car and motorcycle. HCII2019, Communications in Computer and Information Science (CCIS), 1034, Springer International Publishing, Cham, 2019, pp. 404−411.

J. Wang, W. Wang, P. Hansen, Y. Li, F. You, The situation awareness and usability research of different HUD HMI design in driving while using adaptive cruise control, HCI International 2020 − Late Breaking Papers: Digital Human Modeling and Ergonomics, Mobility and Intelligent Environments., Vol. 12429, Springer, Cham, 2020, pp. 236−248.

F. You, Y. Li, R. Schroeter, J. Friedrich, J. Wang, Using eye-tracking to help design HUD-based safety indicators for lane changes, in: The 9th International Conference, 2017.

F. You, Y. Wang, J. Wang, X. Zhu, P. Hansen, Take-over requests analysis in conditional automated driving and driver visual research under encountering road hazard of highway, Advances in Human Factors and SystemsInteraction, Springer International Publishing, Cham, 2018.

F. You, Y. Yang, M. Fu, J. Yang, X. Luo, L. Li, et al., Icon design recommendations for central consoles of intelligent vehicles, in: T. Ahram, R. Taiar, V. Gremeaux-Bader, K. Aminian (Eds.), Human Interaction, Emerging Technologies and Future Applications II. IHIET 2020. Advances in Intelligent Systems and Computing, vol 1152, Springer International Publishing, Cham, 2020.

PART 4

Research on interaction design of on-board robots

CHAPTER 16

Research hotspots and trends of social robot interaction design: a bibliometric analysis

Jianmin Wang[1], Yongkang Chen[2], Siguang Huo[2], Liya Mai[2] and Fusheng Jia[2]

[1]Car Interaction Design Lab, College of Arts and Media, Tongji University, Shanghai, P.R. China [2]College of Design and Innovation, Tongji University, Shanghai, P.R. China

Chapter Outline

16.1 Introduction 203
16.2 Research design 204
    16.2.1 Data sources 204
    16.2.2 Research methods 205
16.3 Bibliometric results and analysis 205
    16.3.1 Trend analysis of annual outputs of social robot interaction design literature 205
    16.3.2 Research hotspots of social robot interaction design 207
    16.3.3 Evolution and trend of social robot interaction design research hotspots 210
    16.3.4 Knowledge base of SR-human-robot interaction research 211
    16.3.5 Distribution of social robot interaction design literature sources 215
    16.3.6 High-impact countries and research institutions 216
    16.3.7 High-impact author analysis 217
16.4 Discussion 218
    16.4.1 More realistic research scenarios 218
    16.4.2 Effective measurement of research indicators 218
    16.4.3 Longitudinal trial 219
    16.4.4 More specific design strategies 219
    16.4.5 Uncovering the psychological cognitive mechanisms behind user behavior ... 219
16.5 Conclusions 219
References 220
Further reading 220

16.1 Introduction

The social robot is a type of robotics technology that interacts with users in a natural and intuitive way by using the same social norms as humans and typically is in an anthropomorphic form. The transformation of robots from service tools to social partners will profoundly impact economic and social dimensions. Therefore to fully integrate them into the home and everyday environment, careful consideration of the future roles of robots and ensuring their perfect fulfillment of these roles through human-robot interaction (HRI) design is of paramount importance [1–3]. Social robot interaction design (SRID) is the key to creating a positive user experience, establishing emotional connections, and adding value, which requires comprehensive research into the social behaviors, gestures, body movements, role design, and voice, among other aspects of robots, in a systematic manner.

In the past decade, research in SRID has experienced rapid growth, leading to a substantial body of literature on the subject. Such a significant growth in the literature requires new approaches to review and analyze trends within knowledge domains. In addition, SRID research exhibits a strong cross-disciplinary nature with a complex and diverse knowledge structure. Relying solely on traditional literature review methods makes it challenging to comprehensively grasp the current state of SRID research and the evolution of research hotspots. Bibliometrics is an effective quantitative method for visualizing and analyzing the existing published literature. It enables the extraction of regular patterns and underlying insights from a large volume of papers and is now widely employed in numerous scientific research fields. Therefore this paper will use bibliometrics to address the following five research questions to provide a comprehensive overview of SRID:

RQ1: What are the main focuses of SRID's current research hotspots?

RQ2: How has the research content in SRID evolved over the past decade in terms of temporal development? What are the overall trends?

RQ3: What are the primary knowledge foundations and classical theories of SRID research?

RQ4: Which scholars and institutions worldwide are the research subjects of SRID? What are the collaborative relationships between them?

16.2 Research design

16.2.1 Data sources

Web of Science (WOS) is the most comprehensive English language database in the world, and the literature in the WOS Core Collection has undergone peer review and rigorous scrutiny by publishing journals, so it is considered to possess greater disciplinary representativeness than other databases. To ensure the validity and reliability of the research data, this study chooses the WOS Core Collection as the data source. Our search terms combine social robot and interaction design to filter and obtain literature relevant to HRI design for social robots. Since there are many different expressions of relevant keywords in the English context, we list as many expressions as possible and connect these keywords with the Boolean operator symbol OR. The specific search strategy is TS = (social robot OR social robotic OR social robots OR assistive robot OR conservational robot OR chatbot OR chat robot OR embodied conversational agent) and (HRI design OR interaction design OR collaborative design OR interactive design OR HRI design).

To avoid the loss of interdisciplinary literature, the citation index sources were set to "All" (SSCI, SCI, A&HCI, CPCI-S, CPCI-SSH, BKCI-S, BKCI-SSH, and ESCI, among others), and the sources were not streamlined. Setting the time span as nearly 10 years (January 1, 2014 to July 25, 2023), a total of 3379 initial documents were retrieved. As this study primarily focuses on empirical research, 179 review papers were excluded from the analysis. Subsequently, we reviewed the abstract of each paper to determine whether it met the criteria of this study. Among them, 761 papers were excluded due to their deviation from the research topic. Finally, 23 non-English papers were also excluded. As a result, 2416 relevant papers (from 2014 to 2023) were retained, and these were exported in TXT plain text format for bibliometric analysis. Among them, 1339 were journal articles, and 1077 were conference papers.

16.2.2 Research methods

Bibliometrics is widely recognized as one of the primary methods for analyzing literature reviews and has become a mainstream approach in scientific policy and research management in recent years. Commonly used bibliometric software includes CiteSpace, VOSviewer, HistCite, and BibExcel. Among them, CiteSpace and VOSviewer are both bibliometric analysis tools running on JAVA programs that can effectively establish the mapping relationship between the knowledge units of the literature and visualize the macrostructure of the knowledge through knowledge mapping. CiteSpace provides various vital indicators, such as keyword burst and time zone, which can visually identify the development trend of a specific research area and the evolution process of knowledge. VOSviewer provides text mining functions to build knowledge maps based on metrics such as cooccurrence, bibliographic coupling, and cocited references. Our study used two bibliometric visualization tools, VOSviewer (V1.6.19) and CiteSpace (V6.2.R4), to conduct a comprehensive analysis of SRID research, explore the knowledge structure and development trend of SRID research, and further discuss possible future research priorities. Fig. 16.1 illustrates the research framework of this study.

16.3 Bibliometric results and analysis

16.3.1 Trend analysis of annual outputs of social robot interaction design literature

Statistical patterns of change in the output of academic literature over time can effectively assess the accumulation of knowledge, research dynamics, and maturity level of a discipline. The annual distribution curve of the literature output in SRID is shown in Fig 16.2. Based on the trend of literature output in the last decade, the entire development of SRID research can be categorized into three main stages: the initial stage (2013–16), the developmental stage (2017–19), and the booming stage (2020–23). As shown in Fig. 16.2, the literature production of SRID research was relatively low during the initial stage (2013–16), with an average annual publication volume of 115 papers per year, and the annual publication count did not exceed 150 papers. Starting in 2017, this field entered a stage of rapid development, with an average annual publication volume of 246 papers published per year and a rapid growth rate. From 2020 to 2023, with the development and

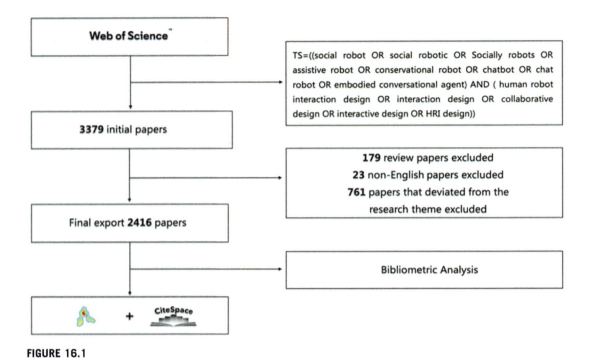

FIGURE 16.1

Research framework.

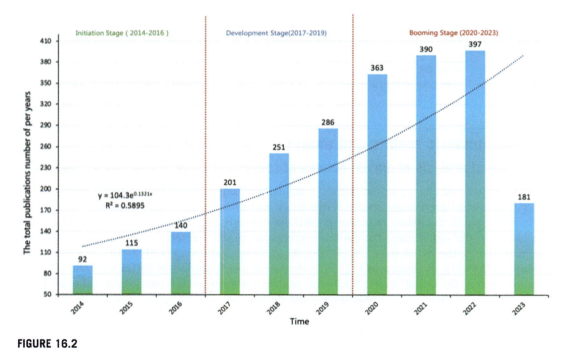

FIGURE 16.2

A distribution map of the annual publication volume of social robot interaction design.

popularization of technologies such as artificial intelligence, affective computing, and cloud computing, the literature output in this field began to sharply increase. By 2023, the average annual number of publications was as high as 332 per year, the growth rate was more stable, and SRID research entered a prosperous stage.

16.3.2 Research hotspots of social robot interaction design

A paper's keywords serve as a high-level summary of the authors' research results, typically encompassing research objects, perspectives, and methods. High-frequency cooccurring keywords reflect the long-standing research hotspots of SRID. The 2416 documents within the search contained a total of 6889 keywords. To ensure the consistency of keyword expressions and the clarity of homonyms, keywords were standardized and synonyms were merged based on suggestions from Frey and Dueck, following discussions among three senior professors and the research team. For example, "customer experience" was replaced with "user experience," "autism spectrum disorders" was replaced with "autism spectrum disorder," and "older adults," "elderly people," "aging," and so on, were replaced with "elderly." Importing the WOS data into VOSviewer, we set a cooccurrence frequency threshold of more than eight to ensure the quality of the plots. After merging synonymous keywords, the resulting keyword cooccurrence clustering diagram for SRID research is shown in Fig. 16.3A. There are 288 keyword nodes and 9560 connecting lines in the figure, and keywords with the same color in the graph are the same clusters, which form a total of five major clusters. That is, the current research of SRID mainly focuses on #1 the study of human-robot relationships in social robots, #2 research on the emotional design of social robots, #3 research on social robots for children's psychotherapy, #4 research on companion robots for elderly rehabilitation, and #5 research on educational social robots. Fig. 16.3B is a keyword cooccurrence cluster-time overlay diagram reflecting the temporal distribution of these five SRID research clusters. Table 16.1

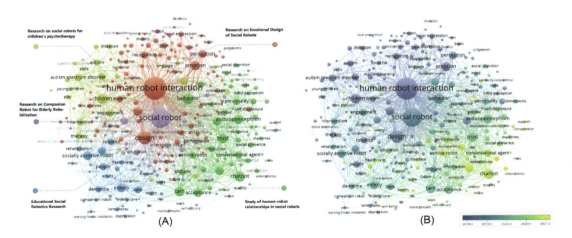

FIGURE 16.3

(A and B) Keyword cooccurrence clustering.

provides detailed information about the SRID research clusters, and the following sections outline the specific research content within these clusters.

Cluster #1: The study of human-robot relationships in social robots, which consists of 101 keywords, including mainly acceptance, anthropomorphism, attitude, trust, privacy, self-disclosure, satisfaction, technology acceptance model, social presence, warmth, transparency, roles, experience, and other keywords. The human-robot relationship is an essential topic in SRID research and is an emerging research theme (see Fig. 16.3B). Furthermore, as robots become highly intelligent, people will naturally and unconsciously apply the social rules of interpersonal communication to HRIs, even if the robots are not humans. Scholars such as Nass use the "computer is a social actor (CASA) paradigm" to describe this phenomenon. The anthropomorphic design and autonomous actions of social robots make their behaviors particularly easy to perceive as social behaviors. The research content of the human-robot relationship mainly includes human-robot trust (trust), robot acceptance (acceptance), user self-disclosure willingness (self-disclosure), user satisfaction (satisfaction), and user privacy concerns (privacy). This direction of research is mainly based on the user's psychological cognition, through the use of the model research method (structural equation model) and user experiments to conduct quantitative empirical research, to identify the influencing factors affecting human-robot relationships, as well as the influence mechanism. From the large number of research results accumulated in the current literature, the influencing factors (antecedent variables) of the human-robot relationship can be broadly categorized into three dimensions: human, robot, and environment. For example, the keywords anthropomorphism, roles, appearance, transparency, social presence, and warmth appearing in Cluster #1 are antecedent variables belonging to the Robot dimension. Combined with the characteristics of high-frequency keywords, the research focus of Cluster #1 can be further summarized as building a more harmonious social human-machine relationship through the interaction design of social robots.

Table 16.1 Specific information on social robot interaction design's five research clusters.

Cluster	Cluster label	Average time	Keywords number	Keywords
1(green)	Study of human-robot relationships in social robots	2021	101	Acceptance; anthropomorphism; attitude; trust; privacy; self-disclosure; technology acceptance model; warmth; transparency
2(red)	Research on emotional design of social robots	2019	60	Affective computing; affective robotics; emotion; facial expression; body language; EEG; gaze; gesture recognition
3(yellow)	Research on social robots for children's psychotherapy	2019	51	Autism; therapy intervention; children; anxiety; training; social skills; spectrum disorder; games; speech
4(blue)	Research on companion robots for elderly rehabilitation	2020	48	Elderly; health care; loneliness; mental health; assisted therapy; codesign; depression; user-centered design
5(purple)	Educational social robotics research	2020	28	Educational robot; performance; knowledge; teacher; classroom; child-robot interaction; empathy; interaction design

Cluster #2: Research on the emotional design of social robots. This cluster comprises 60 cluster members, primarily including keywords such as affective computing, affective robotics, emotion, emotion recognition, facial expression, face, body language, EEG, brain, eye contact, gaze, gesture recognition, and more. Emotional design for social robots refers to imbuing robots with emotions, enabling them to recognize better, understand, and respond to users' emotions through emotional expressions. A social robot with an effective emotional design can enhance the interactive experience of HRI, improve user satisfaction, and facilitate a more natural and enjoyable communication process. Specifically, emotional recognition in social robots refers to enabling robots to recognize and understand human emotional expressions, allowing the robots to perceive and appropriately respond to users' emotions during interactions. The emotional expression design for social robots primarily includes two categories: verbal cues and nonverbal cues. Verbal cues encompass the tone, intonation, and conversational style of social robots' speech, while nonverbal cues encompass facial expressions, gestures, body movements, and gaze, among others. For example, robots can use gestures, natural language, and even eye contact to engage in positive emotional interactions with users. As shown in Fig. 16.3B, the average occurrence time of keywords in this cluster is relatively early, which belongs to the research basis of human-computer interaction design of social robots. From the perspective of comprehensive keyword characteristics, this cluster primarily delves into the impact of social cues and emotional factors on HRI, focusing on the recognition and expression of social robot emotions and their integration into specific designs and evaluations.

Cluster #3: Research on social robots for children's psychotherapy is composed of 51 keywords such as autism, therapy intervention, children, anxiety, training, social skills, spectrum disorder, games, and speech. This cluster focuses on the specific impact of the interaction design of social robots on the effectiveness of children's psychotherapy and emotional experience. The psychotherapy social robot provides targeted interventions by analyzing children's psychological needs and emotional states, and is specifically designed to alleviate negative emotions such as anxiety and depression. A large number of studies have demonstrated that social robots have advantages in children's psychotherapy compared with other therapeutic methods, such as avoiding the psychological barriers that may arise from personal emotions and social stigma in traditional psychotherapy scenarios, not being affected by time and location restrictions, not being limited by the number of psychologists and higher acceptance and enjoyment in interactions. From the high-frequency keywords in the cluster, social robots are also often used in psychotherapy scenarios, such as in therapy for children with autism spectrum disorders and social skills training. Typical psychotherapy social robots include Woebot, Wysa, Joyable, and so on. These robots are used in more than 130 countries around the world and have been shown to be effective through psychological evaluations. Nonetheless, this research direction is still in its early stages. Future investigations must address fundamental research questions related to children's perception, cognition, behavior, and psychotherapy, among other aspects. Furthermore, many current studies lack rigorous quantitative statistical analysis, and there is a relatively limited number of child participants in user experiments.

Cluster #4: Research on companion robots for elderly rehabilitation mainly contains 48 keywords such as elderly, health care, loneliness, mental health, assisted therapy, codesign, depression, user-centered design, and others, and at the same time presents a very inclusive perspective. This cluster adopts an inclusive perspective and primarily focuses on the research of interaction design for rehabilitation companion robots. These robots aim to support the independent living of elderly individuals in their homes, meeting their real-world needs, promoting physical and mental health, and preserving their dignity, privacy, and independence. Numerous studies have shown that assisting elderly individuals with physical and cognitive training through social robots has a significant positive impact on their cognitive function, overall well-being, and quality of life. For instance, it leads to a reduction in diseases such as diabetes, Alzheimer's disease, and cardiovascular conditions among the elderly population. In this research scenario, the social robot mainly acts as an exercise partner or trainer to provide social interaction and companionship for the elderly. To maximize user enthusiasm and engagement, robots should be user-centered, adapting their interaction strategies based on the personalized needs, actual abilities, and daily habits of the elderly.

Cluster #5: Educational social robotics research contains a total of 28 cluster members, mainly containing the keywords educational robot, performance, knowledge, teacher, classroom, child-robot interaction, cognition, empathy, interaction design, and other keywords. Educational social robots are designed to provide educational and learning support by interacting naturally with students, developing logical thinking skills, language expression abilities, teamwork abilities, creativity, and more. In learning environments, these robots typically play roles such as tutoring teachers, learning partners, and information providers. The research topics mainly include user interface design, dialog language design, emotional interaction design, and personalized learning support, among other aspects, and involve interdisciplinary knowledge such as education, psychology, and computer science. By combining the characteristics of high-frequency keywords, the research hotspots in Cluster #5 can be succinctly summarized as the investigation of design strategies for the interaction between social robots and student users. The aim is to provide effective learning support and create a more enjoyable user experience.

16.3.3 Evolution and trend of social robot interaction design research hotspots

To better understand the evolution and development trends of SRID research hotspots, we conducted a statistical analysis of the average appearance time of keywords, resulting in Fig. 16.4 (keyword: time zone) and Fig. 16.5 (burst keywords' emergence). The time zone map intuitively reflects the frequency size and first appearance time of keywords within the search range and is often used by scholars to identify the previous trend of research topics. Fig. 16.5 lists the top 30 keywords and their emergence intensity in different periods, where the darker parts characterize the intensity of keyword bursts in the literature and the years in which they were relatively prominent. The burst keywords represent the level of attention given to research topics in different periods and can also reflect the changing trends and past research hotspots in the field. The time zone and burst keywords combine keywords with the time dimension for analysis. By combining the insights from the time zone and burst keywords, we aim to obtain more robust research results.

In Fig. 16.4, it is evident that early SRID research focused on appearance design for social robots, with keywords such as anthropomorphism, uncanny valley, humanoid robots, and facial expressions. Subsequently, the research started to turn to social cues of emotional design, such as gestures, body language, affective computing, face, eyes, and so on. The subsequent research focused more on specific application scenarios, such as psychotherapy, rehabilitation assistance, and companion learning. The keywords after 2020 are human-robot collaboration, machine learning, deep learning, natural language processing, satisfaction, adoption, individual differences, participatory design, and more.

The top 30 burst keywords in Fig. 16.5 provide further insights, revealing the transition from early keywords such as facial expressions and uncanny valley to emotion recognition, and onwards to affective computing, artificial intelligence, and participatory design. This trend underscores the shift from embodied interaction design research to human-robot relationship research. Combining the insights from the time zone analysis, burst keywords, and cooccurrence cluster map, it is

16.3 Bibliometric results and analysis   211

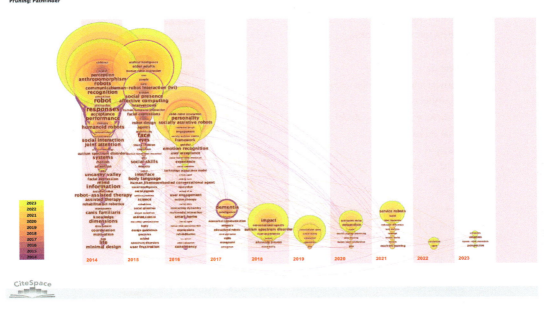

FIGURE 16.4

Keywords time zone view of social robot interaction design.

evident that the future of SRID research will be centered around several key areas. These include human-robot collaboration, affective computing, artificial intelligence, human-robot trust, user self-disclosure, emotion recognition and expression, and interaction design evaluation. This synthesis of research trends underscores the anticipated focus on advancing the understanding of how social robots can effectively collaborate with humans, integrate emotions, leverage AI capabilities, foster trust, facilitate user self-disclosure, and enhance interaction design evaluation.

16.3.4 Knowledge base of SR-human-robot interaction research

References establish a cocitation relationship when they are cited together in the same documents. The higher the cocitation strength between two papers, the more significant the correlation between them. According to the statistics, 75,022 valid references from 44,524 scholars were cited in the 2416 papers within the search scope. Based on VOSviewer, we extracted references with a citation frequency of no less than 20 in 2014–23. We generated a reference cocitation cluster network of 300 references and 19,944 cocitation relationships, as shown in Fig. 16.6. Consistent with the results of the keyword cooccurrence cluster, the reference cocitation cluster still forms five clusters: #1 the study of human-robot relationships in social robots, #2 research on the emotional design of

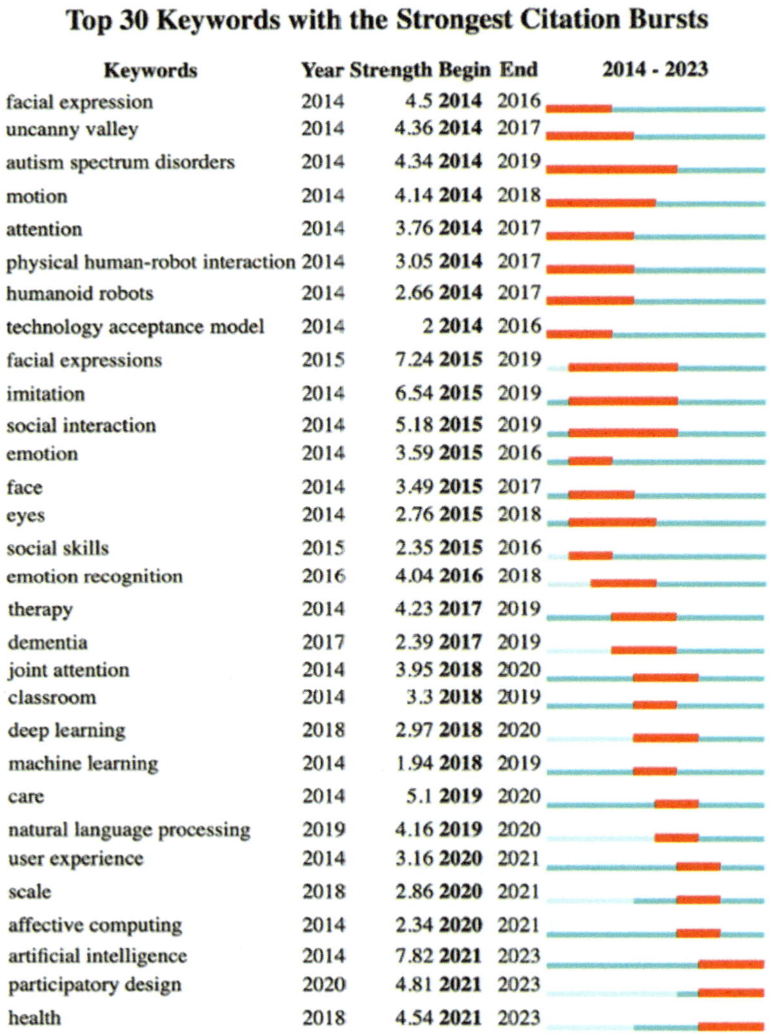

Top 30 Keywords with the Strongest Citation Bursts

Keywords	Year	Strength	Begin	End	2014 - 2023
facial expression	2014	4.5	2014	2016	
uncanny valley	2014	4.36	2014	2017	
autism spectrum disorders	2014	4.34	2014	2019	
motion	2014	4.14	2014	2018	
attention	2014	3.76	2014	2017	
physical human-robot interaction	2014	3.05	2014	2017	
humanoid robots	2014	2.66	2014	2017	
technology acceptance model	2014	2	2014	2016	
facial expressions	2015	7.24	2015	2019	
imitation	2014	6.54	2015	2019	
social interaction	2014	5.18	2015	2019	
emotion	2014	3.59	2015	2016	
face	2014	3.49	2015	2017	
eyes	2014	2.76	2015	2018	
social skills	2015	2.35	2015	2016	
emotion recognition	2016	4.04	2016	2018	
therapy	2014	4.23	2017	2019	
dementia	2017	2.39	2017	2019	
joint attention	2014	3.95	2018	2020	
classroom	2014	3.3	2018	2019	
deep learning	2018	2.97	2018	2020	
machine learning	2014	1.94	2018	2019	
care	2014	5.1	2019	2020	
natural language processing	2019	4.16	2019	2020	
user experience	2014	3.16	2020	2021	
scale	2018	2.86	2020	2021	
affective computing	2014	2.34	2020	2021	
artificial intelligence	2014	7.82	2021	2023	
participatory design	2020	4.81	2021	2023	
health	2018	4.54	2021	2023	

FIGURE 16.5

Keywords burst.

social robots, #3 research on social robots for Children's psychotherapy, #4 research on companion robots for elderly rehabilitation, and #5 research on educational social robots. The papers in these five clusters constitute the most critical knowledge foundation in the field of SRID, connecting the research contents of different disciplines.

In Cluster #1 (study of human-robot relationships in social robots), the article "Measurement Instruments for the Anthropomorphism, Animacy, Likeability, Perceived Intelligence, and Perceived

16.3 Bibliometric results and analysis

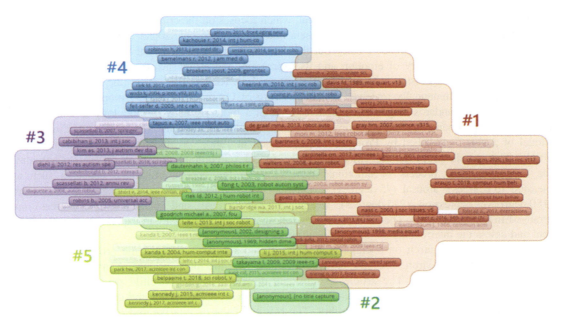

FIGURE 16.6

Reference cocitation cluster network.

Safety of Robots," published by C. Bartneck et al. in 2009 in the *International Journal of Social Robotics*, received the most attention, with a total of 201 citations in our cocitation network. In contrast to traditional reviews, this study conducted a systematic literature review on anthropomorphism, animacy, likeability, perceived intelligence, and perceived safety, which are the five key concepts of social robot design. It also examined measurement methodologies and refined the results into five semantic difference scales with high internal consistency. This work serves as the conceptual basis for studying human-robot relationships in social robots. The second most cited paper is "Anthropomorphism and the Social Robot" by B.R. Duffy, published in *Robotics and Autonomous Systems* in 2003. The paper delves into the mechanisms underlying anthropomorphism and introduces the critical social features necessary for robots to engage in social interactions. It also discusses how anthropomorphic design can make it easier for users to comprehend and justify their actions. In addition, the paper "Machines and Mindlessness: Social Responses to Computers," published by Clifford Nass et al. in 2000, has also attracted much attention. This paper further validates the CASA paradigm by reviewing a series of experimental studies, proving that the depth of social response is related to the "personality" of the computer, and explains the psychological cognitive process behind this phenomenon. This paper represents the second generation of research within the CASA paradigm, which focuses on explaining the social responses generated in HRIs to advance further research on human-robot relationships.

Cluster #2 (research on the emotional design of social robots) consists mainly of classic literature related to the emotional design of social robots. At the center of the cluster is "Emotion and Sociable

Humanoid Robots," a 2003 paper by C. Breazeal, which focuses on the critical role of emotional expressive behavior in social interactions between users and anthropomorphic social robots. The paper takes the Kismet robot as an example to carry out the development and verification work, puts forward the key principles in the theory of emotion and its expression, and establishes the scientific basis for the emotion model and expression behavior of social robots. L.D. Riek's 2012 article (the second most cited at 65 citations) systematically reviewed 54 HRI Wizard of Oz experiments and proposed new experimental reporting standards to guide researchers to adopt a more rigorous methodology to conduct and report HRI experiments and obtain more convincing research conclusions. The article "A Circumplex Model of Affect," published by Professor James Russell in 1980, was the third most frequently cited article. This article proposes the emotional space model (circumplex model of affect) to visualize the mechanism and process of emotional generation. In this model, emotional concepts such as pleasure (0 degrees), arousal (45 degrees), excitement (90 degrees), distress (135 degrees), displeasure (180 degrees), depression (225 degrees), sleepiness (270 degrees), and relaxation (315 degrees) are arranged in a circular order. The model provides follow-up researchers with an expressive way to represent and assess the interrelatedness of emotional dimensions.

In Cluster #3 (research on social robots for children's psychotherapy), the article "Keepon: A Playful Robot for Research, Therapy, and Entertainment" is the most influential and is at the center of the cocitation network. The article designs and develops a Keepon child psychotherapy robot and conducts longitudinal field observations of typical preschool children and children with autism spectrum disorder using the modified robot. The results of the qualitative and quantitative analyses suggest that the Keepon robot can help children understand socially meaningful information and motivate them to share it. These findings also emphasize the significance of interaction style in establishing human-robot relationships. Second is the 2005 paper by B. Robins et al. in Universal Access in the Information Society, which describes a long-term longitudinal study of four children with autism spectrum disorder. Quantitative analyses showed an increase in the duration of predefined behaviors in the later experiments, while qualitative analyses indicated a further increase in social interaction skills (imitation, turn-taking, and role-switching) and communicative competence demonstrated by children in an interactive setting. The findings also confirm the necessity and benefits of long-term longitudinal studies of HRI. Subsequently, a paper by E.S. Kim et al. was published in the *Journal of Autism and Developmental Disorders* in 2013. This paper examined the effects of interacting with three different objects: (1) an adult, (2) a touchscreen computer game, and (3) a social robot, on interaction outcomes in 4–12-year-old children with autism spectrum disorders ($N = 24$). The results indicated that children said more words when the interaction partner was a robot compared with a human or computer game partner, and the study makes a strong case for the great potential of social robots in developing social skills and psychotherapy.

In Cluster #4 (research on companion robots for elderly rehabilitation), the most highly cited paper is "Assessing Acceptance of Assistive Social Agent Technology by Older Adults: The Almere Model." The article extended the technology acceptance model and the unified theory of acceptance and use of technology model to test the intention of older users to use assistive robots, with the model incorporating function-related variables (such as perceived usefulness and perceived ease of use) and social interaction-related variables to explain the use intention. Three different elderly robots were tested using controlled experiments and longitudinal data, and the results showed strong support for the model, accounting for 59%–79% of the variance in intention to use and 49%–59% of the variance in actual use. In a 2007 paper by K. Wada et al., two therapeutic robots were introduced into an elderly care facility for a long-term follow-up experiment to investigate the psychological and social impact of robots. The results showed that interaction with the robots increased their social interactions. In addition, physiological measurements showed that the subjects' vital organ responses to stress improved after the introduction of the robots. The 2013 paper by J. Fasola et al. discusses the design methodology and implementation details of a social assistive robot (SAR) system to help older adults participate in physical activity, including insights from psychological research into intrinsic motivation, and proposes five clear SAR-based therapy design principles for interventions. A multisession user study of the elderly ($N = 33$) also demonstrated the effectiveness of the SAR exercise system in terms of factors such as fun and social attractiveness.

In Cluster #5 (research on educational social robots), the review paper "Social Robots for Education: A Review" by T. Belpaeme et al. was the most influential. This paper presents three critical studies by answering the following questions: (1) What is the effect of robot tutors in achieving learning outcomes? (2) How do the appearance and behavior of robots affect learning outcomes? (3) What are the potential roles of robots in educational settings? The paper provides a comprehensive overview of the application of educational social robots. The second most highly cited paper in Cluster #5 is titled "Interactive Robots as Social Partners and Peer Tutors for Children: A Field Trial." The paper reports an 18-day field experiment in a Japanese elementary school to investigate how educational robots establish relationships with children. The experimental results suggest that robots may be more successful in establishing common ground and influence when children already have some initial proficiency or interest in English. These results suggest that the design of interactive robots should have something in common with their users, providing both social and technical challenges. In addition, C. Bartneck's 2004 paper "A Design-Centered Framework for Social Human-Robot Interaction" has also attracted much attention. The paper proposes a framework for categorizing attributes of social robots such as form, morphology, social norms, autonomy, and interactivity, and provides extensive guidelines for designing social robots.

16.3.5 Distribution of social robot interaction design literature sources

Analyzing the citations and the number of papers of the literature sources can help researchers understand which journals/conferences are the sources that contribute the most to the SRID research field. According to the statistics, the 2416 valid documents within the scope of retrieval were published in 909 journals and conferences. Table 16.2 lists the top 10 high-impact journals/conferences in terms of total citations in the last decade, which account for 17.136% of the total number of publications. Among them, the *International Journal of Social Robotics* is the journal with the most papers in this research field, containing 184 papers, and it also has the highest total citations, reflecting its significant influence in the field of HRI design for social robots. Next is *Computers in Human Behavior*, with 39 publications and a total of 1268 citations. Ranking third is *Frontiers in Psychology*, with 34 publications and 764 total citations. These journals/conferences come from different research fields, such as robotics, psychology, computer science, management, geriatrics, and more. It can be seen that SRID is a key research topic with solid interdisciplinary characteristics, involving knowledge from multiple subject areas. According to the disciplinary statistical analysis of the WOS system, the 2416 documents are related to 83 major disciplines, and the top ten disciplines in terms of the number of publications are computer science, robotics,

Table 16.2 Distribution of SR-human-robot interaction literature sources (top 10).

Ranking	Literature source	Publisher	Citations	Number of publications	Average number of citations
1	International Journal of Social Robotics	SPRINGER	4010	184	21.793
2	Computers in Human Behavior	ELSEVIER	1268	39	32.512
3	Frontiers in Psychology	FRONTIERS	764	34	22.470
4	International Journal of Human-Computer Studies	ELSEVIER	734	22	33.363
5	ACM/IEEE International Conference on Human-Robot Interaction (HRI2015)	ACM	563	12	46.916
6	International Journal of Contemporary Hospitality Management	EMERALD	544	7	77.714
7	Journal of the American Medical Directors Association	ELSEVIER	440	4	110.034
8	ACM/IEEE International Conference on Human-Robot Interaction (HRI2018)	ACM	435	13	33.461
9	Frontiers in Robotics and AI	FRONTIERS	388	69	5.623
10	ACM/IEEE International Conference on Human-Robot Interaction (HRI2019)	ACM	363	30	12.021

engineering, automated control systems, psychology, business, economics, education, neuroscience, sociology, and telecommunications. These disciplinary subjects are important research areas for SRID, providing both theoretical foundations and methodological tools for SRID studies. Furthermore, SRID also provides valuable development opportunities for advancements in these disciplines. In addition, journals such as the *International Journal of Human-Computer Studies*, *ACM Transactions on Human-Robot Interaction*, and *Interaction Studies* also have a significant influence on SRID research.

16.3.6 High-impact countries and research institutions

As seen in Fig. 16.7A, a total of 80 countries/regions around the world have contributed to the research field of SRID, and three cooperative communities have been formed in terms of cooperation, dominated by the United States, Germany, and Spain. Among these, the United States is the most influential country, with the highest number of publications and total citations (642 publications, 9398 total citations), Followed by Germany (247 articles and 2747 citations), the United Kingdom (216 articles and 3719 citations), China (198 articles and 1647 citations), Japan (196 articles and 1791 citations), Italy (186 articles and 1884 citations), New Zealand (179 articles and 2535 citations), Australia (114 articles and 1288 citations), Canada (112 articles and 1491 citations),

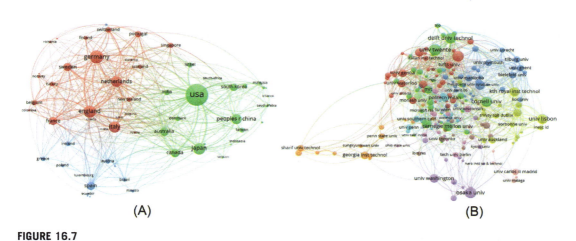

FIGURE 16.7

(A) National cooperation network. (B) Institutional cooperation network.

and Spain (109 articles and 958 citations). The publications of the top 10 countries accounted for 63.286% of the total number of publications, and their total citation counts are all above 100, which is an essential source of output for SRID research worldwide.

A total of 1959 research organizations worldwide have conducted SRID-related research in the last decade. Running VOSviewer and selecting organizations to set the node threshold to five yields a collaborative network of institutions with an annual node count of 218, as shown in Fig. 16.7B. In terms of total citations, the top five research institutions were Stanford University (876 citations), followed by the Massachusetts Institute of Technology (MIT) (781 citations), University of Auckland (761 citations), Hertfordshire University (707 citations), and the University of Twente (682 citations). In terms of publications, the top five research institutions are the University of Twente (41 papers), followed by Osaka University (39 papers), Eindhoven University of Technology (38 papers), Cornell University (36 papers), and Delft University of Technology (35 papers).

16.3.7 High-impact author analysis

Through author cocitations, we can study the more active scholars in the field of SRID research around the world. A total of 7053 authors within the search scope have contributed to the research field of SRID. Table 16.3 lists the top five authors with high literature production and their respective total citations. Among them, Professor Paiva, Ana from the Polytechnic University of Lisbon published the most papers, with an H index of 43 and a total of 476 citations, and is ranked first with 30 publications within the search scope. Professor Ana Paiva is followed by Hiroshi Ishiguro (Osaka University) with 29 papers and a total of 369 citations, Kerstin Dautenhahn (University of Waterloo) with 26 papers and 599 citations, Selma Sabanovic (Indiana University) with 24 papers and 237 citations, and Guy Hoffman (Cornell University) with 21 papers and 483 citations. In addition, among the high-yield authors, Professor Broadbent, Elizabeth of the University of Auckland had the highest number of total citations, with 13 papers published within the search scope and a total of 729

Table 16.3 Top five high-yield authors.

Ranking	Author	Institution	Number of published papers	Number of citations	H index
1	Paiva, Ana	Polytechnic University of Lisbon	30	476	43
2	Ishiguro, Hiroshi	Osaka University	29	369	54
3	Dautenhahn, Kerstin	University of Waterloo	26	599	56
4	Sabanovic, Selma	Indiana University	24	237	32
5	Guy Hoffman	Cornell University	21	483	33

citations, followed by Breazeal, Cynthia (16 papers and 654 total citations), Dautenhahn, Kerstin (26 papers and 599 total citations), Mutlu, Bilge (18 papers and 550 total citations), and Guy Hoffman (21 papers and 483 total citations). These high-impact authors are considered pioneers and significant contributors to SRH-RID research.

16.4 Discussion

In the previous section, we identified the hot topics in the field of SRID research, classic foundational literature, research collaboration networks, and so on. In this section, we will discuss the existing challenges and future research directions.

16.4.1 More realistic research scenarios

Current research is still dominated by laboratory studies, lacking research in real and complex natural interaction scenarios, and the results obtained in laboratory environments may have limitations in terms of real-world applications. Research conducted in close-to-real-world contexts can help obtain more meaningful results and provide deeper insights into human behavior and robot interactions.

16.4.2 Effective measurement of research indicators

Current research presents difficulties and challenges in obtaining reliable data to evaluate human-social robot interactions. For example, the commonly used questionnaires are not well suited to young children, and the measurement of some variables is easily affected by the state of the subject and the external environment, such as trust and users' emotional state. Therefore there are still some challenges in how to effectively and objectively measure the true level of such latent variables and improve the reliability, validity, and replicability of the data. This may require larger sample sizes, a more diverse set of approaches for measuring variables, and richer data types.

16.4.3 Longitudinal trial

Current research is mainly based on short-term experiments, while interactions between humans and social robots often occur over the long term. Short-term experiments cannot capture the long-term effects and how HRI changes over time. For example, in the case of social robots used for psychotherapy and online education, the evaluation of learning and psychotherapy effects is a long-term process. Future research could conduct more long-term longitudinal trials to provide more comprehensive data to support the conclusions, thus providing insights into the phenomenon and its mechanisms changing over time.

16.4.4 More specific design strategies

Although the interaction design of social robots has accumulated a significant number of research findings, the current research is relatively scattered and has not yet formed a comprehensive design framework. Taking the research on psychotherapy social robots as an example, the current research focuses on the acceptance, effectiveness, and feasibility of robot intervention, while the basic research related to the robot interaction design and the user's emotional experience is relatively lacking, and the mapping mechanism between the interaction design factors and the healing effect is unclear. Therefore in the future, it is necessary to study further the specific design features and specific design strategies of SRID, especially how the combination effect of different interaction design features will affect the efficiency of human-computer interaction and emotional experience.

16.4.5 Uncovering the psychological cognitive mechanisms behind user behavior

There is a close interplay between a user's psychological cognitive process, their experience, and their behavior. Understanding the psychological cognitive process of users can help design social robots that better meet user needs and elicit positive emotions. The current empirical research usually only gives general conclusions and lacks empirical studies that explain how the interaction design of social robots affects user perception and user behavioral responses from the perspective of user psychology and cognition. Therefore this will also be a future research direction.

16.5 Conclusions

This paper provides a comprehensive review of the literature related to SRID research in the last decade based on bibliometrics. The main contributions include:

1. Identifying the critical research hotspots and emerging trends in this field over the past decade.
2. identifying classic literature, leading countries, journals, and authors, shaping the knowledge structure of the field.
3. Discussing the challenges that the current SRID research is facing in terms of more authentic research scenarios, effective metrics for research indicators, longitudinal experimental design, more specific design strategies, and mining the mechanisms behind interaction behaviors.

The paper also presents potential research directions and opportunities. These findings can help scholars in this field gain a better understanding of the research structure and latest trends in SRID, and provide a solid foundation for further development.

References

[1] S. Khan, C. Germak, Reframing HRI design opportunities for social robots: lessons learnt from a service robotics case study approach using UX for HRI, Future Internet 10 (2018) 101.
[2] I. Deutsch, H. Erel, M. Paz, G. Hoffman, O. Zuckerman, Home robotic devices for older adults: opportunities and concerns, Comput. Hum. Behav. 98 (2019) 122–133.
[3] D. Yang, Y.-J. Chae, D. Kim, Y. Lim, D.H. Kim, C. Kim, et al., Effects of social behaviors of robots in privacy-sensitive situations, Int. J. Soc. Robot. (2022) 1–14.

Further reading

D.O. Aghimien, C.O. Aigbavboa, A.E. Oke, W.D. Thwala, Mapping out research focus for robotics and automation research in construction-related studies: a bibliometric, Approach. J. Eng. Des. Technol. 18 (2020) 1063–1079.

N. Akalin, A. Kristoffersson, A. Loutfi, Do you feel safe with your robot? Factors influencing perceived safety in human-robot interaction based on subjective and objective measures, Int. J. Hum.-Comput. Stud. 158 (2022) 102744.

B. Alenljung, J. Lindblom, R. Andreasson, T. Ziemke, User experience in social human-robot interaction, Rapid Automation: Concepts, Methodologies, Tools, and Applications, IGI Global, Hershey, PA, USA, 2019, pp. 1468–1490.

R. Arora, Validation of an sor model for situation, enduring, and response components of involvement, J. Mark. Res. 19 (1982) 505–516.

P. Asgharian, A.M. Panchea, F. Ferland, A review on the use of mobile service robots in elderly care, Robotics 11 (2022) 127.

F. Babel, J. Kraus, L. Miller, M. Kraus, N. Wagner, W. Minker, et al., Small talk with a robot? The impact of dialog content, talk initiative, and gaze behavior of a social robot on trust, acceptance, and proximity, Int. J. Soc. Robot. 13 (2021) 1485–1498.

S. Bacci, B. Bertaccini, A. Petrucci, Insights from the co-authorship network of the italian academic statisticians, Scientometrics 128 (2023) 4269–4303.

C. Bartneck, J. Forlizzi, A design-centred framework for social human-robot interaction. In Proceedings of the RO-MAN 2013th IEEE InternationalWorkshop on Robot and Human Interactive Communication (IEEE Catalog No. 04TH8759), Kurashiki, Japan, 22 September 2004; pp. 591–594.

C. Bartneck, D. Kulić, E. Croft, S. Zoghbi, Measurement instruments for the anthropomorphism, animacy, likeability, perceived intelligence, and perceived safety of robots, Int. J. Soc. Robot. 1 (2009) 71–81.

G.A. Bekey, Current trends in robotics: technology and ethics, Robot Ethics: The Ethical and Social Implications of Robotics, MIT Press, Cambridge, MA, USA, 2012, pp. 17–34.

T. Belpaeme, J. Kennedy, A. Ramachandran, B. Scassellati, F. Tanaka, Social robots for education: a review, Sci. Robot. 3 (2018). eaat5954.

E. Bendig, B. Erb, L. Schulze-Thuesing, H. Baumeister, The next generation: chatbots in clinical psychology and psychotherapy to foster mental health—a scoping review, Verhaltenstherapie 29 (2019) 266–280.

I. Benke, U. Gnewuch, A. Maedche, Understanding the impact of control levels over emotion-aware chatbots, Comput. Hum. Behav. 129 (2022) 107122.

C. Breazeal, Emotion and sociable humanoid robots, Int. J. Hum. Comput. Stud. 59 (2003) 119−155.

E. Broadbent, R. Tamagawa, A. Patience, B. Knock, N. Kerse, K. Day, et al., Attitudes towards health-care robots in a retirement village, Australas. J. Ageing 31 (2012) 115−120.

C. Chen, Citespace Ii: detecting and visualizing emerging trends and transient patterns in scientific literature, J. Am. Soc. Inf. Sci. Technol. 57 (2006) 359−377.

E. Chew, U.S. Khan, P.H. Lee, Designing a novel robot activist model for interactive child rights education, Int. J. Soc. Robot. 13 (2021) 1641−1655.

W.-C. Cho, K.Y. Lee, S.-B. Yang, What makes you feel attached to smartwatches? The stimulus−organism−response (S−O−R) perspectives, Inf. Technol. People 32 (2019) 319−343.

L. Christoforakos, A. Gallucci, T. Surmava-Große, D. Ullrich, S. Diefenbach, Can robots earn our trust the same way humans do? A systematic exploration of competence, warmth, and anthropomorphism as determinants of trust development in HRI, Front. Robot. AI 8 (2021) 79.

H. Chung, H. Kang, S. Jun, Verbal anthropomorphism design of social robots: investigating users' privacy perception, Comput. Hum. Behav. 142 (2023) 107640.

K. Dautenhahn, Socially intelligent robots: dimensions of human-robot interaction, Philos. Trans. R. Soc. B Biol. Sci. 362 (2007) 679−704.

D. David, S.-A. Matu, O.A. David, Robot-based psychotherapy: concepts development, state of the art, and new directions, Int. J. Cogn. Ther. 7 (2014) 192−210.

D.P. Davison, F.M. Wijnen, J. van der Meij, D. Reidsma, V. Evers, Designing a social robot to support children's inquiry learning: a contextual analysis of children working together at school, Int. J. Soc. Robot. 12 (2020) 883−907.

G.B. De'Aira, J. Xu, Y.-P. Chen, A. Howard, The effect of robot vs. human corrective feedback on children's intrinsic motivation, in: Proceedings of the 2019 14th ACM/IEEE International Conference on Human−Robot Interaction (HRI), Daegu, Republic of Korea, 11−14 March 2019, pp. 638−639.

J. deWit, P. Vogt, E. Krahmer, The design and observed effects of robot-performed manual gestures: a systematic review, ACM Trans. Hum.-Robot Interact. 12 (2023) 1.

Y. Dong, D. Li, Gender representation in textbooks: a bibliometric study, Scientometrics, 128, 2023, pp. 5969−6001.

X. Dou, C.-F. Wu, K.-C. Lin, S. Gan, T.-M. Tseng, Effects of different types of social robot voices on affective evaluations in different application fields, Int. J. Soc. Robot. 13 (2021) 615−628.

B.R. Duffy, Anthropomorphism and the social robot, Robot. Auton. Syst. 42 (2003) 177−190.

F. Eyssel, D. Kuchenbrandt, Social categorization of social robots: anthropomorphism as a function of robot group membership, Br. J. Soc. Psychol. 51 (2012) 724−731.

J. Fasola, M.J. Matarić, A socially assistive robot exercise coach for the elderly, J. Hum.-Robot Interact. 2 (2013) 3−32.

J. Fasola, M.J. Mataric, Using socially assistive human-robot interaction to motivate physical exercise for older adults, Proc. IEEE 100 (2012) 2512−2526.

B.J. Frey, D. Dueck, Clustering by passing messages between data points, Science 315 (2007) 972−976.

J.C. Giger, N. Piçarra, P. Alves-Oliveira, R. Oliveira, P. Arriaga, Humanization of robots: is it really such a good idea? Hum. Behav. Emerg. Technol. 1 (2019) 111−123.

T. Gnambs, M. Appel, Are robots becoming unpopular? changes in attitudes towards autonomous robotic systems in Europe, Comput. Hum. Behav. 93 (2019) 53−61.

P.H. Gobster, Mining the LANDscape: Themes and Trends over 40 Years of Landscape and Urban Planning, Volume 126, Elsevier, Amsterdam, The Netherlands, 2014, pp. 21−30.

I. Guemghar, P. de Oliveira Padilha, A. Abdel-Baki, D. Jutras-Aswad, J. Paquette, M.-P. Pomey, Social robot interventions in mental health care and their outcomes, barriers, and facilitators: scoping review, JMIR Ment. Health 9 (2022) e36094.

P.A. Hancock, T.T. Kessler, A.D. Kaplan, J.C. Brill, J.L. Szalma, Evolving trust in robots: specification through sequential and comparative meta-analyses, Hum. Factors 63 (2021) 1196–1229.

M. Heerink, B. Kröse, V. Evers, B. Wielinga, Assessing acceptance of assistive social agent technology by older adults: the almere model, Int. J. Soc. Robot. 2 (2010) 361–375.

M. Holohan, A. Fiske, "Like I'm talking to a real person": exploring the meaning of transference for the use and design of AI-based applications in psychotherapy, Front. Psychol. 12 (2021) 720476.

T. Kanda, T. Hirano, D. Eaton, H. Ishiguro, Interactive robots as social partners and peer tutors for children: a field trial, Hum. Comput. Interact. 19 (2004) 61–84.

E.S. Kim, L.D. Berkovits, E.P. Bernier, D. Leyzberg, F. Shic, R. Paul, et al., Social robots as embedded reinforcers of social behavior in children with autism, J. Autism Dev. Disord. 43 (2013) 1038–1049.

Y.W. Kim, D.Y. Kim, Y.G. Ji, Complexity in in-vehicle touchscreen interaction: a literature review and conceptual framework, in: Proceedings of the HCI in Mobility, Transport, and Automotive Systems. Automated Driving and In-Vehicle Experience Design: Second International Conference, MobiTAS 2020, Held as Part of the 22nd HCI International Conference, HCII 2020, Copenhagen, Denmark, 19–24 July 2020, Proceedings, Part I 22. Springer: Berlin/Heidelberg, Germany, 2020, pp. 289–297.

R. Kirby, J. Forlizzi, R. Simmons, Affective social robots, Robot. Auton. Syst. 58 (2010) 322–332.

J. Kleinberg, Bursty and hierarchical structure in streams, Data Min. Knowl. Discov. 7 (2003) 373–397.

E.A. Konijn, J.F. Hoorn, Robot tutor and pupils' educational ability: teaching the times tables, Comput. Educ. 157 (2020) 103970.

D. Kontogiorgos, A. Pereira, O. Andersson, M. Koivisto, E. Gonzalez Rabal, V. Vartiainen, et al. The effects of anthropomorphism and non-verbal social behaviour in virtual assistants, in: Proceedings of the 19th ACM International Conference on Intelligent Virtual Agents, Paris, France, 2–5 July 2019, pp. 133–140.

H. Kozima, M.P. Michalowski, C. Nakagawa, Keepon: a playful robot for research, therapy, and entertainment, Int. J. Soc. Robot. 1 (2009) 3–18.

P. Kumari, How does interactivity impact user engagement over mobile bookkeeping applications? J. Glob. Inf. Manag. JGIM, 30, 2021, pp. 1–16.

K.M. Lee, W. Peng, S.-A. Jin, C. Yan, Can robots manifest personality?: An empirical test of personality recognition, social responses, and social presence in human-robot interaction, J. Commun. 56 (2006) 754–772.

S. Lemaignan, C.E. Edmunds, E. Senft, T. Belpaeme, The Pinsoro dataset: supporting the data-driven study of child-child and child-robot social dynamics, PLoS ONE 13 (2018) e0205999.

J. Li, F. Goerlandt, G. Reniers, B. Zhang, Sam Mannan and his scientific publications: a life in process safety research, J. Loss Prev. Process. Ind. 66 (2020) 104140.

X. Li, Y. Sung, Anthropomorphism brings us closer: the mediating role of psychological distance in user-AI assistant interactions, Comput. Hum. Behav. 118 (2021) 106680.

S.X. Liu, Q. Shen, J. Hancock, Can a social robot be too warm or too competent? older chinese adults' perceptions of social robots and vulnerabilities, Comput. Hum. Behav. 125 (2021) 106942.

C. Lutz, A. Tamò, A. Guzman, Communicating with robots: antalyzing the interaction between healthcare robots and humans with regards to privacy, Human-Machine Communication: Rethinking Communication, Technology, and Ourselves, Peter Lang, New York, NY, USA, 2018, pp. 145–165.

C. Lutz, A. Tamó-Larrieux, The robot privacy paradox: understanding how privacy concerns shape intentions to use social robots, Hum.-Mach. Commun. 1 (2020) 87–111.

N. McCartney, A.L. Hicks, J. Martin, C.E. Webber, A longitudinal trial of weight training in the elderly: continued improvements in year 2, J. Gerontol. Ser. A Biol. Sci. Med. Sci. 51 (1996) B425–B433.

A. Mehrabian, J.A. Russell, An Approach to Environmental Psychology, MIT Press, Cambridge, MA, USA, 1974.

M. Mende, M.L. Scott, J. van Doorn, D. Grewal, I. Shanks, Service robots rising: how humanoid robots influence service experiences and elicit compensatory consumer responses, J. Mark. Res. 56 (2019) 535–556.

P. Mongeon, A. Paul-Hus, The journal coverage of web of science and scopus: a comparative analysis, Scientometrics 106 (2016) 213–228.

L. Morillo-Mendez, M.G. Schrooten, A. Loutfi, O.M. Mozos, Age-related differences in the perception of robotic referential gaze in Human-robot interaction, Int. J. Soc. Robot. (2022) 1–13.

S. Naneva, M. Sarda Gou, T.L. Webb, T.J. Prescott, A systematic review of attitudes, anxiety, acceptance, and trust towards social robots, Int. J. Soc. Robot. 12 (2020) 1179–1201.

C. Nass, Y. Moon, Machines and mindlessness: social responses to computers, J. Soc. Issues 56 (2000) 81–103.

J. Park, H. Choi, Y. Jung, Users' cognitive and affective response to the risk to privacy from a smart speaker, Int. J. Hum. Comput. Interact. 37 (2021) 759–771.

L.S. Pauw, D.A. Sauter, G.A. van Kleef, G.M. Lucas, J. Gratch, A.H. Fischer, The avatar will see you now: support from a virtual human provides socio-emotional benefits, Comput. Hum. Behav. 136 (2022) 107368.

B.A. Pickut, W. Van Hecke, E. Kerckhofs, P. Mariën, S. Vanneste, P. Cras, et al., Mindfulness based intervention in parkinson's disease leads to structural brain changes on MRI: a randomized controlled longitudinal trial, Clin. Neurol. Neurosurg. 115 (2013) 2419–2425.

B. Reeves, C. Nass, The Media Equation: How People Treat Computers, Television, and New Media Like Real People, Volume 10, Cambridge University Press, Cambridge, UK, 1996, p. 236605.

N. Reich-Stiebert, F. Eyssel, C. Hohnemann, Exploring university students' preferences for educational robot design by means of a user-centered design approach, Int. J. Soc. Robot. 12 (2020) 227–237.

M. Rheu, J.Y. Shin, W. Peng, J. Huh-Yoo, Systematic review: trust-building factors and implications for conversational agent design, Int. J. Hum. −Comput. Interact. 37 (2021) 81–96.

S. Riches, L. Azevedo, A. Vora, I. Kaleva, L. Taylor, P. Guan, et al., Therapeutic engagement in robot-assisted psychological interventions: a systematic review, Clin. Psychol. Psychother. 29 (2022) 857–873.

L.D. Riek, Wizard of Oz studies in Hri: a systematic review and new reporting guidelines, J. Hum.-Robot Interact. 1 (2012) 119–136.

B. Robins, K. Dautenhahn, R.T. Boekhorst, A. Billard, Robotic assistants in therapy and education of children with autism: can a small humanoid robot help encourage social interaction skills? Univers. Access. Inf. Soc. 4 (2005) 105–120.

M. Rueben, A.M. Aroyo, C. Lutz, J. Schmölz, P. Van Cleynenbreugel, A. Corti, et al. Themes and research directions in privacy-sensitive robotics, in: Proceedings of the 2018 IEEE Workshop on Advanced Robotics and Its Social Impacts (ARSO), Madrid, Spain, 1–5 October 2018, pp. 77–84.

J.A. Russell, A circumplex model of affect, J. Personal. Soc. Psychol. 39 (1980) 1161.

B. Schäfermeier, J. Hirth, T. Hanika, Research topic flows in co-authorship networks, Scientometrics 128 (2023) 5051–5078.

F.L. Schmidt, J.E. Hunter, Measurement error in psychological research: lessons from 26 research scenarios, Psychol. Methods 1 (1996) 199.

S.R. Schweinberger, M. Pohl, P. Winkler, Autistic traits, personality, and evaluations of humanoid robots by young and older adults, Comput. Hum. Behav. 106 (2020) 106256.

K. Smeds, F. Wolters, M. Rung, Estimation of signal-to-noise ratios in realistic sound scenarios, J. Am. Acad. Audiol. 26 (2015) 183–196.

Y. Song, D. Tao, Y. Luximon, In robot we trust? the effect of emotional expressions and contextual cues on anthropomorphic trustworthiness, Appl. Ergon. 109 (2023) 103967.

C. Theodoraki, L.-P. Dana, A. Caputo, Building sustainable entrepreneurial ecosystems: a holistic approach, J. Bus. Res. 140 (2022) 346–360.

N.J. Van Eck, L. Waltman, Software survey: vosviewer, a computer program for bibliometric mapping, Scientometrics 84 (2010) 523–538.

M.M. Van Pinxteren, R.W. Wetzels, J. Rüger, M. Pluymaekers, M. Wetzels, Trust in humanoid robots: implications for services marketing, J. Serv. Mark. 33 (2019) 507–518.

C.L. van Straten, J. Peter, R. Kühne, A. Barco, On sharing and caring: investigating the effects of a robot's self-disclosure and question-asking on children's robot perceptions and child-robot relationship formation, Comput. Hum. Behav. 129 (2022) 107135.

K. Wada, T. Shibata, Living with seal robots—its sociopsychological and physiological influences on the elderly at a care house, IEEE Trans. Robot. 23 (2007) 972–980.

Z. Wang, J. Chen, J. Chen, H. Chen, Identifying interdisciplinary topics and their evolution based on bertopic, Scientometrics (2023) 1–26.

Z. Wang, J. Wang, Do the emotions evoked by interface design factors affect the user's intention to continue using the smartwatch? The mediating role of quality perceptions, Int. J. Hum. Comput. Interact. 39 (2023) 546–561.

Y.-H. Wu, J. Wrobel, M. Cornuet, H. Kervé, S. Damnée, A.-S. Rigaud, Acceptance of an assistive robot in older adults: a mixed-method study of Human-robot interaction over a 1-month period in the living lab setting, Clin. Interv. Aging 9 (2014) 801.

A. Zhang, P.-L.P. Rau, Tools or peers? Impacts of anthropomorphism level and social role on emotional attachment and disclosure tendency towards intelligent agents, Comput. Hum. Behav. 138 (2023) 107415.

Y. Zhang, W. Song, Z. Tan, H. Zhu, Y. Wang, C.M. Lam, et al., Could social robots facilitate children with autism spectrum disorders in learning distrust and deception? Comput. Hum. Behav. 98 (2019) 140–149.

CHAPTER 17

Experimental study on abstract expression of human-robot emotional communication

Jianmin Wang[1], Yuxi Wang[1], Yujia Liu[1], Tianyang Yue[1], Chengji Wang[1], Weiguang Yang[1], Preben Hansen[2] and Fang You[1]

[1]Car Interaction Design Lab, College of Arts and Media, Tongji University, Shanghai, P.R. China [2]Department of Computer and Systems Sciences, Stockholm University, Stockholm, Sweden

Chapter Outline

- 17.1 Introduction 225
- 17.2 Literature review 226
    - 17.2.1 Development of emotion expression design in virtual images 226
    - 17.2.2 Facial emotion evaluation model 227
- 17.3 Emotion expression design of virtual image 228
- 17.4 Evaluation of facial expression design of virtual image 229
- 17.5 Method 233
    - 17.5.1 Participants 233
    - 17.5.2 Design of experiment 233
    - 17.5.3 Procedure 235
    - 17.5.4 Results 236
- 17.6 Conclusions 239
- References 239
- Further reading 240

17.1 Introduction

Facial expressions play an important role in understanding people's intentions and emotions toward each other in the process of communication. This paper studies general expressions for the Chinese population, using them as the main participants of the design and experiment. The demand for virtual images has been further increased from simple facial animation to more intelligent emotion expressions; these emotion expressions bring changes in the human emotional state and further influence human behavior and emotional state. Therefore emotion expression features have become the focus of research, but few researchers have studied emotion recognition and emotion expression in robots.

In this research, based on the present facial expression design of virtual images in intelligent products, we studied how to reasonably extract the abstract expression features of virtual images and correspond the pleasure−arousal−dominance (PAD) emotion values to virtual images for better interaction and communication with people. For example, the expression design of service robots can improve the effect of human-robot communication and enhance user experience (UX). In this study, the virtual image was designed to better enable humans to communicate with robots and allow humans to understand abstract emotion expression more widely. In addition, a quantitative evaluation method of virtual images based on the 3D PAD emotion model was proposed, which can help designers better design and iterate the emotion expression of virtual image facial expressions and improve the effectiveness of human-robot communication. Based on this research, the study performed the following tasks [1−4]:

1. In the first part of this paper, we introduce the application scenarios of existing virtual images in detail and overview the development of facial expressions and a 3D (PAD) emotion model.
2. In the second part of this paper, we summarize, discuss, and design the facial emotion features of virtual images.
3. In the third part of this paper, we use the PAD emotion model to conduct experimental evaluation and analysis to verify whether the facial expression design of avatars conforms to human cognition.

17.2 Literature review

Since the prevalence of virtual assistants (VAs), the cooperation between human beings and VAs has gradually increased. VAs based on emotion design affect human life and behavior. The emotion expression function of VAs also plays an important role in many application scenarios in real life. To effectively improve UX and human-robot interaction (HRI), VAs can use nonverbal behaviors, especially facial expressions, to express emotions and play an important role in enhancing human-computer interaction (HCI).

17.2.1 Development of emotion expression design in virtual images

In the research and exploration of the emotion expression design of virtual images, the facial design of virtual images serves as an important factor that influences user behavior and response. Han studied three robot facial design styles—iconic, cartoon, and realistic—and showed that iconic or cartoon-like faces are preferred to realistic robot faces. In Tawaki's research, it was proposed that when communicating with people, the robot's face conveys the most important and richest information. The author simplified facial expressions into face elements, which are two eyes and a mouth and are geometrically transformed to form facial expressions. Albrecht I et al. proposed that the facial expressions of VAs can enhance the expressiveness of voice reminders. For example, in the aspect of intelligent vehicle interaction, the vehicle VA can not only reduce the driver's distraction but also enhance the deep understanding of the information conveyed. However, the current research direction of VAs is more on how to help users complete certain tasks and less on how to facilitate emotional communication between humans and robots.

17.2.2 Facial emotion evaluation model

Facial expression and its relationship with emotion have been widely studied. The existing emotion description models mainly include the following types: dimensional emotion model, discrete emotion model, and classification model. For instance, in 1971 Ekman and Friesen used the discrete emotion model to express different emotions and put forward six basic emotion types: anger, disgust, fear, sadness, happiness, and surprise. These six emotions can be combined, and various compound emotions such as depression, distress, and anxiety can be derived from them.

In contrast to the discrete emotion model, the bipolar pleasure-arousal scale of Mehrabian and Russell is widely used. Russell developed the ring model of emotion classification in 1989 and argued that people's emotions can be divided into two dimensions: happiness and intensity. Plutchik designed the emotional wheel model. He suggested that people's emotions have three dimensions: strength, similarity, and polarity. This model emphasizes that when emotions show different strengths, they will influence each other to produce similar or opposite emotions. Wundt first proposed the 3D theory of emotion, which includes dimensions such as pleasant-unpleasant, excited-calm, and tension-relaxation. Furthermore, Osgood suggested three dimensions to describe the emotional experience, which are evaluation, potency, and activity. The changes in each dimension are continuous. In 1974 Mehrabian and Russell put forward the 3D PAD emotional model with the highest recognition so far in which the three dimensions: P (pleasure) indicates the positive and negative state of an individual's emotions; A (arousal) represents the individual's neurophysiological activation level; and D (dominance) signifies the individual's control over the situation.

The PAD emotion model has been used in various research studies on emotion calculation. Becker used PAD space to quantify the emotional calculation of virtual humans and used coordinates or vector Q (P, A, D) to represent the emotion state. As shown in Fig. 17.1, the origin of the coordinates indicates that the emotion is in a calm state. Therefore we can quickly find the basic

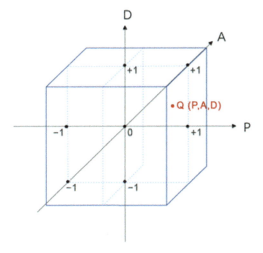

FIGURE 17.1

Pleasure—arousal—dominance 3D emotion space.

emotional points and some indistinguishable emotional points in the 3D PAD emotional space, such as sadness $-P-A-D$ and shame $-P+A-D$, which are divided into different emotional spaces by different arousal degrees. In this paper, the 3D emotion space with the best comprehensive performance was selected to define emotions.

The models above only study emotions from a qualitative perspective. In other words, based on the levels of pleasantness and arousal, it is not possible to conclude whether a participant felt either fascinated or proud, and thus this article uses the PAD 3D emotion space model to study emotion values, integrally and comprehensively define the position of emotions in the space, and conduct quantitative analysis.

17.3 Emotion expression design of virtual image

In the process of interaction between people and VAs, facial expression is an effective way of expressing emotions. Mehrabian provided a formula: emotion expression = 7% language communication +38% pronunciation and intonation expression +55% facial expression. It shows that most people convey emotional information through facial expressions in nonverbal communication, and facial emotion expression is an important communication channel in the process of face-to-face interaction.

DiSalvo et al. conducted a study on the influence of facial features and the size of virtual images on human-like perception. They suggested that the facial expression design of virtual images should balance three factors, which are human-ness, robot-ness, and product-ness. The "human-like nature" allows users to interact with Vas more intuitively; the "robot-ness" is the expectation of a virtual image's cognitive ability; and the "product-ness" allows people to regard the virtual image as a tool or equipment. Manning introduced the triangle design space for cartoon faces, which means that the more iconic the face of the avatar, such as a simplified emoji, the more it can enlarge the meaning of facial emotions and make people focus on information instead of media. At the same time, the more symbolic the avatar's face is, the more people it can represent, the stronger its credibility, and the more it can avoid the influence of the uncanny valley theory. If a robot looks too much like a human, it can be scary and even offensive. Designing animals and robots with human characteristics while maintaining their nonhuman appearance is much safer than using more anthropomorphic animals or entities. For designers, the safest combination seems to have an obvious nonhuman appearance but can express emotions like humans.

The eyes are the organs that can directly reflect people's emotions in facial expressions. Tawaki also explored ways to simplify faces, showing that two eyes and one mouth are important components of facial expression and can effectively convey emotion. The gaze direction of the eyes can convey the visual focus of the VA and show its behavioral motivation. The emotions of virtual images are divided into two types: positive emotions and negative emotions. When the concave face of the eyes is upward, it means that the virtual image is filled with positive emotions, such as happiness, gratitude, and pride. When the eyes of negative emotions concave upward, it indicates that the virtual image is in a negative emotion, such as shame, sadness, or anger. When the eyes are slightly narrowed, it indicates that the avatar is in a mild and relaxed state. Mouth changes are one of the criteria for judging emotion expression. When the corners of the mouth turn up, they show positive emotions, such as joy, hope, gratitude, and admiration. A larger upward radian represents more pleasure. Otherwise, when the mouth turns down, it shows negative emotions, such as sadness, shame, hostility, and other emotions.

House et al. found that eyebrows made great contributions to the importance of perceiving users' emotions. Ekman found that eyebrow movement is related to communication, and the type of facial emotion expression action depends on the dynamics of environmental prompts to a great extent. It can be seen that eyebrows cannot be ignored in the design of facial emotion expression.

Emoticons are auxiliary elements in visual symbols. On the one hand, they can express the emotions of virtual images more accurately through hand gestures, items, and so on. On the other hand, they can convey emotions that cannot be accurately transmitted to users visually, thus satisfying their functions and creating a relaxed and pleasant atmosphere. In the usage of emojis, users of different ages, countries, and cultural backgrounds will have the same understanding of the meaning of the emoji, thus achieving unity of cognition. Robots can use a combination of emoticons and facial expressions to achieve a higher degree of anthropomorphism.

We adopted and extended the PAD emotional dimension method to construct the emotions expressed by facial expressions. Emotions are divided into categories according to the eight spatial principles of PAD. Young's research involved simplified and exaggerated facial expressions and hand gestures that could provide powerful expressive mechanisms for robots. In this study, we refer to the mapping relationship between expression parameters and facial expression features in the animation Tom and Jerry and Disney classic animated characters, and it mainly includes four parts: eye feature, mouth feature, eyebrow feature, and emoticon feature. The iconic design elements that best meet the avatar expression are extracted. Emojis' visual emotional symbols are referenced in the design of emoticons, including facial expressions, hand gestures, and functional symbols. Table 17.1 analyzes and summarizes facial emotion features.

17.4 Evaluation of facial expression design of virtual image

To verify the rationality of the design and evaluate whether the expression design of the virtual image accords with the user's cognition, we need to measure the emotional changes of the users when they see the design. There are many measurement methods, such as the PAD semantic difference scale in the self-report measurement method, SAM self-emotion measurement method, PrEmo method, and so on. In this study, a nonverbal self-assessment manikin (SAM) was used to evaluate three dimensions: pleasure, arousal, and dominance. The SAM model has five human figures in each dimension. There are scoring points for each human figure. As shown in Fig. 17.2, the scoring range of each dimension is from 1 to 9, with 1 representing the lowest and 9 representing the highest.

Bradley and Lang demonstrated that when rating the dimensions of a group of pictures, although the conclusions of using the semantic difference scale and SAM self-emotion evaluation scale were almost identical, the participants passed the cartoon picture information in the SAM measurement. Therefore we can assume that the score of the SAM is the same as the value of PAD. Instead of text information, describing the emotion type with nontext characteristics can eliminate systemic errors caused by language semantic differences.

Table 17.1 Summary table of emotion expression features of facial design expressions.

Emotional space dimension	Emotion	Eyebrow	Eye	Mouth	Emoticons
Exuberant	Love	Eyebrow level	Eyes wide open, Notch down	Mouth up	Love
	Joy	Eyebrow level/Raised eyebrow level	Slightly squinted, Notch down	Corners of the mouth face up slightly	Flush
	Happiness	Eyebrow level/Raise eyebrows	Half-squinted eyes, Notch down	Grin open	Blushing/Hands raised
	Pride	Raised eyebrow level	Eyes wide open, Notch down	Corners of the mouth substantially upward	Stars/V gestures/ medals
	Gratification	Eyebrow level, Slightly larger spacing	Slightly squinted, Notch down	Corners of the mouth up	Blushing
Dependent	Hope	The two eyebrows face down at the ends, like "/\" [a]	Slightly squinted, Notch down	Corners of mouth slightly upward	Stars/hands closed
	Liking	The two eyebrows face down at the ends, like "/\"	Slightly squinted, Notch down	Corners of the mouth face up slightly	Blushing/Love
	Gratitude	Eyebrow level, Slightly larger spacing	Slightly squinted, Notch down	Corners of the mouth face up slightly	Blushing/Love/ Hands closed
	Admiration	Eyebrow level/Raised eyebrow level	Slightly squinted, Notch down	Corners of the mouth face up slightly	Great gestures/fist gestures
Relaxed	Relief	The two eyebrows face down at the ends, like "/\"	Slightly squinted, Notch up	Corners of the mouth face up slightly	Blushing
	Satisfaction	Eyebrow level, Raised eyebrow level	Slightly squinted, Notch down	Corners of the mouth up	Blushing/give the OK sign
Docile	Mildness	The two eyebrows face down at the ends, like "/\"	Slightly squinted, Notch down	Corners of the mouth face up slightly	Blushing

	Eyebrow	Eyes	Mouth	Other
Bored	Raise both ends of the eyebrows, Raise and lower eyebrows	Eyes squinted, Notch up	Corners of the mouth are tilted down	Hold up head with hand
Resentment	Eyebrows down, like "/\"	Slightly squinted, Notch up	Corners of the mouth face down slightly	Tears, Spread your hands
Pity	Raise the ends of the eyebrows, Spacing becomes smaller	Half-squinted eyes, Notch up	Corners of the mouth face down sharply	Tears Wipe away tears
Distress	Eyebrows down, like "/\"	Wide-eyed/slanted, Notch up	Corners of the mouth face down sharply	Hold up head with hand
Fear	Eyebrows down, like "/\", Spacing becomes smaller	Half-squinted eyes, Notch up	Corners of the mouth face down slightly	Blushing
Shame	Eyebrows down, like "/\", Spacing becomes smaller	Half-squinted eyes, Notch up	Corners of the mouth face down slightly	Hold your forehead
Remorse	Eyebrows down, like "/\", Spacing becomes smaller	Half-squinted eyes, Notch up	Corners of the mouth face down sharply	Sigh
Disappointment	Raised eyebrows like "\/"	Wide-eyed/slanted, Notch up	Mouth wide open	Hold your face with both hands
Fear	Eyebrows raised at both ends	Half-squinted eyes, Notch up	Mouth wide open	Letters and symbols
Anxious	Eyebrows slanted and raised	Wide-eyed, slanted, Notch up	Corners of the mouth face down sharply	Letters and symbols
Reproach	Eyebrows slanted and raised	Wide-eyed, Notch up	Corners of the mouth face down sharply	Angry symbol
Disliking	Eyebrows slanted and raised	Wide-eyed, Notch up	Corners of the mouth face down sharply	Flames of fury
Hostile				
Anger				
Hate				

[a] A Chinese character figuratively describes the eyebrow.

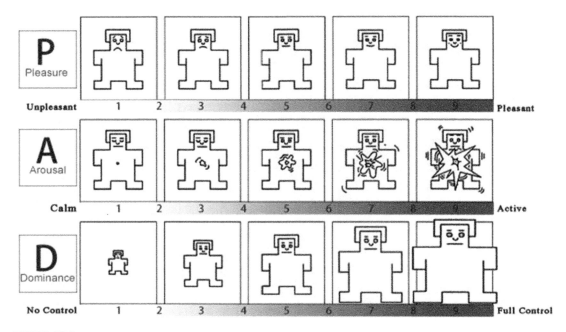

FIGURE 17.2

Self-assessment manikin scale. The numbers in the figure represent the scoring range of each dimension, with 1 representing the lowest and 9 representing the highest.

Through SAM evaluation, the average scores of the three dimensions of emotion words and emotion expressions were obtained. Psychological experiments show that the intimacy degree between the emotional state of the participants and basic emotions is the emotional tendency of the minimum distance called PAD, whose values of basic emotions are obtained through two rounds of experiments, which can effectively evaluate the relationship between the user's emotional tendency and the degree of inclination. At any coordinate position Q (P, A, D) in PAD's 3D space, the corresponding emotional state is Q. The PAD's emotional tendency is the distance between the 3D emotional space and the coordinated position between the emotional state of the participants and their basic emotions. The smaller the distance value, the higher the degree of the participant's emotional tendency will be. The formula for calculating the coordinate distance of emotional space is as follows (z is a positive integer):

$$Li = \sqrt{(P-pi)^2 + (A-ai)^2 + (D-di)^2} (i \in z) \quad (17.1)$$

Li is the shortest distance between emotional state and basic emotions in 3D emotional space; *P*, *A*, and *D* are the values of emotional state *Q* in 3D emotional space; *pi*, *ai*, and *di* are the coordinate values of basic emotional types *Q*i. According to Formula (17.1), the distance values between the emotional state of the participants and basic emotions are calculated, which are denoted as L1, L2, L3, and so on. For example, *Q* = Pride (0.68, 0.61, 0.58), *Q*i = Pride (0.72, 0.57, 0.55) and their minimum distance value *Li* is calculated as 0.06 by the formula.

Through the data collection of the PAD value of the emotion word, the position of each emotion word in the PAD space is obtained, the position of the PAD value of the emotion word is used as the basic emotion, and the minimum distance between the PAD values of the two experimental expression designs is calculated. If the minimum distance of the expression design corresponding to the emotion word is less than or equal to other expression designs, it shows that the expression design conforms to the human's emotional cognition. Otherwise, it shows that the expressions do not conform to human emotion perception and need to be adjusted and iterated.

17.5 Method

The study involved two experiments. The first experiment was divided into the emotion measurement task of emotion words and the emotion measurement task of expression. The second experiment only included the emotion measurement task of expression. Based on the results of the first experiment, we conducted a design iteration and proposed a design solution. The second experiment served as supplementary verification of design guidelines and specifications. The experiments used the Ortony, Clore, and Collins (OCC) model to extract rules from emotion cognition to generate emotion words. Then the numerical calculation of the PAD dimensional space was performed for the emotion word and expression design, and the numerical relationship between emotion words and expression design was determined and verified through a quantitative study. PAD emotion space mapping was used as a bridge to match human perceptions of expression design with emotion words to ensure the accuracy of abstract expression design.

17.5.1 Participants

The initial task of the first experiment recruited 153 student volunteers, including 54 male participants and 99 female participants, whose ages ranged from 20 to 26, with an average age of 23.71. In the second task, 33 college students volunteered to participate in this experiment, including 9 male users and 24 female users, with an average age of 23.1.

The second experiment recruited 50 college students, including 26 male participants and 24 female participants, with an average age of 24.02. Participants in both experiments were required to be free of color blindness and color weakness.

17.5.2 Design of experiment

To reduce external interference, the test was carried out indoors. People with cultural and geographical differences interpret facial expressions differently. The research and experiment were conducted in China; thus we defined the Chinese people's understanding of emotions by scoring the emotion words with PAD. In the emotion measurement task of emotion words in the first experiment, we translated the emotion words of PAD into Chinese—English bilingual languages. Participants scored the emotion words to obtain the accurate value of the emotion words in the emotion space, which was used as the benchmark for the mapping of the emotions in the affective space.

In the emotion measurement task of facial expressions in the first experiment, we designed facial expressions and used regular reasoning about events, objects, and attitudes to obtain the emotion word set in the OCC model of emotions. For commercial confidentiality reasons, the design used in this article was slightly adjusted in the thickness of the brush strokes. Facial expression designs 1–20, as shown in Fig. 17.3, represent different emotions: happiness, hate, satisfaction, gratitude, reproach, distress, pride, fear, mildness, pity, boredom, shame, disappointment, hope, resentment, love, gloating, anger, relief, and admiration. Before the start of the experiment, the participants were randomly divided into groups, with about 16 people in each group. After the experiment started, the expressions were presented to the participants in random order, and the participants scored the expressions on the SAM scale.

We modified and iterated the first-edition expression design based on the results of the first experiment. The expression iteration design is shown in Fig. 17.4; the numbers in Fig. 17.4 represent the same emotions as in Fig. 17.3. The expression and emotion measurement task design of the second experiment is the same as the first one.

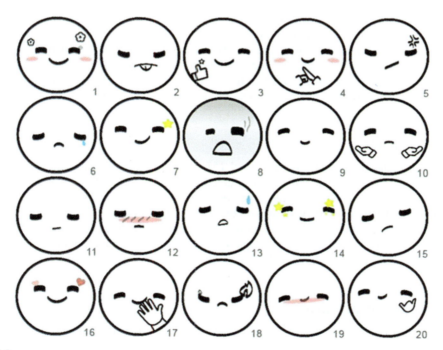

FIGURE 17.3

First edition expression design. (The numbers in the figure mark the expressions, as described in paragraph 2 of Section 17.5.2.)

17.5 Method   235

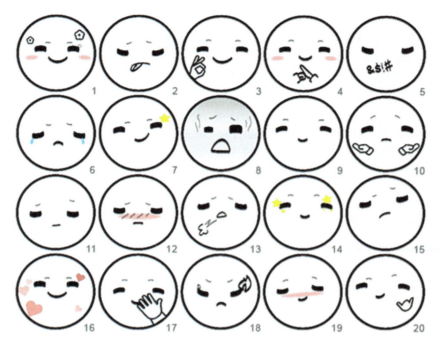

FIGURE 17.4

Iterative expression design. (The numbers in the figure mark the expressions, as described in paragraph 2 of Section 17.5.2).

17.5.3 Procedure

Before the start of the experiment, the experimenter introduced the purpose and tasks of the experiment and handed out the basic information form and the experiment informed consent form to the participants. Participants studied the meaning of the three emotional dimensions of PAD, the distribution in the 3D space, and the scoring method of the scale and then officially started the experiment after understanding and adapting.

At the beginning of the experiment, after the participants heard the "ding" prompt, the emotion words and expression designs were presented on the screen in front of the participants in random order. In the normative study of psychology, it takes 21 seconds for the participants to complete the evaluation of a picture. Thus each emotion word or expression design was displayed for 6 seconds. Participants had to watch the screen during the entire display process. After 6 seconds, the screen was switched to a white screen. Participants responded to the display on the screen at a constant 15-second interval. The expression design or emotion words were scored on three dimensions, and the subjective feelings of the heart when they saw these emotion words were evaluated. After scoring all expression designs or emotion words, the researcher conducted in-depth interviews with the participants and selected specific ambiguous expression designs or emotion words for investigation.

17.5.4 Results

In this study, 20 emotion words and emotion expressions were scored on the SAM scale, and a total of 137 valid results were recovered in the first task of the first experiment. In the second task of the first experiment, 33 valid results were recovered, and in the second task of the second experiment, 50 valid results were recovered.

According to the PAD nine-point scale, the emotional grade deviation was divided into 0−1 emotional value ranges, which were light <0.25, 0.25 ≤ mild <0.5, 0.5 ≤ moderate <0.75, and 0.75 ≤ severe <1. The PAD value was analyzed correspondingly with the expression elements. Table 17.2 shows the data with emotional offset in the first round of the relationship between the emotion words and the emotion expressions corresponding to the minimum distance, which is the

Table 17.2 Minimum distance between the first round of emotion words and emotion expressions.

	Pride	Love	Admiration	Gratitude	Resentment	Hope	Relief	Satisfaction	Mildness	Shame	Happiness
Reproach	1.29	1.22	0.89	0.91	0.40	1.16	0.74	1.13	0.62	0.62	0.40
Pride	0.51	0.46	0.52	0.49	1.21	0.42	0.31	0.28	0.51	1.40	0.38
Love	0.57	0.43	0.23	0.20	1.13	0.40	0.46	0.40	0.55	1.30	0.42
Admiration	0.48	0.40	0.43	0.41	1.18	0.35	0.38	0.27	0.55	1.37	0.34
Gratitude	0.38	0.30	0.46	0.43	1.28	0.25	0.47	0.18	0.65	1.46	0.23
Resentment	1.58	1.49	1.05	1.07	0.48	1.44	0.93	1.37	0.61	0.59	1.44
Hope	0.50	0.39	0.34	0.31	1.17	0.35	0.42	0.31	0.55	1.35	0.35
Relief	1.02	0.92	0.54	0.55	0.85	0.89	0.30	0.90	0.11	0.75	0.92
Satisfaction	0.67	0.55	0.29	0.28	1.03	0.52	0.32	0.46	0.40	1.21	0.52
Fear	1.46	1.34	0.90	0.92	0.37	1.64	1.06	1.34	0.92	0.44	1.33
Disappointment	1.79	1.69	1.19	1.22	0.29	0.45	1.13	1.61	0.92	0.25	1.65
Mildness	0.61	0.50	0.29	0.27	1.05	0.45	0.38	0.43	0.48	1.22	0.46
Pity	1.63	1.55	1.10	1.13	0.39	1.49	0.91	1.43	0.70	0.49	1.50
Distress	1.78	1.68	1.20	1.23	0.20	1.63	1.17	1.61	0.97	0.12	1.65
Anger	1.47	1.44	1.21	1.22	0.67	1.36	1.17	1.40	1.10	0.82	1.39
Shame	0.79	0.66	0.26	0.27	0.95	0.63	0.35	0.58	0.35	1.11	0.64
Happiness	0.38	0.27	0.41	0.38	1.28	0.23	0.51	0.21	0.66	1.46	0.23
Hate	1.40	1.34	1.01	1.02	0.40	1.28	0.79	1.23	0.65	0.61	1.28
Boredom	1.61	1.52	1.08	1.07	0.58	1.47	0.83	1.38	0.61	0.68	1.46
Gloating	0.77	0.69	0.44	0.43	0.88	0.63	0.30	0.58	0.34	1.07	0.65

PAD value of the emotion words displayed horizontally and the PAD value of the emotion expressions displayed vertically. The yellow marked part refers to the minimum distance between the emotion expression and the corresponding emotion word, while the green one indicates that the distance between other emotion expressions and the given emotion word is less than the distance between the corresponding emotion expressions. The high frequency of green annotation indicates that the emotion expression needs to be modified (slight deviation is normal).

Most expressions in the first edition of facial design in the second task have emotion offset. For example, the minimum distance between the emotion expression pride and its corresponding emotion word is 0.51. However, the minimum distance between the emotion expressions of admiration, gratitude, hope, and happiness are 0.48, 0.38, 0.50, and 0.38, respectively, which are all less than 0.51, indicating that the emotion expression of pride needs to be modified. During the interview, some participants mentioned that they could not understand the meaning of the expression; thus we considered the iterative design of these expressions. Having explored the PAD value of the pride emotion expression, the arousal degree of the expression is not high. Combined with the interview content, we made some adjustments to enlarge the eyes in iterative design. Thus adjusting the inclination of the expression is helpful to emphasize the expression of the pride emotion. Other emotion expressions listed in the table were also analyzed according to the process above, and subsequently, their optimization direction was obtained, which is not described here.

In the second experiment of this study, the eyebrow elements were added based on the first round of experiments to further strengthen the embodiment of different emotions in the virtual image. We adopted and expanded the PAD emotional dimension method to construct the emotions expressed by facial expressions and analyzed and summarized the features. Table 17.1 concludes emotions in the eight emotional dimensions, including eyebrows, eyes, mouth, and auxiliary elements. At the same time, the three visual information dimensions proposed in the previous research, namely, the inclination of the mouth, the opening of the face, and the inclination of the eyes, are used to iterate the facial expression design combined with the teaching value of PAD. For instance, in the positive emotional space, the eyebrows of the avatar are in a horizontal state; the eyes are slightly narrowed or widened; the shape of the eye notch is downward; and the corners of the mouth are turned upward. The auxiliary elements such as blush and love are integrated. In the boring emotional space, the eyebrows of the virtual image are lifted at both ends; the whole eyebrows are turned down; the distance between the eyebrows is reduced; the eyes are narrowed or squinted; the notch is up; and the corners of the mouth are slightly curled down. The auxiliary elements and symbols such as tears can help convey emotions better.

Comparing the results of the iterative expression experiment with those of the first edition, it is found that the emotional offset of the iterative emotion expression design is significantly reduced compared with that of the first edition. However, there is still a slight emotional offset between some emotion expressions, as shown in Table 17.3.

The element of eyebrows was added to the iterative expression design. Two raised eyebrows are common facial expressions. When the eyebrows are raised, the virtual images have higher arousal and dominance such as satisfaction and pride. When the eyebrows are down, it means that the pleasure and arousal are low and the mood is in a neutral or negative state such as pity and relief. When the eyebrows are high and low, the virtual image is between positive emotion and negative emotion, and the meaning of love and resentment is very different. When the distance between two eyebrows becomes smaller, most of them represent negative emotions. In addition, some of the

Table 17.3 Minimum distance between second-round emotion words and emotion expression.

	Admiration	Resentment	Hope	Satisfaction	Disappointment	Mildness	Pity
Reproach	1.29	0.85	1.34	1.35	1.19	1.09	1.14
Pride	0.78	1.61	0.21	0.31	1.95	1.04	1.85
Love	0.62	1.51	0.06	0.26	1.85	0.95	1.75
Admiration	0.43	1.18	0.32	0.27	1.49	0.58	1.39
Gratitude	0.31	1.14	0.35	0.35	1.45	0.58	1.35
Resentment	1.21	0.39	1.64	1.59	0.18	0.86	0.09
Hope	0.59	1.31	0.24	0.24	1.64	0.75	1.54
Relief	0.52	0.87	0.83	0.82	0.97	0.09	0.88
Satisfaction	0.59	1.31	0.24	0.24	1.64	0.75	1.54
Fear	1.14	0.55	1.67	1.58	0.89	1.21	0.86
Disappointment	1.23	0.43	0.93	1.62	0.17	0.88	0.10
Mildness	0.58	0.68	0.93	0.86	0.90	0.19	0.80
Pity	1.21	0.39	1.64	1.59	0.18	0.86	0.09
Distress	1.47	0.50	1.91	1.87	0.13	1.17	0.23
Anger	1.41	0.69	1.57	1.56	0.96	1.15	0.93
Shame	1.05	0.19	1.50	1.51	0.43	0.92	0.38
Happyiness	0.53	1.40	0.07	0.20	1.73	0.83	1.63
Hate	1.01	0.40	1.28	1.23	0.67	0.65	0.51
Boredom	1.08	0.58	1.47	1.38	0.50	0.61	0.42
Gloating	0.44	0.88	0.63	0.58	1.18	0.34	1.08

hand gestures were changed in the second experiment. For example, the hand gesture satisfaction was modified, and the minimum distance value in the second experiment was significantly smaller than the minimum distance value in the first experiment.

The experimental results show that it is normal for a small number of emotions to have a slight emotional deviation. For example, the emotional offset value of admiration and gratitude is less than the minimum distance. The difference between these two emotion expressions lies in the different hand gestures and the subtle changes in eyes and eyebrows. After analysis and comparison of Q (0.48, 0.34, −0.13) of admiration and Q (0.50, 0.35, −0.10) of gratitude, the PAD values of these two emotions are in the same emotional space and very close. Thus the emotional offset difference between admiration and gratitude is 0.12.

The results of the task in the second experiment show that, after the second task of design iteration optimization, the number of offsets of emotional expressions and emotion words is significantly smaller than that of the first task. In addition, the iterative version of the design is significantly optimized based on the first task. The calculation method of PAD space combined with the minimum distance can effectively measure the numerical difference between expressions in this research to obtain the expressions that are most in line with human emotional cognition. The larger the inclination of the eyes, eyebrows, and mouth, the greater the degree of awakening. Incorporating blush and emojis can effectively enhance pleasure and arousal, which is because the nonsingle color of the face can improve pleasure and dominance. Eyebrows are an indispensable element of facial expressions that can improve the arousal of users. Thus adding appropriate semantic emoticons is very helpful to improve dominance. In addition, the adjustment of hand gestures can increase the user's perception of emotion expressions.

17.6 Conclusions

In this study, the virtual image was used as a carrier to design the facial expressions of the emotion, and a quantitative evaluation method for virtual images based on the PAD 3D emotion model was introduced. The focus of this study was whether it can guide designers to effectively enhance human-robot communication. The results show that using the quantitative evaluation method, namely, the PAD emotion value and emotion tendency obtained by evaluating the emotion words and emotion expressions, the PAD emotion value and the image design are mapped symmetrically and can determine whether the expression design conforms to human emotion cognition. The regular features and expression design principles of the VA's facial design expressions can be used to improve the design. Based on the features of facial elements, the forms of visual presentation can produce many variations. Both experiments found that auxiliary elements such as hand gestures can effectively increase the emotional recognition of virtual images.

The research on the abstract expression design of the virtual image is part of our work on HRI interaction. Multichannel interactions, such as voice interaction and hand gesture interaction, will be added to test the overall emotion interaction.

References

[1] G. Iannizzotto, L.L. Bello, A. Nucita, G.M. Grasso, A vision and speech enabled, customizable, virtual assistant for smart environments, in: Proceedings of the 2018 11th International Conference on Human System Interaction (HSI), Gdansk, Poland, 4–6 July 2018, pp. 50–56.
[2] S. Blairy, P. Herrera, U. Hess, Mimicry and the judgment of emotional facial expressions, J. Nonverbal Behav. 23 (1999) 5–41.
[3] K. Itoh, H. Miwa, Y. Nukariya, K. Imanishi, D. Takeda, M. Saito, et al. Development of face robot to express the facial features, in: Proceedings of the RO-MAN 2004, 13th IEEE International Workshop on Robot and Human Interactive Communication, Kurashiki, Japan, 22 September 2004, pp. 347–352.
[4] Z. Liu, M. Wu, W. Cao, L. Chen, J. Xu, R. Zhang, et al., A facial expression emotion recognition based human-robot interaction system, IEEE/CAA J. Autom. Sin. 4 (2017) 668–676.

Further reading

I. Albrecht, J. Haber, K. Kahler, et al. "May I talk to you?:-)"-facial animation from text, in: 10th Pacific Conference on Computer Graphics and Applications, IEEE, 2002, pp. 77-86.

C. Becker-Asano, I. Wachsmuth, Affective computing with primary and secondary emotions in a virtual human, Auton. Agents Multi-Agent Syst. 20 (2009) 32–49.

M.M. Bradley, P.J. Lang, Measuring emotion: the self-assessment manikin and the semantic differential, J. Behav. Ther. Exp. Psychiatry 25 (1994) 49–59.

M. Davis, K.J. Davis, M.M. Dunagan, Communication without words, 3rd ed., Scientific Papers and Presentations, 14, Elsevier Inc., Cham, Switzerland, 2012, pp. 149–159. ISBN 978-0-12-384727-0.

DiSalvo, C.; Gemperle, F.; Forlizzi, J.; Kiesler, S. All robots are not created equal: The design and perception of humanoid robot heads, in Proceedings of the 4th Conference on Designing Interactive Systems: Processes, Practices, Methods, and Techniques, London, UK, 22 June 2002; pp. 321–326.

Ekman, P. About brows: emotional and conversational signals. Human Ethology; Routledge: London, UK, 1979.

P. Ekman, W.V. Friesen, Constants across cultures in the face and emotion, J. Pers. Soc. Psychol. 17 (1971) 124–129.

H.A. Elfenbein, N. Ambady, Universals and cultural differences in recognizing emotions, Curr. Dir. Psychol. Sci. 12 (2003) 159–164.

N. Foen, Exploring the Human-Car Bond Through an Affective Intelligent Driving Agent (AIDA), Massachusetts Institute of Technology, Cambridge, MA, USA, 2012.

Gebhard, P. ALMA: a layered model of affect, in: Proceedings of the Fourth International Joint Conference on Autonomous Agents and Multiagent Systems, Utrecht, The Netherlands, 25 July 2005, pp. 29–36.

Han, J.; Kang, S.; Song, S. The design of monitor-based faces for robot-assisted language learning, in Proceedings of the 2013 IEEE RO-MAN, Gyeongju, Korea, 26–29 August 2013; pp. 356–357.

House, D.; Beskow, J.; Granström, B. Timing and interaction of visual cues for prominence in audiovisual speech perception, in Proceedings of the INTERSPEECH, Aalborg, Denmark, 3–7 September 2001; pp. 387–390.

N. Jiang, R. Li, C. Liu, Application of PAD emotion model in user emotion experience assessment, Packag. Eng. (2020) 1–9 (In Chinese).

Koda, T.; Sano, T.; Ruttkay, Z. From cartoons to robotspart 2: facial regions as cues to recognize emotions, in: Proceedings of the 6th ACM/IEEE International Conference on Human-Robot Interaction (HRI), New York, NY, USA, 6 March 2011, pp. 169–170.

Laurans, G.; Desmet, P. Introducing PrEmo2: new directions for the non-verbal measurement of emotion in design, in: Proceedings of the 8th International Conference on Design and Emotion, London, UK, 11–14 September 2012.

X. Li, X. Fu, G. Deng, Preliminary trial of chinese simplified pad emotion scale in beijing university students, Chin. J. Ment. Health 22 (2008) 327–329 (In Chinese).

Y. Liu, S.S. Jang, The effects of dining atmospherics: an extended Mehrabian–Russell model, Int. J. Hosp. Manag. 28 (2009) 494–503.

A. Manning, Understanding comics: the invisible art, IEEE Trans. Dependable Secur. Comput. 41 (1998) 66–69.

A. Mehrabian, Pleasure-arousal-dominance: a general framework for describing and measuring individual differences in temperament, Curr. Psychol. 14 (1996) 261–292.

M. Mori, Bukimi no Tani [The Uncanny Valley], Energy 7 (1970) 33–35.

A. Ortony, G.L. Clore, A.J. Collins, The Cognitive Structure of Emotions, Cambridge University Press (CUP), Cambridge, UK, 1988. ISBN 9780511571299.

C.E. Osgood, Dimensionality of the semantic space for communication via facial expressions, Scand. J. Psychol. 7 (1966) 1–30.

X. Qi, W. Wang, L. Guo, M. Li, X. Zhang, R. Wei, Building a Plutchik's wheel inspired affective model for social robots, J. Bionic Eng. 16 (2019) 209–221.

J.A. Russell, M. Lewicka, T. Niit, A cross-cultural study of a circumplex model of affect. J. Pers. Soc. Psychol. 57 (1989) 848–856.

Schneider, E.; Wang, Y.; Yang, S. Exploring the Uncanny Valley with Japanese video game characters, in: Proceedings of the DiGRA Conference, Tokyo, Japan, 28 September 2007.

Z. Shao, Visual humor representation of cartoon character design—taking (Tom and Jerry) as an example, Decorate 4 (2016) 138–189 (In Chinese).

Steptoe, W.; Steed, A. High-fidelity avatar eye-representation, in: Proceedings of the 2008 IEEE Virtual Reality Conference, Reno, NV, USA, 8–12 March 2008, pp. 111–114.

Tawaki, M.; Kanaya, I.; Yamamoto, K. Cross-cultural design of facial expressions for humanoid robots, in Proceedings of the 2020 Nicograph International (NicoInt), Tokyo, Japan, 5–6 June 2020; p. 98.

Williams, K.J.; Peters, J.C.; Breazeal, C.L. Towards leveraging the driver's mobile device for an intelligent, sociable in-car robotic assistant, in: Proceedings of the 2013 IEEE Intelligent Vehicles Symposium (IV), Gold Coast, QLD, Australia, 23–26 June 2013, pp. 369–376.

W. Wundt, Outline of psychology, J. Neurol. Psychopathol. 1–5 (1924) 184.

Young, J.E.; Xin, M.; Sharlin, E. Robot expressionism through cartooning, in: Proceedings of the ACM/IEEE International Conference on Human-Robot Interaction, Arlington, VA, USA, 10–12 March 2007, pp. 309–316.

T. Zhang, Research on Affective Speech Based on 3D Affective Model of PAD, Taiyuan University of Technology, Taiyuan, China, 2018 (In Chinese).

CHAPTER 18

Self-assessment emotion tool: nonverbal measurement tool of user's emotional experience

Jianmin Wang[1], Yujia Liu[1,2], Yuxi Wang[1], Jinjing Mao[1], Tianyang Yue[1] and Fang You[1]

[1]Car Interaction Design Lab, College of Arts and Media, Tongji University, Shanghai, P.R. China [2]College of Design and Innovation, Tongji University, Shanghai, P.R. China

Chapter Outline

18.1 Introduction 243
18.2 Related works 244
    18.2.1 Emotion and emotional space 244
    18.2.2 Nonverbal emotion measurement tool 245
18.3 Design of self-assessment emotion tool 246
    18.3.1 Emotion set 246
    18.3.2 Emotion design of self-assessment emotion tool 247
18.4 Validation of self-assessment emotion tool images 250
    18.4.1 Pilot study 250
    18.4.2 Validation study 253
    18.4.3 Summary of results 255
18.5 Conclusions and discussion 257
    18.5.1 Methods of application 257
    18.5.2 Conclusions 257
References 257
Further reading 258

18.1 Introduction

In the domain of product interaction and visual design, users undergo different emotional experience processes regarding man-made objects carefully constructed by designers. In roughly the last 40 years, there has been a considerable amount of research on the experience of human-computer interaction. The main focus of user experience research is on the emotional response to human-computer interaction.

Emotions could exert a broad impact on the formation of human-computer interactions, the communication of the interaction, and the evaluation of the object of the interaction. To assess the impact of this important factor, researchers have used a variety of tools in academic and practice-based design research. The methods used for such investigations are usually validated emotion-measurement instruments from the field of experimental psychology (e.g.). The disadvantage of these methods is that they are not always well-suited to the highly interactive nature of digital media. Most methods are applied after the experiment, providing a measure of the overall experience. Self-reported measurement instruments have become an effective means of designing studies and surveys because of their advantages in terms of low cost and efficient collection. This study describes a nonverbal text-based self-assessment emotion tool (SAET), which was developed through the Ortony, Clore, and Collins (OCC) model based on the well-established derived rules for emotion perception, and the validation of the numerical computation of the emotion set and image pleasure-arousal-dominance (PAD) dimension space, as well as the dual validation of image and textual ideation, which was conducted to finally explore the resulting measurement tool with comprehensive coverage of emotion categories and high image-recognition.

18.2 Related works

18.2.1 Emotion and emotional space

Emotions are feelings that encompass the entirety of the human organism, incorporating its senses, mentality, and spirit. Emotions have different meanings and ways of being understood in different disciplines. The focus of research involves many aspects, which can be mainly categorized into two overall concerns: the study of the formation mechanism of emotions, and the study of the classification of types of emotions and methods of measurement.

One of the most important research orientations in the study of the formation mechanism of emotion is the far-reaching cognitive appraisal theory, which is based on previous psychological studies of emotion. Its basic idea is to direct environmental influences from objective stimuli to cognitive appraisal and to push physiological influences from the arousal row activity of the autonomic nervous system to the higher cognitive activity of the cerebral cortex. The representative theoretical model is the OCC model proposed by Ortony et al. [1], which expresses emotions in terms of a series of cognitively derived conditions.

In this model, emotions are assumed to be constituted by events (happy or not), actions (satisfied or not), objects (liked or not), and dispositional (positive or negative) responses to situations. It is inferred inductively through different cognitive conditions, and approximately 22 emotion types are specified, including the basic construct rules used to generate these emotion types, providing a taxonomy of emotions and giving the underlying reasoning process. For example, "Worry about not being needed by others" or "Will be forgotten" is an unpleasant thing, "Worry" indicates that things will be unhappy if they happen, and "Will be forgotten" indicates that the outcome is related to oneself. According to the generation rules of the emotion "Fear" in the OCC model, the sentence will be identified as having the emotion of fear and worry.

In contrast to the emotion generation rules of the OCC model, the other model uses an explicit multidimensional space to express emotions. One of the main ideas of emotion classification research is to organize the various types of emotions relationally through different quantities (or dimensions). Dimensionality is a characteristic of emotions, and different theories of emotions have proposed a wide variety of dimensional divisions, all of which are not only useful for the understanding of emotions but, more importantly, facilitates the basis for establishing tools for measuring emotions.

One of the representative studies is the PAD model of emotion proposed by Mehrabian [2], which consists of Pleasure (positive and negative characteristics of an individual's emotional state), Arousal (the individual's level of neurophysiological activation), and Dominance (the individual's state of dominance over situations and others) The three dimensions are independent of each other. $+P$ and $-P$ denote positivity and negativity, $+A$ and $-A$ denote high and low activation, and $+D$ and $-D$ are used for active and passive states. It has been shown that the PAD three-dimensional spatial model can effectively measure and explain human affective states. Gebhard [3] proposed a layered model of affect (ALMA), which introduces OCC emotion sets in the PAD space and assigns spatial coordinate values from -1 to $+1$ to these affective sets. This approach makes it possible to manipulate the computational aspects of affect, to achieve quantitative expressions of specific affects, and to combine the cognitive generation rules of affect.

18.2.2 Nonverbal emotion measurement tool

Affective evaluation is the basis of affective design research. In design practice, affective measurement can be used to define the user experience of a product in the early stages of design, evaluate the design or compare different prototypes in the middle stages, and assess the type of affective expression of a product in the later stages. The main methods of emotion measurement are self-reporting, physiological measurement, and behavioral observation among which self-reporting is an effective tool for designing studies and surveys due to its low cost and ease of large-scale information collection. Therefore the development of self-reported forms of product-styling emotion measurement tools and design methods based on them have had positive implications for both affective design research and product development applications.

The self-reported measurement of emotions contains two representations: a verbal report questionnaire and a nonverbal text report questionnaire. The former asks participants to indicate which words best match their feelings in reporting their current state, while the latter uses images or animations to represent emotions for participants to choose. Nonverbal text-based emotion measurement instruments do not convey emotions through words, and some studies have shown that such approaches are largely consensual across cultures.

Bradley and Lang's [4] self-assessment manikin (SAM) is a three-dimensional self-report instrument of emotions based on the PAD model. SAM uses abstract two-dimensional characters to represent different levels of pleasure, arousal, and dominance. A nine-point scale is used to describe the scores of each dimension. Research has shown that SAM can be used as an alternative to verbal report-based PAD scales in the form of nonverbal text, and has been used in emotion-assessment studies in a variety of domains such as home and advertising. Desmet et al. [5] proposed a product emotion measure (PrEmo) constructed by collecting a large number of design students' emotional words about product appearance to construct 18 emotions to assess product appearance, and it has been widely used in industrial fields such as product design and automotive exterior styling. Later, in 2012, PrEmo2 was iterated to improve the theoretical aspects of the emotion set and the design of the two-dimensional images to better fit the meaning of the emotion words. The year 2017 saw the iteration of a self-reporting tool for the emotions of 14 animated characters. Pic-a-mood is another scale that measures two dimensions of pleasure-arousal proposed by Vastenburg et al. [6], which has three persona designs for male, female, and robotic characters and is now used in research in the areas of airport experience, personal devices, and interactive media. Among the many nonverbal text measurement tools, the issue of recognition of certain images of emotions remains to be addressed. Studies have shown that positive-active class expressions are more difficult to recognize compared with negative emotions. Furthermore, the issue of the number of emotion sets is still underexplored, and further research is needed to cover more categories while ensuring a concise and clear number of emotions in the tool. As the field matures, it is increasingly important to develop tools with a more comprehensive theoretical basis and validation.

18.3 Design of self-assessment emotion tool

As discussed in the previous section, the nonverbal emotion measurement tool expresses emotion through images or animations, rather than words, so it is largely consensual across cultures and provides a more viable solution than verbal reports. Given the systematic academic and quantitative accuracy of affect-measurement tools, however, many factors need to be taken into account in the design of nonverbal emotion tools. First, the use of an OCC model deriving rules from emotion cognition yields the emotion set of the tool, which gives it a firm theoretical foundation. Second, the numerical calculation of the dimensional space PAD is performed on the emotion set and expression design to verify the numerical relationship between emotion words and expression design, which renders the tool quantitatively accurate. Finally, the accuracy of the cognitive understanding of emotion words and expression designs can be ensured by directly matching the semantic meaning of the expression designs with the semantic meaning of the emotion words. SAET was developed with the above features and should be easy to use (i.e., in-process measurement), understandable (i.e., not too demanding for the participants) during operation, and should have the possibility of cross-cultural use. Because animations require time to play in their entirety, which would severely disrupt the participant's interaction with the product or interface under evaluation, they are not applicable as content for the tool. For these reasons, the development of a new set of visual emotion measurement tools is necessary.

18.3.1 Emotion set

To make the emotion set more comprehensive and cognitively meaningful, SAET used 22 emotion words generated by the emotion generation rules in the OCC model and added two additional emotions: Boredom and Mildness, from other emotion measurement tools. Thus 24 emotion types were formed.

Since the OCC model uses cognitive evaluation rules to derive emotions, different objects, events, and other factors may yield similar emotion words. Therefore it is necessary to rate these 24 emotion words on the SAM scale at the first stage to derive the PAD value of each emotion word and then conduct relevance analysis. If two or more emotions are correlated in terms of values in the dimensional space and the words have very similar meanings, only one of them needs to be retained for the sake of simplicity of the measurement tool.

The PAD values of emotion words were evaluated among college students and the general population, and 178 results were collected through both online and offline methods, of which 153 were valid. The normal distribution test was conducted on the data results, and it was found that the data did not conform to the normal distribution. As such Spearman correlation analysis was used to make a two-by-two comparison between emotions. If two emotions were correlated at the same time in all three dimensions of pleasure, arousal, and dominance, they were marked with orange color blocks, as shown in Fig. 18.1 (those that did not correlate with other words are not shown). The results show that the four groups of emotions, gratification and satisfaction, joy and happy-for, remorse and pity, and distress and fears confirmed — are highly correlated, and close in Chinese semantics. Therefore one of the four groups was excluded, and the other 20 emotions were retained to form the emotion set of SAET. The SAM scale scores of 1–9 were converted to an interval of -1 to $+1$, and the final scores are shown in Table 18.1.

18.3.2 Emotion design of self-assessment emotion tool

Image design is generally classified as anthropomorphic, symbolic, and abstract. Symbolic design tends to make it easier for people to project their identities onto the robot. Hence a symbolic design style was used to allow users to better experience their emotions vicariously. Dynamic expressions are more expressive than static expressions, but static images were chosen to make the emotion measurement tool simpler and to cause as little visual fatigue as possible. The image was designed around a highly expressive face, containing eyes and a mouth as elements.

The eye is an important organ in facial expressions that reflect emotions. The direction of eye gaze can convey the focus of facial attention and also show the motivation of a person's behavior. When a person is in a positive emotion, the concave side of the eye usually faces down, indicating emotions such as happiness, gratitude, and pride; in a negative emotion, the concave side of the eye faces up, indicating that the avatar is experiencing an emotion such as shame, sadness, or anger. When the eyes are slightly narrowed, it indicates that the avatar is in a state of gentleness and relaxation. The change in the mouth is also one of the criteria for judging the expression of emotion. When the corners of the mouth are curved upward, the emotions are mostly positive emotions

FIGURE 18.1

Relevance of emotion words in the three dimensions of pleasure-arousal-dominance.

Table 18.1 Coordinate values of 20 emotion words in pleasure-arousal-dominance three-dimensional emotion space.

	Reproach	Pride	Love	Admiration	Gratitude	Resentment	Hope	Relief	Satisfaction	Fear
P	−0.50	0.72	0.73	0.48	0.50	−0.53	0.65	0.30	0.66	−0.64
A	0.25	0.57	0.55	0.34	0.35	0.09	0.55	−0.07	0.33	0.45
D	0.43	0.55	0.36	−0.13	−0.10	−0.35	0.38	0.26	0.45	−0.60
	Disappointment	Mildness	Pity	Distress	Anger	Shame	Happy-For	Hate	Boredom	Gloating
P	−0.65	0.18	−0.57	−0.75	−0.74	−0.62	0.67	−0.58	−0.22	0.12
A	−0.28	−0.15	−0.25	−0.27	0.36	0.08	0.49	0.12	−0.49	0.13
D	−0.54	0.08	−0.49	−0.55	0.26	−0.55	0.43	0.07	−0.24	0.03

such as joy, hopefulness, gratitude, and admiration. The greater the curve of the mouth upward, the higher the pleasantness; while in sadness, shame, hostility, and other emotions, the corners of the mouth face downward. Emoticons, as visual symbolic auxiliary elements, express the emotions of characters more precisely through gestures, props, and other elements on the one hand, and convey emotions that cannot be accurately conveyed to users by words through visuals on the other hand, creating a relaxed and pleasant atmosphere while satisfying the function. In the use of emoticons, users of different ages, countries, and cultural backgrounds will all have the same understanding of the meaning of emoticons and reach a unity of cognition.

Fig. 18.2 shows that many emotions have certain correlations in people's cognition. To effectively distinguish various types of emotions, the facial expressions were designed in multicolor and integrated with emojis. After referring to the mapping results between the expression parameters of the animated Tom and Jerry and the classic Disney animation characters and the facial expression features, we extracted iconic design elements that best fit the virtual image expressions and drew 20 emotion images. The emoji design of the virtual image refers to the visual emotion symbols of emojis, including facial expressions, hand gestures, and emoticons. Due to the specific application of the design in the experiment, which involves cooperation with industry content, here we ensure the size of the PAD value and the relationship between the positive and negative of the premise to retain the facial features of the eyebrows, eyes, and mouth to show the emotion design, as shown in Fig. 18.2.

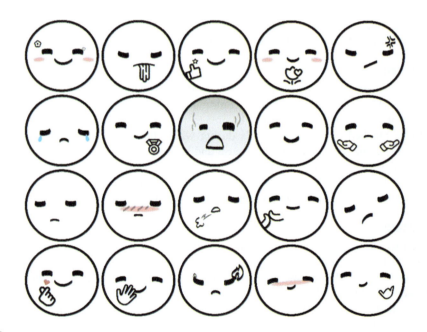

FIGURE 18.2

Twenty emotions of expression design. The emotions are (from *left* to *right*, *top* to *bottom*): happy-for, hate, satisfaction, gratitude, reproach, distress, pride, fear, mildness, pity, boredom, shame, disappointment, hope, resentment, love, gloating, anger, relief, and admiration.

18.4 Validation of self-assessment emotion tool images

A validation experiment was conducted to assess the recognizability of the SAET images, divided into a pilot study and a validation study. Participants were asked to rate the image by SAM to determine what score they thought the image was for each of pleasure, arousal, and dominance. For the validation study, they were also asked to choose the best emotion word for the image and indicate which emotion they thought the SAET image showed.

18.4.1 Pilot study

Prior to the validation study, a pilot study was conducted. Its goal was to obtain the accuracy of the initially designed SAET images and iterate the image expression design from the results.

18.4.1.1 Participants

The participants were masters and undergraduate students from Tongji University in China who were enrolled in a user research and interaction design course. A total of 9 male and 24 female students participated ($N = 33$). Their ages ranged from 20 to 26 years ($M = 22.09$, $SD = 2.3$).

18.4.1.2 Apparatus

Participants used their cell phones to scan a QR code to enter and fill out the questionnaire. The SEAT images were played on the large screen in the conference room in randomized order.

18.4.1.3 Procedure

Participants were instructed to score the expression design images on a three-dimensional (PAD) scale using the SAM scale. The experiment would automatically play the image for 6 seconds, and the participants were required to observe the image for that period. Subsequently, the playback screen would be blacked out for 10 seconds during which the participants would rate the images and then move on to the next image with a pause and black screen for the same duration, and so on until the material was all played out. In the experiment, the order of the images before and after was disrupted to balance the order effect. Participants judged independently throughout, without having mutual conversations with the experimenter or other people.

18.4.1.4 Results

To verify whether the expression design is in line with people's perceptions, the design drawings need to be evaluated. The emotion images need to be converted to the interval from -1 to $+1$ according to a scale of 1–9, and the degree of the emotional tendency can be obtained by analyzing the distance relationship between the emotion images and the word PAD coordinates. The image with the smallest distance from the word is considered to be closer to the emotional tendency of that word, and the coordinate distances in the emotional space can be obtained by calculating the Euclidean distance algorithm by calculating:

$$L_n = \sqrt{(p_n - p'_n)^2 + (a_n - a'_n)^2 + (d_n - d'_n)^2}, n = [1, 20], n \in Z \qquad (18.1)$$

where L_n is the coordinate distance between the static image and the 20 emotions in 3D space. P, A, and D denote the coordinate values of the measured emotional state e in the emotion space P_n, a_n, d_n; e_n denotes the emotional word; and p'_n, a'_n, d'_n are the coordinates of the corresponding figure f_n. According to the coordinate values of formula (1), the distance between each picture and text can be calculated and labeled as $L_1, L_2, ..., L_{20}$. If the static image f_n (p'_n, a'_n, d'_n) calculates L_5 as the minimum of L_n, then the emotional tendency of that static image corresponds to the fifth term of the textual emotion.

The results of the pilot study on the Euclidean distance of SAET images and emotion words are shown in Table 18.2. The yellow blocks indicate the distance between the emotion design and the corresponding text, and the green blocks indicate that the distance of the emotion from the text is smaller than the original design of the emotion. For example, the distance of the original design of emotion Pride to the word Pride is 0.54 (yellow), but the emoticons Admiration, Gratitude, Happy-for, and Gloating are 0.52, 0.43, 0.45, and 0.33, respectively, from the word Pride in the text, that is, smaller than 0.54.

18.4.1.5 Discussion

The image designs with emotions that have green color blocks in each vertical column of Table 18.2 are considered to require modification and iteration. The PAD values of words can effectively guide the design element characteristics of expressions. For example, the PAD values for anger are (-0.74, 0.36, 0.26), which shows that people's emotional tendency for anger is strongly unpleasant, moderately activated, and moderately low dominance, and the design direction can be eyes open like an inverted eight, eyebrows furrowed, and mouth tightly closed and pulled down. In contrast, the PAD values of sadness were (-0.75, -0.27, -0.55), it can be seen that people's emotional tendency toward sadness is strongly unpleasant, moderately low activated, and moderately low dominated. The designed direction can be eyes slightly squinted, eyebrows slightly furrowed, eyes spaced wider, and mouth slightly closed and slightly pulled down. In addition, in interviews with participants, it was found that many expressions struggle to represent differences in emotion relying only on the eyes and mouth, suggesting the inclusion of another key feature, the eyebrows. In a study by Ekman, eyebrow movement was found to be associated with communication and the type of facial emotion-expressive movements depended to a large extent on the ambient eyebrow dynamics. This shows that eyebrows are an element that cannot be ignored in the design of facial emotion expression, so the design iteration adds the element of eyebrows to the first round of experiments to further enhance the embodiment of different emotions in the images.

The iterative design of the image refers to the PAD value of the initial text, which on the one hand adds eyebrows and modifies emojis to the facial feature elements on the other hand, such as red blush, yellow stars, and hand gestures. For example, the "satisfied" expression is expressed through the "OK" gesture, while "proud" is expressed through the one-sided rise of the mouth and the addition of yellow stars at the end of the eye. Finally, according to the relationship between the size of PAD values, positive and negative shapes, and facial features, the design of 20 emotion expressions was modified, as shown in Fig. 18.3.

252 Chapter 18 Self-assessment emotion tool

Table 18.2 Pleasure-arousal-dominance Euclidean distance between 20 images and 20 emotional words in the pilot study.

		Repr oach	Pride	Love	Admir ation	Grati tude	Resent ment	Hope	Relief	Satisfa ction	Fear	Disappoi ntment	Mild ness	Pity	Distr ess	Anger	Sha me	Happy-For	Hate	Bore dom	Gloa ting
Image	Reproach	0.27	1.27	1.02	0.90	0.82	0.47	0.78	0.72	0.98	0.54	0.63	0.74	0.42	0.92	0.50	0.64	0.47	0.38	0.69	0.49
	Pride	1.07	0.54	0.24	0.17	0.27	1.17	0.36	0.56	0.12	1.27	1.38	0.72	1.13	1.48	1.15	1.41	0.21	1.18	1.12	0.47
	Love	0.99	0.64	0.24	0.16	0.12	1.15	0.20	0.61	0.31	1.15	1.34	0.75	1.12	1.40	1.16	1.31	0.25	1.17	1.12	0.42
	Admiration	1.06	0.52	0.16	0.06	0.21	1.14	0.32	0.60	0.18	1.23	1.38	0.76	1.14	1.45	1.12	1.38	0.13	1.18	1.15	0.45
	Gratitude	1.14	0.43	0.07	0.09	0.28	1.21	0.39	0.70	0.20	1.31	1.48	0.85	1.24	1.55	1.19	1.47	0.04	1.28	1.25	0.55
	Resentment	0.72	1.59	1.26	1.13	0.99	0.94	0.90	0.64	1.14	0.67	0.33	0.55	0.17	0.67	0.98	0.58	1.24	0.30	0.19	0.68
	Hope	1.09	0.56	0.16	0.06	0.15	1.16	0.26	0.61	0.22	1.20	1.38	0.76	1.14	1.44	1.15	1.36	0.16	1.19	1.15	0.44
	Relief	0.91	1.06	0.69	0.57	0.41	1.08	0.32	0.15	0.74	0.60	0.83	0.25	0.63	0.84	0.76	0.76	0.73	0.66	0.75	0.23
	Satisfaction	0.60	0.72	0.33	0.21	0.05	1.09	0.11	0.46	0.29	1.08	1.22	0.61	0.99	1.31	1.09	1.21	0.32	1.05	0.98	0.29
	Fear	0.97	1.46	1.18	1.08	0.98	0.58	1.11	1.01	1.22	0.31	0.75	1.03	0.68	0.52	0.68	0.47	1.16	0.65	0.93	0.71
	Disappointment	0.84	1.81	1.48	1.35	1.20	0.90	0.19	0.96	1.41	0.43	0.20	0.89	0.33	0.36	0.98	0.23	1.45	0.74	0.52	0.89
	Mildness	0.98	0.66	0.28	0.16	0.11	1.05	0.19	0.55	0.30	1.08	1.26	0.69	1.03	1.32	1.05	1.23	0.27	1.07	1.05	0.32
	Pity	0.66	1.64	1.32	1.19	1.05	0.87	0.97	0.74	1.22	0.59	0.23	0.66	0.09	0.55	0.93	0.48	1.29	0.21	0.30	0.73
	Distress	0.68	1.79	1.48	1.35	1.22	0.79	1.14	1.03	1.43	0.31	0.28	0.97	0.39	0.22	0.88	0.11	1.45	0.36	0.64	0.90
	Anger	0.31	1.42	1.28	1.19	1.18	0.06	1.17	1.21	1.32	0.69	0.97	1.26	0.85	0.75	0.10	0.85	1.24	0.75	1.18	0.91
	Shame	0.97	0.85	0.45	0.34	0.15	1.07	0.06	0.41	0.40	0.99	1.13	0.54	0.91	1.13	1.09	1.12	0.45	0.97	0.89	0.24
	Happy-for	1.17	0.45	0.05	0.09	0.26	1.22	0.37	0.71	0.24	0.99	1.49	0.87	1.26	1.49	1.21	1.47	0.07	1.30	1.26	0.55
	Hate	1.14	1.37	1.13	1.02	0.95	0.34	0.91	0.87	1.11	1.10	0.63	0.89	0.56	0.77	0.39	0.59	1.09	0.39	0.78	0.62
	Boredom	0.70	1.24	0.90	0.77	0.99	0.90	0.55	0.33	0.78	0.74	0.66	0.32	0.43	0.86	0.92	0.77	0.88	0.52	0.42	0.34
	Gloating	1.22	0.33	0.07	0.19	0.38	1.26	0.49	0.81	0.30	1.39	1.58	0.97	1.35	1.59	1.24	1.56	0.29	1.38	1.36	0.42

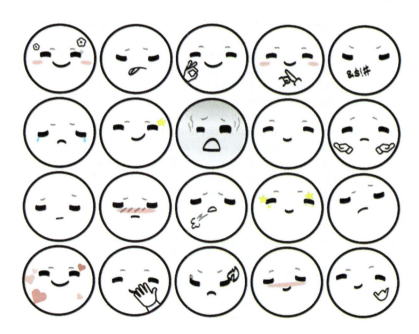

FIGURE 18.3

Iterative image design of 20 emotions. The emotions are (from *left* to *right*, *top* to *bottom*): happy-for, hate, satisfaction, gratitude, reproach, distress, pride, fear, mildness, pity, boredom, shame, disappointment, hope, resentment, love, gloating, anger, relief, and admiration.

18.4.2 Validation study

Participants were asked to rate the image by SAM in the validation study, but in contrast to the pilot study, they also needed to select the emotion word that best matched the image and indicate which emotion they thought the SAET image showed.

18.4.2.1 Participants

We conducted a validation study of the images offline with a population of college students and faculty members in a total of two rounds, and a total of 50 valid results were collected. A total of 26 male and 24 female Chinese participants took part in the study ($N = 50$). The age of the participants ranged from 19 to 33 years (M = 23.9, SD = 3.05).

18.4.2.2 Apparatus

Similarly, participants used their cell phones to scan a QR code to access and fill out the questionnaire. SEAT images were played in randomized order on a large screen in the conference room.

18.4.2.3 Procedure

The validation study was divided into two rounds. In the first round, SAM was used to score the expression design drawings in three dimensions (PAD), which was consistent with the steps of the pilot study. In the experiment, the image would automatically play and stay for 6 seconds, during which time the participant would observe it. Then the playback interface will be blacked out for 10 seconds, during which time the participant would score the image before moving on to the next, and so on until the material was played out. The first round thus had the same experimental scheme as the pilot.

After the first round of the experiment, a 5-minute break was taken to enter the second round, which required the selection of the closest Chinese emotion word for the design drawing. The second round of the experiment was the same, with the images automatically played on a large screen in randomized order and then blacked out after 6 seconds, except that the questionnaire filled in by the participants was replaced by multiple-choice questions with emotional words. The participant was given 10 seconds to select the emotional meaning he or she thought the image represented. During this process, participants judged independently and did not have mutual conversations with the experimenter or other people.

18.4.2.4 Results

The results of the first round of this validation study on the Euclidean distance between SAET images and the PAD of emotion words are shown in Table 18.3. The results show that after the design iterations, the Euclidean distances of PAD values for most of the images corresponding to words are the smallest, except for Admiration, Hope, Disappointment, and Mildness. For example, the distance between the word Admiration and the image Gratitude is 0.31, while the distance to the image Admiration is 0.43, which is thus smaller than the distance to the image Admiration. Similarly, the distance between the word Resentment and the image Shame is 0.13, which is smaller than the distance to the image Resentment, which is 0.37; the word Resentment is smaller than the distance to the image Shame. The distance between the word Hope and the image Love is 0.06, respectively, which is slightly smaller than the distance (0.07) to the image Hope. The distances

254 Chapter 18 Self-assessment emotion tool

Table 18.3 Pleasure-arousal-dominance Euclidean distance between 20 images and 20 emotional words in validation study.

		Repr oach	Pride	Love	Admir ation	Grati tude	Resent ment	Hope	Relief	Satisfa ction	Fear	Disappoi ntment	Mild ness	Pity	Distr ess	Anger	Sha me	Happy-For	Hate	Bore dom	Gloa ting
Image	Reproach	0.26	1.42	1.42	1.29	1.29	0.85	1.34	1.12	1.35	1.02	1.19	1.09	1.14	1.38	0.18	1.03	0.85	0.48	1.22	0.94
	Pride	1.24	0.06	0.23	0.78	0.74	1.61	0.21	0.84	0.31	1.77	1.95	1.04	1.85	2.02	1.48	1.79	0.19	1.45	1.64	0.92
	Love	1.13	0.16	0.07	0.62	0.58	1.51	0.06	0.78	0.26	1.66	1.85	0.95	1.75	1.92	1.45	1.69	0.10	1.39	1.55	0.82
	Admiration	0.95	0.46	0.38	0.43	0.40	1.18	0.32	0.42	0.27	1.40	1.49	0.58	1.39	1.57	1.18	1.37	0.32	1.06	1.17	0.46
	Gratitude	0.98	0.51	0.39	0.31	0.29	1.14	0.35	0.46	0.35	1.34	1.45	0.58	1.35	1.53	1.19	1.31	0.36	1.06	1.15	0.42
	Resentment	0.84	1.58	1.49	1.04	1.07	0.37	1.44	0.86	1.37	0.82	0.40	0.65	0.31	0.48	0.87	0.50	1.44	0.52	0.29	0.67
	Hope	1.15	0.24	0.12	0.52	0.48	1.40	0.07	0.67	0.21	1.57	1.73	0.83	1.63	1.81	1.37	1.58	0.10	1.29	1.43	0.70
	Relief	0.88	0.96	0.87	0.52	0.53	0.87	0.83	0.25	0.82	0.89	0.97	0.09	0.88	1.04	0.87	0.82	0.85	0.65	0.78	0.17
	Satisfaction	0.97	0.40	0.32	0.44	0.41	1.27	0.28	0.45	0.19	1.49	1.57	0.62	1.47	1.65	1.28	1.45	0.26	1.16	1.23	0.53
	Fear	1.10	1.63	1.52	1.06	1.09	0.38	1.67	1.25	1.52	0.20	0.72	1.07	0.68	0.73	0.83	1.51	0.67	0.99	0.88
	Disappointment	1.05	1.82	1.72	1.23	1.26	0.43	0.93	1.11	1.62	0.80	0.17	0.88	0.10	0.27	1.03	0.42	1.68	1.05	0.39	0.91
	Mildness	0.79	1.08	0.98	0.58	0.59	0.68	0.93	0.38	0.86	1.03	0.90	0.19	0.80	0.98	0.95	0.86	0.93	0.67	0.56	0.19
	Pity	1.04	1.85	1.75	1.27	1.29	0.39	1.70	1.15	1.65	0.75	0.11	0.93	0.04	0.20	0.99	0.37	1.71	0.66	0.46	0.94
	Distress	1.19	2.06	1.96	1.47	1.50	0.50	1.91	1.38	1.87	0.74	0.13	1.17	0.23	0.05	1.08	0.39	1.92	0.80	0.68	1.16
	Anger	0.40	1.55	1.53	1.31	1.32	0.69	1.45	1.19	1.47	0.82	1.03	1.12	0.99	1.02	0.09	0.84	1.47	0.39	1.15	0.96
	Shame	0.84	1.55	1.44	0.96	0.99	0.13	1.39	0.99	1.39	0.46	0.48	0.80	0.41	0.53	0.78	0.26	1.41	0.50	0.65	0.68
	Happy-for	1.14	0.23	0.13	0.53	0.49	1.40	0.07	0.66	0.20	0.46	1.73	0.83	1.63	1.78	1.36	1.58	0.09	1.28	1.42	0.70
	Hate	1.14	1.35	1.29	0.97	0.98	0.38	1.22	0.84	1.21	1.30	0.75	0.72	0.62	0.90	0.41	0.60	1.24	0.15	0.76	0.58
	Boredom	0.97	1.61	1.52	1.08	1.07	0.58	1.47	0.83	1.38	1.01	0.50	0.61	0.42	0.60	1.04	0.68	1.46	0.69	0.09	0.72
	Gloating	0.74	0.77	0.69	0.44	0.43	0.88	0.63	0.30	0.58	1.15	1.18	0.34	1.08	1.25	0.95	1.07	0.65	0.78	0.88	0.09

between the word Disappointment and the images Pity and Distress are 0.11 and 0.13, respectively, which is slightly smaller than the distance to the image Disappointment. The distance between the word Disappointment and the images Pity and Distress is 0.11 and 0.13, respectively, which is slightly smaller than the distance of 0.17 to the image Disappointment; the distance between the word Mildness and the image Relief is 0.09, and thus smaller than the distance of 0.19 to the image Mildness. Although the Euclidean distances between these four emotions and the image are not the smallest, they are small compared with other distances.

The accuracy results of the second round of evaluation experiments for image selection of emotion words are shown in Table 18.4. The results show that all the expression designs have the highest accuracy rates, with individual accuracy rates ranging from 40% to 92%, which indicates that all of them represent the corresponding emotional word in terms of the emotional perception of the image designs. Pride, Love, Satisfaction, Fear, Distress, Anger, Shame, and Gloating—these eight emotion images are highly accurate and do not produce ambiguities with other emotions, especially Love, Distress, and Anger. In particular, the accuracy of the three emotion images of Love, Distress, and Anger reached over 90%.

In addition, there are four types of emotions that are easy to confuse: Reproach and Anger, Disappointment and Mildness, Mildness and Distress, and Pity and Disappointment. They are all at the same level of pleasantness, and there is no confusion between positive and negative emotions.

Table 18.4 Validation experiments for cognitive interpretation of selected emotion words.

Image	Hit rate	Other options	Hit rate
Reproach	52%	Anger	38%
Pride	74%	None	None
Love	92%	None	None
Admiration	54%	Satisfaction	28%
Gratitude	44%	Hope	18%
Resentment	40%	Hate	24%
Hope	40%	Admiration	20%
Relief	42%	Mildness	34%
Satisfaction	72%	None	None
Fear	82%	None	None
Disappointment	50%	Distress	34%
Mildness	44%	Relief	38%
Pity	42%	Disappointment	36%
Distress	92%	None	None
Anger	92%	None	None
Shame	72%	None	None
Happy-for	50%	Love	24%
Hate	54%	Resentment	14%
Boredom	48%	Disappointment	20%
Gloating	82%	None	None

Note: The "hit rate" is the percentage of correct responses for each image. "Other options" is the most common incorrect response.

18.4.3 Summary of results

The results of the pilot and the validation study show that SAET's image design is effective. Each designer's image design for emotion yields different results. The deterministic design found through the pilot study differs somewhat from the results generally perceived by the public, and the analysis from the perspective of the PAD values provides designers with a very clear baseline. This is also reflected in the results of the validation study, where the image design has improved to a great extent.

Furthermore, although the results of the first round of the validation study showed that not every emotion had the smallest Euclidean distance, with the exception of resentment and shame, these emotion groups were significantly correlated to their text (Fig. 18.4), so it is possible that the PAD values for the expression designs were less distant from the other significantly correlated texts. The results of the second round evaluation for the selection of emotion words for the images indicate that the emotion set is largely representative of the textual emotion meaning of the corresponding words in terms of the meaning represented by the expressions. As the composition of the emotion

256 Chapter 18 Self-assessment emotion tool

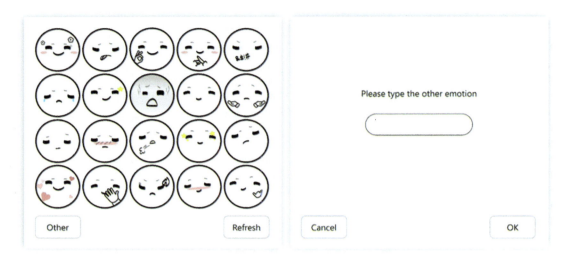

FIGURE 18.4

Screenshot of the self-assessment emotion tool system.

set itself is derived from cognitive rules, the correlations between emotion words are greatly enhanced when the richness of the emotion set is ensured, meaning that some emotion words are naturally confused with other emotion words. Therefore a combination of the PAD minimum distance from the emotion words can be a good way to obtain an expression design that matches the words, while ensuring the highest correctness of the design.

At the same time, we observed how the male and female groups scored the expressions. For the most part, there were no significant differences between them, except for some obvious differences in the ratings of dominance. We found that for females, the score on the dominance dimension is higher in the positive emotions than that of the males among these differential scores, for example, in the emotion images happy-for and satisfaction, while the males had greater scores than women in the dominance dimension of ratings on negative emotions—for example, in the emotion images Pity and Anger. This may be because women perceive positive emotions as more likely to affect others, while men perceive negative emotions as more likely to affect others. However, there is little data to support this conclusion, and there are exceptions—for example, women have a higher dominance score for Resentment compared with men. This therefore cannot be taken as a definitive rule.

Combining the experimental results from the pilot study and the validation study, it was found that the image designs in SAET after the iterations were effective in conveying the emotional meaning of the text, and the vast majority of them corresponded to the PAD values, which also indicates a good grasp of the intensity of the emotion.

18.5 Conclusions and discussion

18.5.1 Methods of application

SAET is a graphical self-reporting tool (Fig. 18.4) that can be disseminated via a web-based questionnaire, currently available in both Chinese and English. The SAET interface is arranged in a four-by-five matrix with randomized expressions. When a participant makes a choice by clicking on an expression, the interface will be refreshed automatically. Participants can also manually click the refresh button at the bottom right to rearrange the emotions. Besides this, the participant is left with the possibility of making other choices such as clicking on the "Other" button in the bottom left corner to jump to the screen for entering the name of another type of emotion.

18.5.2 Conclusions

There are still areas of improvement for the self-assessment emotion measurement tool, such as a broader group of participants and a more concise emotion set. However, overall the SAET is a theoretically sound, methodologically innovative, and practically meaningful instrument based on a well-established derived rule model of affective cognition and validated with dimensional spatial numerical calculations for the measurement of experienced emotions. It is also a visual tool that can compensate for the limitations of users in expressing their emotions in words and give them visual cues, which can be used both in the first stage of user research on emotional experiences and in the later stage of emotional evaluation of products or services. This three-dimensional approach to evaluating the emotion space can be used as a quantitative way to assess the design of expressions, providing a reference for the design of emotional experiences and expressions. The construction of a self-reported form of emotion measurement tool and emotion-assessment method has positive implications for both the accumulation of academic research on emotion design and the practical application of user-experience design.

References

[1] A. Ortony, G.L. Clore, A. Collins, The Cognitive Structure of Emotions, Cambridge University Press, Cambridge, MA, USA, 1988.
[2] A. Mehrabian, Pleasure-arousal-dominance: a general framework for describing and measuring individual differences in temperament, Curr. Psychol. 14 (1996) 261−292.
[3] Gebhard, P. ALMA: a layered model of affect, in: Proceedings of the 4th International Joint Conference on Autonomous Agents and Multiagent Systems (AAMAS 2005), Utrecht, The Netherlands, 25−29 July 2005.
[4] M.M. Bradley, P.J. Lang, Measuring emotion: the self-assessment manikin and the semantic differential, J. Behav. Ther. Exp. Psych. 25 (1994) 49−59.
[5] P. Desmet, P. Hekkert, J. Jacobs, When a car makes you smile: development and application of an instrument to measure prod-uct emotions, Adv. Consum. Res. 27 (2000) 111−117.

[6] M. Vastenburg, N. Romero Herrera, D. Van Bel, P. Desmet, PMRI: development of a pictorial mood reporting instrument, CHI '11 Extended Abstracts on Human Factors in Computing Systems, Association for Computing Machinery, Vancouver, BC, Canada, 2011, pp. 2155–2160.

Further reading

J.N. Bassili, Emotion recognition: the role of facial movement and the relative importance of upper and lower areas of the face, J. Pers. Soc. Psychol. 37 (1979) 2049–2058.

M. Blow, K. Dautenhahn, A. Appleby, C.L. Nehaniv, D.C. Lee, Perception of robot smiles and dimensions for human-robot interaction design, in: Proceedings of the ROMAN 2006—The 15th IEEE International Symposium on Robot and Human Interactive Communication, Hatfield, UK, 6–8 September 2006, pp. 469–474.

P. Desmet, G. Laurans, Developing 14 animated characters for non-verbal self-report of categorical emotions, J. Des. Res. 15 (2017) 214–233.

P. Ekman, About Brows: Emotional and Conversational Signals, Cambridge University Press, Cambridge, UK, 1979.

J. Forlizzi, K. Battarbee, Understanding experience in interactive systems, in: Proceedings of the 5th Conference on Designing Interactive Systems: Processes, Practices, Methods, and Techniques, Cambridge, MA, USA, 1–4 August 2004, Association for Computing Machinery, New York, NY, USA, 2004, pp. 261–268.

M. Hassenzahl, S. Diefenbach, A. Göritz, Needs, affect, and interactive products—facets of user experience, Interact. Comput. 22 (2010) 353–362.

G. Huisman, M. Van Hout, E. Van Dijk, T. Van Der Geest, D. Heylen, LEMtool: measuring emotions in visual interfaces, in: Proceedings of the SIGCHI Conference on Human Factors in Computing Systems, Paris, France, 27 April–2May 2013; Association for Computing Machinery, New York, NY, USA, 2013, pp. 351–360.

P.R. Kleinginna, A.M. Kleinginna, A categorized list of emotion definitions, with suggestions for a consensual definition, Motiv. Emot. 5 (1981) 345–379.

H. Kobayashi, S. Suzuki, H. Takahashi, Automatic extraction of facial organs and recognition of facial expressions, in: Proceedings of the 8th IEEE International Workshop on Robot and Human Interaction, RO-MAN '99 (Cat. No.99TH8483), Pisa, Italy, 27–29 September 1999.

G.F.G. Laurans, P.M.A. Desmet, Introducing PrEmo2 new directions for the non-verbal measurement of emotion in design, in: Out of Control: Proceedings of the 8th International Conference on Design and Emotion, London, UK, 11–14 September 2012, Central Saint Martins College of the Arts and the Design and Emotion Society: London, UK, 2012.

E.L. Law, V. Roto, M. Hassenzahl, A.P. Vermeeren, J. Kort, Understanding, scoping and defining user experience: a survey approach, in: Proceedings of the SIGCHI Conference on Human Factors in Computing Systems, Boston, MA, USA, 4–9 April 2009, Association for Computing Machinery, New York, NY, USA, 2009, pp. 719–728.

J.A. Russell, A. Mehrabian, Evidence for a three-factor theory of emotions, J. Res. Pers. 11 (1977) 273–294.

Z. Shao, Visual humor representation of cartoon character design—taking tom and jerry as an example, Decorate 4 (2016) 138–139.

M.N. Shiota, J.W. Kalat, Emotion, 2nd ed., Wadsworth Cengage Learning, San Francisco, CA, USA, 2012.

M. Thüring, S. Mahlke, Usability, aesthetics and emotions in human–technology interaction, Int. J. Psychol. 42 (2007) 253–264.

Y. Zhao, A study of audience psychology in emojis, Art. Des. Res. 6 (2016) 46–49.

Robot transparency and anthropomorphic attribute effects on human-robot interactions

CHAPTER 19

Jianmin Wang[1], Yujia Liu[1,2], Tianyang Yue[1], Chengji Wang[1], Jinjing Mao[1], Yuxi Wang[1] and Fang You[1]

[1]Car Interaction Design Lab, College of Arts and Media, Tongji University, Shanghai, P.R. China [2]College of Design and Innovation, Tongji University, Shanghai, P.R. China

Chapter Outline

- 19.1 Introduction 260
- 19.2 Robot-to-human communication 260
  - 19.2.1 Communication transparency 261
  - 19.2.2 Anthropomorphism 261
- 19.3 Human-robot interaction outcomes 262
  - 19.3.1 Safety 262
  - 19.3.2 Usability 262
  - 19.3.3 Workload 262
  - 19.3.4 Trust 263
  - 19.3.5 Affect 263
  - 19.3.6 Current study 264
- 19.4 Methods 265
  - 19.4.1 Participants 265
  - 19.4.2 Design 265
  - 19.4.3 Apparatus and materials 266
  - 19.4.4 Procedure 267
- 19.5 Results 268
  - 19.5.1 Task one: welcome 268
  - 19.5.2 Task two: incoming call 268
  - 19.5.3 Task three: chatting 273
- 19.6 Discussion 275
  - 19.6.1 Transparency 275
  - 19.6.2 Anthropomorphism 276
  - 19.6.3 Human-robot interaction variable relationships 276
  - 19.6.4 Implications for human-robot interaction design 276
  - 19.6.5 Limitations of the current study 277

Human-Machine Interface for Intelligent Vehicles. DOI: https://doi.org/10.1016/B978-0-443-23606-8.00020-8
© 2024 Tongji University Press Co., Ltd. Published by Elsevier Inc. All rights are reserved, including those for text and data mining, AI training, and similar technologies.

19.7 Conclusions and future directions 277
References 277
Further reading 278

19.1 Introduction

The emergence of onboard intelligent robots has enriched the practical application scenarios for robots, enhanced the human-vehicle interaction experience, and improved the overall intelligence level of the intelligent cockpit. Voice communication is one of the main interaction methods to convey information between humans and robots, so robots with human-like voice communication capabilities can provide better services. However, natural voice commands do not fully convey precise information, and humans sometimes prefer uncertain terms, symbols, and concepts.

In the field of onboard robots, a considerable amount of research has proved that anthropomorphic robots provide a better driving experience than traditional forms of interaction with voice and touch screens. For example, Kenton Wiliams et al. [1,2], from MIT published several papers comparing four different types of interactions (smartphone, dynamic robot, static robot, and human passenger) and showed that dynamic robots had a significant effect on reducing the user's cognitive load, improving distractions, and spending more time to get a positive effect. David R. Large et al. published a study that explored how passengers in self-driving cars interact with an onboard conversational agent interface, conducting a comparison of three groups of participants with an anthropomorphic agent interlocutor, a voice command interface, and a traditional touch interface. The results showed that the anthropomorphic agent interlocutor was the most popular interface and significantly improved the enjoyment and sense of control of the journey experience. However, it contained "trust challenges," with participants reporting that they were unable to predict the intentions of the anthropomorphic agent well enough to ensure that they were in control.

Therefore the key to human-robot interactions is how the human communicates with the robot. Maintaining transparency within the team facilitates constructive discussions for a proper understanding and informed decision-making about intelligent machines. Subsequently, it ensures maintaining effective collaboration among team members. At the same time, robots are given more life-like expressions by people because of the anthropomorphic nature of their images. Kontogiorgos, Dimosthenis et al. conducted a comparative evaluation study of social robots with anthropomorphic faces versus smart speakers with only voices. The results showed that anthropomorphic expressions and communicating with nonverbal cues are not always the best options. Anthropomorphic social behavior needs to be balanced with the task. Therefore it is also worth considering whether onboard robots should show anthropomorphic aspects of humanity. This study examined the aspects of transparency and their effect on human-robot outcomes when humans work together with an onboard robot in driving and nondriving situations.

19.2 Robot-to-human communication

One of the important determinants in evaluating whether anthropomorphic robots perform well is maintaining effective communication with humans. With the increasing complexity of the robot system, their ability to communicate grows, resulting in more information exchange. In this paper, we examine aspects of communication-based transparency in anthropomorphic robots in driving and nondriving states and their impact on human-robot team performances, associated cognitive structures, and affective experiences.

19.2.1 Communication transparency

Transparency can be defined as a way of establishing shared intentions and awareness between humans and robotic systems. The key features of transparency in human-robot systems are understanding the robot's purpose, analyzing its actions, and receiving information from them about decision-making and environment-aware analysis processes. Chen et al. regarded transparency as the ability of an intelligent agent (e.g., robot) to effectively communicate with a human being to accurately comprehend its current goals, reasoning, and future state. Good transparency in a team facilitates humans to have a proper understanding and sound judgment of intelligent machines and, subsequently, maintain an effective collaboration. Nevertheless, the situation awareness-based agent transparency (SAT) model proposes three levels of transparency for an autonomous agent, in addition to an explicit sense of the operator's situation awareness (SA): supporting the operator's understanding of the autonomous agent's actions and perceptions of the environmental characteristics (SAT1 Level 1); supporting the operator's understanding of the reasoning behind the autonomous agent's actions and decisions (SAT2 Level 2); and involving the agent's predictions of the outcomes of its reasoning, actions, and any uncertainty associated with the information presented (SAT3 Level 3). Starting with L1, L2, and L3, the analysis and selection of good transparency models and their application in the design of robot interactions will help to maintain transparency between humans and robots. Furthermore, it will ensure the usability of the intelligent system, the safety and positive emotional experience of the human, and a level of trust in the intelligent system.

19.2.2 Anthropomorphism

People tend to automatically rationalize the behavior of robots when they enter the living space of human beings, which is no exception in the automotive space. Apparently, such an anthropomorphic tendency is a powerful driving force for the development of robots. The manifestation of anthropomorphism refers to attributing human motivations, characteristics, and behaviors to inanimate objects, such as speaking technology (as shown in the example described above), and building expectations based on this. Many factors influence robot anthropomorphism, such as movement, verbal communication, emotion, gesture and intelligence, sociocultural background, gender, and group membership. The purpose of robot anthropomorphism research is not only to assess a better experience in undergoing the anthropomorphic tendencies of robots but to also improve the cognitive abilities of intelligent vehicles. People are inspired (to some extent) to believe that this artifact has the ability to think rationally (agency) and feel consciously (experience), which is usually influenced by people's perception of human traits and characteristics, such as voice, social behavior, etc. Passengers of smart driving vehicles that were anthropomorphized (with human names, genders, and voices) rated their cars' cognitive abilities higher than those with the same autonomous driving characteristics but without the associated anthropomorphic cues. The purpose of this paper is to explore the relationship between the anthropomorphic perception of voice and vision in transparency and human-robot cooperative team performances.

19.3 Human-robot interaction outcomes

19.3.1 Safety

In human-computer interactions, especially in the domain of smart cars, the first consideration in human-computer interface design is safety. All displayed information must align with the standards of safe driving, while also complying with relevant laws and regulations. Therefore in the case of an increasing number of electronic devices and richer functions in the vehicle, the National Highway Traffic Safety Administration (NHTSA) expressed concerns about driving safety and issued Visual-Manual NHTSA Driver Distraction Guidelines for In-Vehicle Electronic Devices guidelines (referencing relevant research in Europe and Japan), which are nonbinding and voluntary, with the intent of guiding designs to reduce driver distractions. In this guideline, the distractions that affect a driver's focus are divided into three categories: visual, manual, and cognitive distractions. Visual distractions refer to tasks that require the drivers to take their eyes off the road to obtain visual information; manual distractions to tasks that require the driver to take their hands off the steering wheel and operate equipment, and cognitive distractions to tasks that require the driver to take their attention away from the driving task. As people become more tolerant of electronic devices, there may not be such strict restrictions on them, but it is still necessary for onboard robots to consider safety when interacting with people.

19.3.2 Usability

"Usability is not a quality that exists in any real or absolute meaning. Perhaps it can be best summed up as being a general quality of the appropriateness to a purpose of any particular artifact." John Brooke argues that usability is a fundamental quality that needs to be present for artifacts in general. If its usability is in doubt, especially among robots with concrete images, it is likely that people will stop using it. The feedback from the robot is that it wants to demonstrate different levels of transparency, which may not be consistent with utility, which means transparency is not necessarily proportional to utility. For example, a robot that is too verbose while driving may make it hard for a human to quickly and accurately obtain its intentions, leading to a reduced utility; however, a lack of transparency in some areas may also make the intentions unclear, and the human will make more unnecessary guesses. This study considers whether robots with different levels of information transparency would increase or decrease their utility.

19.3.3 Workload

Workload is a multidimensional concept used to describe the psychological stress or information processing capacity of a person in performing a task that involves mental stress, time pressure, task difficulty, operator ability, effort level, and other factors. Workload is an important construct of HRI because it can greatly affect human-robot team performance. It is, in turn, influenced by a variety of factors, such as task structure, performance requirements, human-machine interface, and human-individual factors such as experience. Transparency is thought of as having the potential to decrease and increase workloads. On the one hand, higher levels of transparency can carry useful information for human decision-making; thus, it can reduce the workload or keep the workload unchanged. For example, robots can inform people about the cause of abnormal conditions on the road, preventing them from getting caught in unnecessary guesses and reducing the amount of cognitive computation people need to perform. On the other hand, the extra information provided by the additional transparency might require extra cognitive resources to process. For example, excessively long decision descriptions and overly complex visual animation switching may lead to an increased brain load for humans due to the requirement to process more information. The impact of transparency on the workload may be influenced by a number of different features, such as text for the voice channel and expression animation for the visual channel. Overall, workload is an important cognitive construct to consider when designing transparent communication, as it is helpful to verify what workload range the transparent design is in for humans.

19.3.4 Trust

In the field of human-computer interactions, trust is considered an important factor in the success of human-computer teams. A seminal paper on trust by Lee et al. identified two fundamental components of human-automation cooperation, which are trust and transparency. Trust is a psychological phenomenon—one's expectation of an outcome or a subjective probability held about the occurrence of a future event. It can also be considered an attitude coming from impressions of the information provided by the system and past experiences of use. Depending on the information, impressions, and usage experiences, users will develop different levels of trust and, thus, different levels of system dependence. Mui proposed a model based on trust in vehicle automation, which contains three trust dimensions: predictability, dependability, and loyalty. Stedmon et al. showed that people exhibit higher trust and performance when interacting with systems that use human voice communication compared with systems that use synthetic voice communication. Therefore trust issues also need to be fully considered in human-robot communication teams.

In intelligent driving, adaptive automation can replace the operator's perceptual capabilities and assist and replace decision-making and action processes. On the one hand, without trust, people may be reluctant to use the system even if it performs well in autonomous driving, which leads to its abandonment; on the other hand, too much trust can lead to misuse, that is, using the system in an unintended way. Under-trust or over-trust caused by these effects can lead to accidents. Therefore calibrated trust allows one to maintain an "appropriate" level of trust, which is necessary to ensure the optimal distribution of functions among team members and collaboration between human-machine teams. As a result, one of the key elements in establishing the appropriate trust is transparency, which indirectly reflects the intelligence of the robot.

19.3.5 Affect

Since human behavior and thought processes are closely related to affects, neglecting the user's emotional state can negatively influence task performance and trust in the system. It is also applicable to the driving environment. Affect has been used to encompass different constructs, including emotion, feeling, or mood. This study takes effect as an evaluation dimension and indicator and shows that affect influences attention, perception, and decision-making. These factors, in turn, influence driving behavior. Jeon, Walker, and Yim concluded that drivers in happy or angry moods have decreased driving performances compared with drivers in normal and fearful moods. Negative potency emotions, such as sadness or depression, have also been shown to reduce one's driving ability. In addition, emotions can be directly related to the driving experience, especially when a person is communicating with a robot while driving (stationary state); this deserves more attention.

19.3.6 Current study

The current study examines the impact of transparency delivered via text messages and team orientation on HRI outcomes (safety, usability, workload, trust, and affect) when working together with an autonomous onboard robot. The conceptual model for the study focuses on how robot communication transparency will influence safety, usability, workload, trust, and affect, ultimately contributing to improved performances. We investigated the differences between two transparency levels for two tasks in the driving state (whether anthropomorphic visual information and sound information are needed) and three transparency levels for one task in the nondriving state (whether anthropomorphic visual information is required for L3, and whether anthropomorphic L1 sound information is required), sent by the onboard robot to humans via voice and vision (Fig. 19.1).

The L1 sound channel information was chosen to represent the anthropomorphic nature of the robot because humans usually emit some onomatopoeic words when they perceive their surroundings in life, and we expect that adding this factor will make the human-robot cooperation a better experience and performance. An intelligent robot generally needs to have L2 understanding and L3 prediction. However, whether all acoustic information is needed with visual information is something to consider; considering the scale of the study, we compared L3 with and without robot expression changes during the stationary task and, while driving, compared robots with and without expression. We expect that robots with expression changes will have better driving experiences and performances.

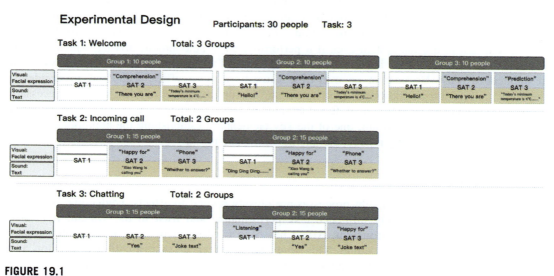

FIGURE 19.1

Different information transparency and anthropomorphic designs for the three tasks.

19.4 Methods

19.4.1 Participants

The study included 30 Chinese participants (25 male and 5 female), in the age range of 22 to 40 years (M = 28.2, SD = 5.83,). They were recruited through questionnaires, which contained demographic information about the participants, 83.8% of whom had university or higher education ($n = 25$), followed by junior college (13.3%, $n = 4$), and high school or below education (3.3%, $n = 1$). They all drove more than 2 to 3 times per week, 70% ($n = 21$) used electric vehicles, and 80% ($n = 24$) had some level of knowledge about onboard robots. Informed consent was obtained from each participant prior to data collection.

19.4.2 Design

There were three tasks in this experiment, and each task was set up with different groups of transparency experiments with different information (Fig. 19.1). Task 1 was a robot under static statues greeting a person who had just boarded a car, and the three groups were set up with different levels of information transparency. When the robot detects and comprehends the person is already in the car, it will send, "Hello! There you are! Good morning." The robot's eyes would look at the driver when it said, "There you are!" and winked with a "comprehension" expression. When the robot predicts that the person needs to know the current temperature and dress advice, it will prompt, "Today's minimum temperature is 4°C; the weather is getting cooler, pay attention to keep warm!" and make a predictive expression for drinking hot tea. The comparison between Group 1 and Group 2 was whether the L1 anthropomorphic "hello" text message was present or not, while Group 3 added the L1 anthropomorphic "hello" text message and the L3 predicted expression. The participants were equally divided into three groups of ten participants each. After completing the post-self-assessment manikin (SAM) affective scale, Task 2 was administered.

Task 2 involved a driving robot alerting the driver of a phone call and asking if the driver wanted to answer. Before the start of the task, the participants were told that they had a friend named Xiao Wang. While the driver was driving smoothly on a straight road at 30 km/h, the operator-controlled phone rang, and the robot said, "Ding Ding Ding, Xiao Wang is calling you, do you want to answer?" When the robot understood the name of the caller and said, "Xiao Wang is calling you," it changed from a smiling foreword expression to "happy-for." When the robot predicted that the person might need to answer the phone, it acted like it was shaking the phone around its ear. The comparison between Group 1 and Group 2 was whether there was an L1 anthropomorphic "Ding Ding Ding" text message. Group 1 and Group 2 of Task 2 had 15 people each and completed the task by filling out the usability, workload, trust, and affect scales, followed by Task 3.

Task 3 involved a person chatting with the robot in a driving state. When the person gave the robot the command to tell a joke, the robot made a listening expression at L1, then understood the person's command, replied, "Yes!," predicted the content of the joke, and made a "happy-for" expression while the robot was broadcasting. The comparison between Group 1 and Group 2 is that Group 1 did not have a visual channel for expressions but only had a robot with a shape that could be interacted with by voice. Group 1 and Group 2 of Task 3 also had 15 participants each, and after completing the task, the participants filled out the scales for usability, workload, trust, and affect.

This study considered the transparency of different information when humans interacted with the robot by increasing or decreasing the feedback information from the robot in the driving and stationary states from both the visual and auditory channels. The experiment was a between-group experiment, with each participant performing one group from each task, and the number of group tests per task was balanced.

19.4.3 Apparatus and materials

The prototype onboard robot "XiaoV" (Fig. 19.2A) is an up-and-down structured robot with a head and base. It has a young female voice and can show dynamic videos. XiaoV can move on two levels of freedom, and its head and face will turn to the driver's seat when interacting with the driver. The feedback from facial expressions and voice output can recommend and describe current events to the drivers. The experimental environment was based on the laboratory's self-developed car simulation simulator system, and the scenario was developed in Unity software, which simulated a real driving environment. The scenario used in the experiment was a two-way two-lane straight road with a large number of oncoming cars in the opposite lane and no overtaking from behind. The simulator was accompanied by a monitoring device, which was used to collect simulator vehicle data and user sweeping behavior data (Fig. 19.2B). All movements of the robot were remotely controlled in real time by the researcher's laptop (The Wizard of Oz).

19.4.3.1 Self-report metrics

The after-scenario questionnaire is a more widely used task-based assessment questionnaire with the benefit of evaluations of three projects: ease of task completion, time required to complete tasks, and satisfaction with support information, which can be used in similar usability studies. The scale options are set from 1 to 7, with higher scores representing higher levels of performance. The driving activity

(A)

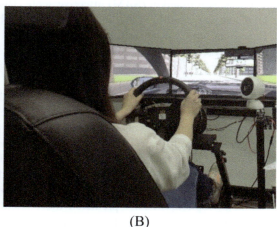
(B)

FIGURE 19.2

(A) Images of the robot "XiaoV." (B) Experiment environment.

load index (DALI) is more appropriate than the other scales for assessing driving-related tasks and contains six dimensions of work—namely, effort of attention, visual demand, auditory demand, temporal demand, interference, and situational stress. The scale options are set from 1 to 10, with higher scores representing a correspondingly higher degree. Mui proposed a model based on trust in vehicle automation, which contains three trust dimensions: predictability, dependability, and loyalty. The present experiment evaluates the trust level from these three dimensions. The scale options are set from 1 to 7, with higher scores representing higher levels. To measure the user (affective) experience, they need to complete SAM, a nonverbal and pictorial assessment technique used for operationalizing the user experience through the constructs of pleasure, arousal, and dominance. The scale options are set from 1 to 9, with higher scores representing a correspondingly higher degree.

19.4.3.2 Performance metrics

In this experiment, the driving data collected in the simulator was used as the evaluation index (Task 1 was performed at rest, without vehicle data). Tasks 2 and Task 3 required the participants to drive smoothly on a straight road at a speed of 30 km/hour. The standard deviation of the vehicle speed represents the driver's longitudinal control ability, and the standard deviation of left lane departure represents the driver's lateral control ability, examining whether the driver is driving smoothly in the current lane in the lateral longitudinal direction. In the results of Task 2 and Task 3, there were no unsafe situations such as lane crossing, red light running, collision, and so on, as well as no unfinished tasks. Based on such a safe condition, the acceleration labeling difference and the standard deviation of left lane departure were used as the main reference basis, with a small standard deviation being regarded as the safer group. The better group was selected by combining the data of other dimensions comprehensively.

In addition, the total sweep frequency and total sweep duration of the robot were extracted from the video recorded by the simulator. Research suggests that a single sweep that is too long (more than 2 seconds) is considered to be a driver distraction and may lead to driving safety hazards. In all task results, the raw data of each group was counted, and no participant had a single sweep time longer than 2 seconds. In addition, the groups with longer and more sweeps showed an increase in the total sweep frequency and total sweep duration at the same time, namely, the total single sweep time did not increase and, consequently, did not pose a threat to driving safety. In the analysis of the results, we used the sweep data as a reference indicator of the driver's interest in the robot, similar to the affective dimension. Therefore the group with a longer and more frequent sweep time was considered in the analysis to have a robot whose performance was more likely to interest the driver in the robot.

19.4.4 Procedure

The entire experiment was conducted on a driving simulator and lasted for about 40 minutes for each participant. Prior to the start of the experiment, the participants were given a short period of simulated driving to familiarize themselves with the simulated driving scenario. After completing the test drive, the participants were required to fill out a basic questionnaire, give informed consent, sign a confidentiality agreement, and then begin the experiment with their full consent to record the video. After listening to the task descriptions, the participants were asked to perform the corresponding tasks and complete the scales required for the tasks, and then they were interviewed in a semistructured manner. The rest was done in the same manner until the experiment was fully completed.

19.5 Results

The data of each experimental group of all the task scenarios is summarized, as shown in Table 19.1. The experimental evaluation refers to several dimensions of the human-robot outcome outlined in Section 19.3 above. Therefore, in addition to Task 1 (stationary state), the experimental data of Tasks 2 and 3 are divided into six aspects: Safety, Usability, Workload, Trust, Sweep, and Affect.

19.5.1 Task one: welcome

19.5.1.1 Sweep

In Task 1, a one-way analysis of variance (ANOVA) was performed on the data from the three groups, whose results showed (Fig. 19.3) that there was a considerable difference in the sweep time between the three groups, $F\ (2,\ 14) = 6.573$, $p = 0.01 < 0.05$. A posthoc multiple comparison revealed that the sweep time was significantly less in Group 1 (M = 1.97, SD = 1.09) than in Group 2 (M = 4.65, SD = 1.38) and Group 3 (M = 4.01, SD = 1.50), with no significant difference between Group 2 and Group 3. In Task 1, there was no significant difference in the number of sweeps among the three groups, $F\ (2,\ 13) = 0.116$, $p = 0.89 > 0.05$. Therefore participants in Group 2 and Group 3 had much longer sweep durations of the robot, which implies that the participants were more interested in the robot performances in Group 2 and Group 3. Additionally, there was no significant difference in the sweep frequency means among the three groups, indicating that the sweep means for the three groups were not markedly different, and the participants in Group 2 and Group 3 had longer single gazes on the robot, that is, the robot's performances were more likely to attract the participants' continuous attention. Since the task was not performed in the case of driving, it can be assumed that Group 2 and Group 3, which elicited longer gazes, had better performances.

19.5.1.2 Affect

According to ANOVA, the participants had higher levels of pleasure score ($F\ (2,20) = 6.529$, $p = 0.007 < 0.05$) and dominance score ($F\ (2,21) = 3.214$, $p = 0.008 < 0.05$). There was no notable difference in activeness ($F\ (2,17) = 6.576$, $p = 0.061$), but it was close to the $p = 0.05$ significance criterion (Fig. 19.4). A posthoc multiple comparison obtained showed that Group 3 was significantly better than Group 1 in terms of the pleasure score ($p = 0.007 < 0.05$) and significantly better than Group 2 in terms of the dominance dimensions ($p = 0.006 < 0.05$).

In conclusion, Group 3, which had the anthropomorphic L1 sound information and the L3 hierarchical visual information of L1, was able to obtain more attention from the participants, as well as more positive affective experiences.

19.5.2 Task two: incoming call

19.5.2.1 Safety

The t-test test showed that the average value of the standard deviation of vehicle speed was higher in Group 1 (M = 0.42, SD = 0.11) than in Group 2 (M = 0.27, SD = 0.28) in Task 3 but was not significantly different ($p = 0.30 > 0.05$), so the drivers were considered to have essentially the same level of longitudinal vehicle control. In contrast, the standard deviation of left lane departure in

19.5 Results

Table 19.1 Summary table of the experimental results.

		Task 1: Welcome			Task 2: Incoming Call		Task 3: Chatting	
		Group 1	Group 2	Group 3	Group 1	Group 2	Group 1	Group 2
Transparency information		L2: voice and vision; L3: voice	L1: voice; L2: voice and vision; L3: voice	L1: voice; L2: voice and vision; L3: voice and vision	L2: voice and vision; L3: voice and vision	L1: voice; L2: voice and vision; L3: voice and vision	L2: voice; L3: voice	L1: vision; L2: voice; L3: voice and vision
1 Safety	Standard Deviation of Vehicle Speed		/		0.42	0.27	1.14	0.90
	Standard Deviation of Left Lane Departure				0.18	0.77	1.09	1.34
2 Usability	Total Means		/		6.70	6.32	6.17	6.45
	Ease of Task Completion				6.61	6.42	5.70	6.27
	Time Required to Complete Tasks				6.79	6.04	6.50	6.69
	Satisfaction with Support Information				6.71	6.50	6.80	6.40
3 Workload	Total Means		/		6.12	5.03	4.45	3.48
	Effort of Attention				7.63	7.27	6.10	4.75
	Visual Demand				5.96	6.00	2.69	1.98
	Auditory Demand				7.08	6.92	7.43	6.24
	Temporal Demand				5.38	4.01	3.54	3.21
	Interference				5.34	3.25	4.67	2.88
	Situational Stress				5.33	2.71	2.25	1.83
4 Trust	Total Means		/		6.43	6.48	6.33	6.35
	Predictability				6.13	6.36	6.08	6.39
	Dependability				6.59	6.57	6.36	6.30
	Loyalty/Desire to continue using				6.57	6.50	6.55	6.35
5 Sweep	Total Sweep Duration(s)	1.973	4.648	4.015	1.671	3.893	1.695	4.439
	Total Sweep Frequency (Fre)	2.50	2.50	2.67	3.00	5.00	1.67	4.00
6 Affect	Pleasure Score	7.37	7.25	7.70	7.83	8.63	7.63	8.45
	Arousal Score	7.60	7.29	6.33	6.93	8.56	7.00	7.79
	Dominance Score	8.50	8.04	8.50	6.93	7.50	7.70	8.59
Conclusion			The better group: Group 3. Reason: L1 anthropomorphic voice information and L3 level visual information can get more attention from the participants and, at the same time, obtain a more positive emotional experience.		The better group: Group 1. Reason: Adding L1 level voice information (Ding Ding Ding), Group 2 can make the participants have more positive emotions and reduce the workload. However, the evaluation of the usability was lower, possibly because the voice of "Ding Ding Ding" made people feel uncomfortable. At the same time, the participants in Group 1 had significantly better vehicle lateral control levels and better safety.		The better group: Group 2. Reason: Compared with Group 1, Group 2 added visual information that can make the participants have more positive emotions. Group 1 communicated with the participants only through voice interactions, which caused the participants' temporal loads to be significantly higher and caused certain interferences. Additionally, the ability of Group 1 to control the vehicle laterally was not as good as Group 2, so the robots with expressions were better than a pure voice robot.	

The red font marks the better data that has been statistically tested. The reasons for choosing a better group are explained at the end of the table.

Group 1 (M = 0.18, SD = 0.24) was significantly lower than in Group 2 (M = 0.77, SD = 0.29) ($p = 0.000 < 0.001$). From the dimension of vehicle safety, it was considered that Group 1, which had a significantly better level of lateral vehicle control, was safer.

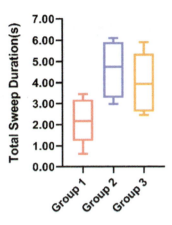

FIGURE 19.3

Individual differences in the total sweep duration between the three groups.

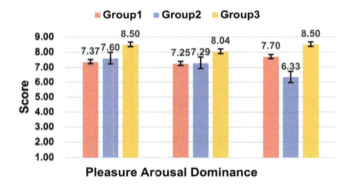

FIGURE 19.4

Ratings of pleasure, arousal, and dominance score between the three groups.

19.5.2.2 Usability

From the usability average scores, it can be concluded that Group 2 ($M = 6.32$, $SD = 0.68$), that is, with L1 level voice messages (Ding Ding Ding), had a lower human evaluation of usability ($p = 0.049 < 0.05$) than Group 1 ($M = 6.70$, $SD = 0.46$). After back testing, satisfaction with time spent was significantly different between the groups ($p = 0.01 < 0.05$), while satisfaction with the level of difficulty ($p = 0.374$) and support information ($p = 0.262$) was not obviously different. It can be concluded that the disfluency of the task progress due to the lengthy voice in Task 3 triggered a lower usability evaluation of the participants in Group 2, and the additional information at the expense of time spent did not improve the usability evaluation in terms of the task difficulty and support information.

19.5.2.3 Workload

From the workload scores (Fig. 19.5), it can be concluded that Group 2, that is, with L1 level voice messages (Ding Ding Ding), participants had a lower workload ($p = 0.002 < 0.01$). In terms of the DALI detail scores, the main reasons were lower interference ($p = 0.003 < 0.01$) and lower situational stress ($p = 0.002 < 0.01$).

19.5.2.4 Trust

Group 1 had similar trust scores (M = 6.43, SD = 0.67) to Group 2 (M = 6.48, SD = 0.51) with no significant difference ($p = 0.519$). The three dimensions of predictability ($p = 0.392$), dependability ($p = 0.954$), and desire to continue using ($p = 0.717$) did not produce distinct differences in terms of each of the trust scores.

19.5.2.5 Sweep

In Task 3, the average sweep time (Fig. 19.6A) was significantly lower in Group 1 (M = 1.671, SD = 0.39) than in Group 2 (M = 3.893, SD = 1.15) by the *t*-test ($p = 0.00 < 0.01$). Similarly, the sweep time means were also significantly lower in Group 1 than in Group 2 ($p = 0.02 < 0.05$) (Fig. 19.6B), indicating that Group 1 enabled participants to focus more on the driving task and was less likely to pose a hazard to driving.

19.5.2.6 Affect

From the pleasure−arousal−dominance (PAD) scores (Fig. 19.7), it can be concluded that Group 2, that is, with L1 voice information (Ding Ding Ding), was able to make participants acquire more positive emotions, with higher levels of pleasure ($p = 0.001 < 0.01$) and activity ($p = 0.001 < 0.01$) and no significant difference in dominance ($p = 0.202$).

In conclusion, the L1 of voice information (Ding Ding Ding) added to Group 2 was able to make participants gain more positive emotions and reduce their workload. At the same time,

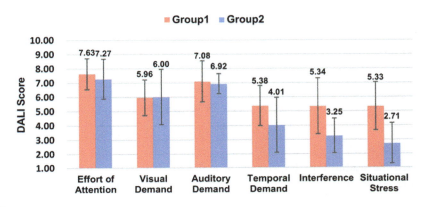

FIGURE 19.5

Ratings of the driving activity load index scale between the two groups.

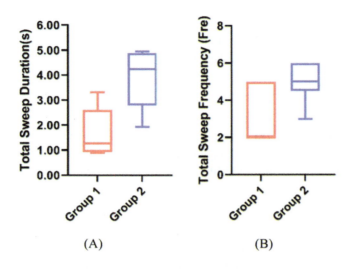

FIGURE 19.6

(A) Individual differences in the total sweep duration between the two groups. (B) Individual differences in the total sweep frequency between the two groups.

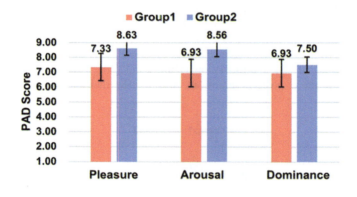

FIGURE 19.7

Ratings of pleasure, arousal, and dominance between the two groups.

participants rated the usability of Experiment 2 as lower. It is probably because the "Ding Ding Ding" voice had no obvious content meaning, which made the participants uncomfortable or did not have sufficient task fluency. Combining the vehicle driving data and the sweeping data, it was concluded that the participants in Group 1 had significantly better lateral control of the vehicle and a higher concentration on driving and had the best safety in terms of vehicle safety dimensions.

19.5.3 Task three: chatting

19.5.3.1 Safety

In Task 4, after the t-test, there was no significant difference in the average standard deviation of vehicle speed in Group 1 (M = 1.14, SD = 0.64) compared with Group 2 (M = 0.90, SD = 0.32). Moreover, the average standard deviation (M = 1.09, SD = 0.59) of Group 1 had no significant difference compared with Group 2 (M = 1.34, SD = 0.72) ($p = 0.06 > 0.05$), so the driver was not considered to be very different in the two sets of vehicle controls.

19.5.3.2 Usability

The participants in Group 2 rated higher overall than Group 1, but there was no significant difference ($p = 0.355$). As seen in the three usability scores, the difficulty satisfaction dimension was significantly different ($p = 0.027 < 0.05$), and the two dimensions of satisfaction with the time required to complete the tasks ($p = 0.327$) and satisfaction with support information ($p = 0.659$) did not produce significant differences. Thus, Group 1, where the robots have no expressions and movements, had lower difficulty satisfaction scores in the human evaluation of their usability, meaning that it would be harder to use robots without expressions and movements.

19.5.3.3 Workload

The overall average value of the workload was lower in Group 2 than in Group 1, but there is no significant difference ($p = 0.495$). As seen on each of the workload scores (Fig. 19.8), Group 2 produced significant variability in both the auditory demand ($p = 0.005 < 0.05$) and interference ($p = 0.000 < 0.001$). That is, expressionless and voiceless robots will increase the auditory demand of the driver and cause stronger interference with the driver.

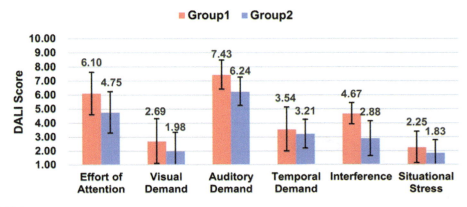

FIGURE 19.8

Ratings of pleasure, arousal, and dominance between the two groups.

19.5.3.4 Trust

There was no significant difference between the two groups in the mean trust score ($p = 0.747$). The dimensions of predictability ($p = 0.281$), dependability ($p = 0.855$), and desire to continue using the robot ($p = 0.422$) did not produce significant differences in the scores of the trustworthiness dimensions.

19.5.3.5 Sweep

By the t-test, the average value of the total sweep duration for Group 1 (M = 1.695, SD = 1.40) was significantly lower ($p = 0.01 < 0.05$) than that of Group 2 (M = 4.439, SD = 1.83) (Fig. 19.9A). Additionally, the mean of the total sweep frequency in Group 1 (M = 1.67, SD = 0.82) was significantly lower ($p = 0.02 < 0.05$) than in Group 2 (M = 4.00, SD = 1.94) (Fig. 19.9B). The results indicated that the participants in Group 2 had a significantly higher sweep duration and sweep frequency than the participants in Group 2, which may be due to the participants' tendency to receive eye contact from the robot with anthropomorphic visual expressions.

19.5.3.6 Affect

From the PAD scores (Fig. 19.10), it can be concluded that Group 2, where the robot has anthropomorphic human expressions, was able to give the participants a higher level of pleasure. The two experimental groups were significant only in terms of the pleasure score ($p = 0.005 < 0.05$), and no notable differences were seen in terms of both the arousal ($p = 0.079$) and dominance scores ($p = 0.066$).

In summary, compared with Group 1, Group 2 has added visual information, as long as the voice information was able to make the participants gain more positive emotions and a correspondingly moderate amount of attention since the participants in Group 1 were communicated via voice

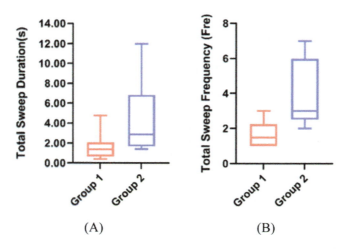

FIGURE 19.9

(A) Individual differences in the total sweep duration between the two groups. (B) Individual differences in the total sweep frequency between the two groups.

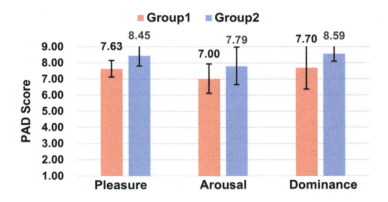

FIGURE 19.10

Ratings of pleasure, arousal, and dominance between the two groups.

interactions, which led to a significantly higher auditory load on the participants and produced some interference. In addition, the participants' ability to control the vehicle laterally was not as good as in Group 2 with expressions, so the robot with expressions was better than the robot without expressions (pure voice).

19.6 Discussion

Effective human resource information is a key component of human-robot team performance. Designing robots for human transparency is critical, particularly to promote the appropriate trust, workload, usability, affect, understanding of the robot, and ultimately, human-robot performance. Despite a growing amount of literature on transparency and how anthropomorphic humans can be applied to intelligent agents, the literature applying transparency principles to robotic communication is still far from adequate. This paper explores the application of the transparency models proposed by Chen et al. to human-robot interactions in smart car scenarios in an attempt to test their impact on human-robot teams.

19.6.1 Transparency

The results of the study on the transparency of robot-human communication showed that, first, it is not that more transparency is better; second, the increase of information is related to the characteristics of the interaction channel. The information of the voice channel can accurately broadcast the intentions that the robot wants to express, but at the same time, it is important to pay attention to the accuracy and rapidity of the expression and the conciseness and clarity of the text content. The voice of the L3 level (predetermination) needs to be shown in a general situation. The L1 (perceptual) voice does not perform well in terms of usability performance and can be removed if the conversation is long. Information from the visual channel enhances the receptivity of the information, but given the fluency of the communication, not every level is suitable for display; otherwise, this would make the change of expression too cumbersome. If the communication is initiated by a human, then the visual information of the L1 listening state is necessary to effectively indicate that the robot is receiving the information from the human. If the communication is not human-initiated, the information can be directed to the L2 comprehension level, which allows the human to quickly accept the information expressed by the robot. Additionally, the visual information of L3 is suggested to be provided to improve the human's pleasure, especially in the stationary state. In summary, the visual and acoustic information of L2 and L3 are suggested to be provided to the human for a good communication result; the visual information of L1 is not good for human-robot communication, except when the human initiates it.

19.6.2 Anthropomorphism

For robot anthropomorphism, the result is that the voice performance in L1 is poor, and the visual performance in L3 is good. The anthropomorphic sound information in L1 may disturb the driver while driving, showing that this will increase the workload because of the disturbance, and the usability of the robot will be reduced accordingly. Despite that, it will make the driver more pleasant. In other words, the premise of not interfering with driving should be met first while pursuing a humanoid emotional performance.

The study also compared the visual channels. The robot with visual expressions was not better than the one without visual expressions in terms of driving performance, but it was better in terms of usability because it was easier to use. Meanwhile, it increased the drivers' positive emotions. Increasing the visual expression of L3 would also enhance the humanoid nature of the robot, gain more attention from the driver, and make it more emotionally pleasing. Therefore the anthropomorphism of sound was not confirmed in this study, but the anthropomorphism of expressions had a good performance.

19.6.3 Human-robot interaction variable relationships

Finally, we observed the relationship between robots conveying information and safety, usability, workload, trust, and affect. In scenarios where the human and robot collaborate on secondary tasks (experimental phone calls, chats, and so on) and people focus their attention on driving with fewer sweeps, the robot usability is better, the workload is lower, and the human has better control of the vehicle. In the stationary state, people have more interest in the robot and a correspondingly better mood in the case of more appropriate sweep times.

19.6.4 Implications for human-robot interaction design

Overall, the results of this study showed the complexity of designing transparent communication with robotically anthropomorphic two-channel audio-visual interactions. Transparent communication that conveys a robot's state may be useful in terms of human performance, but it may also be onerous in terms of the workload if the details of the application (e.g., text mode) are not considered. The text mode needs to be combined with other modes (e.g., graphics), and certain levels of transparency are more appropriate for the text mode, such as L2 and L3, due to the nature of the information available in L3 transparency. However, if the robot is engaging in a passive interaction, then the graphical form is more suitable for L1 transparency. Conversely, if it is engaging in an active interaction, it does not need L1 and goes directly to L2. The use of a graphical form for L3 was not necessary but added to the emotional experience.

The information representation of robots is not a fixed pattern of information transparency; it has different applications in different types of scenarios, and experiments have found that what they have in common remains to be discussed carefully based on the situation. In future robotic systems, these design trade-offs must be considered.

19.6.5 Limitations of the current study

This study has several limitations. The first is that, due to the limitations of the experiment, only three scenarios were set up for the experiment, while the scenarios of the human-robot interactions were complex and diverse. Second, the results of this study were not necessarily fully applicable to all scenarios, such as dangerous driving-type scenarios. Our task scenarios focused on one round of interaction processes, namely, one round of human-robot communication for perception, understanding, and prediction. Additionally, there are many scenarios that require multiple rounds of communication. For example, when the robot helps navigate the whole process, whether it needs to comprehend and anticipate transparent information every time when confirming the navigation location, route, and time remains to be discussed separately.

19.7 Conclusions and future directions

The current research applies an emerging human-robot team transparency model and a simple anthropomorphic approach to the way intelligent robots communicate. It also provides a reference for the design of human-robot interactions that can maintain good performances after empirical evaluations. The results suggest that higher lvels of transparency (L3 voice and vision) can improve human-robot team performance and that lower levels of transparency (L1 voice) need not be demonstrated. The robot's perceptual information when listening to a human (L1 vision) and robot understanding (L2 voice) are the fundamental conditions for communication between the two parties. The current study also showed that the robot's anthropomorphic voice expression did not see a significant enhancement effect, even in terms of user emotion enhancement, an aspect that requires in-depth research and should not be limited to the level of voice interactions.

References

[1] K. Williams, J.A. Flores, J. Peters, Affective robot influence on driver adherence to safety, cognitive load reduction and sociability. In Proceedings of the 6th International Conference on Automotive User Interfaces and Interactive Vehicular Applications, Association for Computing Machinery: Seattle, WA, USA, 17–19 September 2014, pp. 1–8.

[2] K.J. Williams, J.C. Peters, C.L. Breazeal. Towards leveraging the driver's mobile device for an intelligent, sociable in-car robotic assistant. In 2013 IEEE Intelligent Vehicles Symposium (IV); IEEE: Piscataway, NJ, USA, 2013; pp. 369–376.

Further reading

M.M. Bradley, P.J. Lang, Measuring emotion: the self-assessment manikin and the semantic differential, J. Behav. Ther. Exp. Psychiatry 25 (1994) 49–59.

J. Brooke, SUS: a quick and dirty usability scale, Usability Eval. Ind. 189 (1996) 4–7.

J.Y.C. Chen, M.J. Barnes, Human–agent teaming for multirobot control: a review of human factors issues, IEEE Trans. Human-Machine Syst. 44 (2014) 13–29.

J.Y.C. Chen, M.J. Barnes, M. Harper-Sciarini, Supervisory control of multiple robots: human-performance issues and user-interface design, IEEE Trans. Syst. Man. Cybern. Part. C. 41 (2011) 435–454.

J.Y.C. Chen, S.G. Lakhmani, K. Stowers, A.R. Selkowitz, J. Wright, M. Barnes, Situation awareness-based agent transparency and human-autonomy teaming effectiveness, Theor. Issues Ergon. Sci. 19 (2018) 259–282.

B.R. Duffy, Anthropomorphism and the social robot, Robot. Auton. Syst. 42 (2003) 177–190.

C.S. Dula, E.G. Geller, Risky, Aggressive, or emotional driving: addressing the need for consistent communication in research, J. Saf. Res. 34 (2003) 559–566.

F. Eyben, M. Wöllmer, T. Poitschke, B. Schuller, C. Blaschke, B. Färber, et al., Emotion on the road: necessity, acceptance, and feasibility of affective computing in the car, Adv. Human-Computer Interact. 2010 (2010) 1–17.

K. Fischer, How people talk with robots: designing dialogue to reduce user uncertainty, AI Mag. 32 (2011) 31–38.

H.M. Gray, K. Gray, D.M. Wegner, Dimensions of mind perception, Science 315 (2007) 619.

S. Guznov, J. Lyons, M. Pfahler, A. Heironimus, M. Woolley, J. Friedman, et al., Robot transparency and team orientation effects on human-robot teaming, Int. J. Human-Computer Interact. 36 (2020) 650–660.

K.A. Hoff, M. Bashir, Trust in automation: integrating empirical evidence on factors that influence trust, Hum. Factors 57 (2015) 407–434.

M. Jeon, Chapter 1-Emotions and affect in human factors and human-computer interaction: taxonomy, theories, approaches, and methods, in: M. Jeon (Ed.), Emotions and Affect in Human Factors and Human-Computer Interaction, Academic Press, San Diego, CA, USA, 2017, pp. 3–26.

M. Jeon, B.N. Walker, J.-B. Yim, Effects of specific emotions on subjective judgment, driving performance, and perceived workload, Transp. Res. Part. F: Traffic Psychol. Behav. 24 (2014) 197–209.

D. Kontogiorgos, A. Pereira, O. Andersson, M. Andersson, E.G. Rabal, V. Vartiainen, et al., The effects of anthropomorphism and non-verbal social behaviour in virtual assistants. In Proc. 19th ACM Int. Conf. Intelli. Virt. Agent., Paris, France, 2–5 July 2019; Association for Computing Machinery: Paris, France, 2019; pp. 133–140.

D.R. Large, K. Harrington, G. Burnett, J. Luton, P. Thomas, P. Bennett, To please in a pod: employing an anthropomorphic agent-interlocutor to enhance trust and user experience in an autonomous, self-driving vehicle. In Proc. 11th Int. Conf. Autom. User Interf. Inter. Veh. Appl., Utrecht, Association for Computing Machinery: Utrecht, The Netherlands, 21–25 September 2019, pp. 49–59.

J.D. Lee, K.A. See, Trust in automation: designing for appropriate reliance, Hum. Factors 46 (2004) 50.

J.R. Lewis, Psychometric evaluation of an after-scenario questionnaire for computer usability studies: The ASQ, SIGCHI Bull. 23 (1991) 78–81.

J. Maddox, Visual-manual nhtsa driver distraction guidelines for in-vehicle electronic devices, Federal Register, Washington, DC, USA, 2012.

J.E. Mercado, M.A. Rupp, J.Y.C. Chen, M.J. Barnes, D. Barber, K. Procci, Intelligent agent transparency in human-agent teaming for multi-UxV Management, Hum. Factors 58 (2016) 83–93.

B.M. Muir, N. Moray, Trust in automation. Part I. Theoretical Issues in trust and human intervention in automated systems, Ergonomics 37 (1996) 1905–1922.

M.A.V.J. Muthugala, A.G.B.P. Jayasekara, A review of service robots coping with uncertain information in natural language instructions, IEEE Access. 6 (2018) 12913–12928.

C. Nass, I.-M. Jonsson, H. Harris, B. Reaves, J. Endo, S. Brave, et al. Improving automotive safety by pairing driver emotion and car voice emotion. In CHI '05 Extended Abstracts on Human Factors in Computing Systems; Association for Computing Machinery: New York, NY, USA, 2005, pp. 1973–1976.

S. Ososky, D. Schuster, E. Phillips, F.G. Jentsch. Building appropriate trust in human-robot teams. In 2013 AAAI Spring Symposium Series; AAAI: Menlo Park, CA, USA, 2013; pp. 60–65.

R. Parasuraman, P.A. Hancock, O. Olofinboba, Alarm effectiveness in driver-centred collision-warning systems, Ergonomics 40 (1997) 390–399.

R. Parasuraman, V. Riley, Humans and automation: use, misuse, disuse, abuse, Hum. Factors 39 (1997) 230–253.

A. Pauzie, A method to assess the driver mental workload: the driving activity load index (DALI). Intell, Transp. Syst. IET 2 (2009) 315–322.

C. Peter, B. Urban, Emotion in human-computer interaction, in: J. Dill, R. Earnshaw, D. Kasik, J. Vince, P.C. Wong (Eds.), Expanding the Frontiers of Visual Analytics and Visualization, Springer, London, UK, 2012, pp. 239–262.

J.K. Rempel, J.G. Holmes, M.D. Zanna, Trust in close relationships, J. Personal. Soc. Psychol. 49 (1985) 95–112.

T.L. Sanders, T. Wixon, K.E. Schafer, J.Y.C. Chen, P.A. Hancock, The influence of modality and transparency on trust in human-robot interaction. In Proc. IEEE Int. Int-Disc. Conf. CogSIMA, San Antonio, TX, USA, 3–6 March 2014.

A. Selkowitz, S.G. Lakhmani, C.N. Larios, J.Y.C. Chen. Agent transparency and the autonomous squad member. In Proc. Human Fact. Ergono. Soc. Ann. Meet., SAGE: Los Angeles, CA, USA, 2016, Volume 60, pp. 1319–1323.

B. Shneiderman, Designing the user interface strategies for effective human-computer interaction, SIGBIO Newsl. 9 (1987) 6.

A.W. Stedmon, S. Sharples, R. Littlewood, G. Cox, H. Patel, J.R. Wilson, Datalink in air traffic management: Human factors issues in communications, Appl. Ergon. 38 (2007) 473–480.

A. Waytz, J. Heafner, N. Epley, The mind in the machine: anthropomorphism increases trust in an autonomous vehicle, J. Exp. Soc. Psychol. 52 (2014) 113–117.

C.D. Wickens, Multiple resources and performance prediction, Theor. Issues Ergon. Sci. 3 (2002) 159–177.

J. Zlotowski, D. Proudfoot, C. Bartneck. More human than human: does the uncanny curve really matter? In Proc. HRI2013 Workshop on Design of Human Likeness in HRI from Uncanny Valley to Minimal Design, Tokyo, Japan, 3 March 2013; pp. 7–13.

CHAPTER 20

Design of proactive interaction of in-vehicle robots based transparency

Jianmin Wang[1], Tianyang Yue[1], Yujia Liu[1], Yuxi Wang[1], Chengji Wang[1], Fei Yan[2] and Fang You[1]

[1]Car Interaction Design Lab, College of Arts and Media, Tongji University, Shanghai, P.R. China [2]Department of Human Factors, Institute of Psychology and Education; Faculty of Engineering, Computer Science and Psychology, Ulm University, Germany

Chapter Outline

20.1 Introduction 281
20.2 Related works 282
    20.2.1 Proactive interaction 283
    20.2.2 Transparency 283
20.3 Transparency design of proactive interaction 284
    20.3.1 Human-robot interface levels 285
    20.3.2 Transparency design assumptions 285
20.4 Experiment 287
    20.4.1 Participants 287
    20.4.2 Design of experiment 287
    20.4.3 Measurement 289
    20.4.4 Apparatus and materials 289
    20.4.5 Procedure 291
20.5 Results 292
    20.5.1 Telephone task results 292
    20.5.2 Speeding task results 295
20.6 Discussion 297
20.7 Conclusions 299
References 299
Further reading 299

20.1 Introduction

With the rapid development of intelligent vehicles, drivers' requirements for more intelligent assistance from the cockpit have increased. More vehicles are equipped with virtual image voice assistants or vehicle robots with physical entities. The interaction between humans and in-vehicle robots is considered an integration of a complex social and technical system [1], which needs an advanced model to improve safety and trust in autonomous vehicles [2].

Anthropomorphism and proactivity have been widely studied for the future in-vehicle robots. A study by Waytz et al. shows that a more anthropomorphic cockpit can increase human trust and is perceived to have more human-like mental abilities. It also shows that an anthropomorphic robot's voice response can increase trust, pleasure, and dominance of the situation compared with a mechanical voice response. However, there are still concerns about communication barriers for such robots. The accuracy and validity of the output produced by intelligent systems can be problematic because it is difficult for the operator to interpret the output. Part of the reason is that humans have limits to understanding the proactivity of robots.

An important condition for a robot to be able to interact fluently with humans is that the two can share a common cognitive framework and form a coherent mental expectation during the interaction without adding additional learning costs. Therefore to promote a common mental model between human operators and automated systems, in-vehicle robots should be designed according to human cognition to improve communication efficiency and trust. Human cognitive architectures have also been increasingly applied to the intelligent architecture of robots in recent years. Chen and colleagues proposed the situation awareness (SA)-based agent transparency theory to help establish a common mental model between humans and machines, which further generates assistance to the human decision-making process and enhances task performance.

Using cognitive theory to design in-vehicle robots to provide appropriate expression to humans has not been widely studied. This paper applies SA-based agent transparency theory to design in-vehicle robots. The objective of this paper is to explore the appropriate amount of transparency information that can assist human decision-making by conducting experiments. Based on the circumstance that an inappropriate amount of information while driving, especially in critical situations, can affect safety, usability, workload, and trust, the following hypotheses are formulated.

1. The expressed in-vehicle robot's information to drivers needs to be selective.
2. The appropriate amount of transparency information for different driving situations needs to be determined.

To examine these hypotheses, in-vehicle robot transparency design has been developed at three levels: perception, comprehension, and projection based on situation awareness theory (SAT), including channels of voice and visual in the design. Subsequently, experimental evaluation of the transparency design hypothesis has been carried out in selected driving scenarios. After analyzing the experimental results, conclusions of transparency design of in-vehicle robot proactive interaction in different scenarios have been discussed.

20.2 Related works

As human-robot interaction (HRI) becomes more complex, many guidelines and design criteria have been developed depending on the specific applied scenarios. A shared mental model can lead to higher levels of performance; hence, effective team member communication needs to be considered in the design process of proactive robot interaction. Transparency plays an important role in building an understanding of HRI and can contribute to the interaction design of in-vehicle robots.

20.2.1 Proactive interaction

With intelligent systems becoming more complex, researchers have begun focusing on proactivity. Keith Cheverst et al. [3] have argued that a proactive system should be representative of the user and able to initiate behavior without user commands. In proactive interaction, human operators can achieve more supervisory control than active control. However, the change in control mode does not mean that a human's job becomes easier. Highly proactive robots also have a greater influence on human decision-making. More cognitive factors need to be considered in the design. The appropriateness of the quality and quantity of information transfer have to be considered.

Multimodal interaction is also widely used in proactive interaction scenarios. For an anthropomorphic robot, dual-channel interaction between speech and vision is crucial and can significantly affect the mutual understanding between humans and robots. In the context of Industry 4.0, the development of multimodality in the driving environment is also reflected in the research of many different devices, including virtual reality, augmented reality, and robotics. Experiments by Williams et al. also demonstrated that multimodal interaction can be more effective than separate channel interaction in reducing driving workload and distractions as well as in enhancing emotional experience. The importance of multimodal interaction for proactive interaction has also been shown in previous studies.

The core of contemporary research on proactive interaction focuses on the technology domain. Most of them are based on the decision-making of artificial intelligence models to enhance SA, consciousness perception, and emotion perception. Specifically, in the in-vehicle scenario, proactive interaction can help users collect and process information in the environment, therefore reducing the user's information processing burden. A further study proposed human-autonomous team cooperation based on robot initiative to monitor and receive feedback from each other.

Proactive interaction is of high importance due to the development of intelligent interfaces. However, the current work on proactive interaction is focused on the related HRI system in the technical field, which is dedicated to finding solutions from the perspective of artificial intelligence. The explorations of proactive interaction design from the perspective of cognitive theory are sparse.

20.2.2 Transparency

Understanding the reasoning process behind the output of an intelligent system in a dynamic environment is of great importance. Van Dongen et al. found that the perceptibility process of participants to the reasoning process of a decision aid system has a significant impact on their reliance on its recommendations.

In the context of automation, an understanding of the behavior of technical agents is important to ensure good interactions between human and technical agents. This understanding is often referred to as "transparency." Lyons argues that transparency can facilitate optimal calibration between humans and autonomous systems. The design of the appropriate amount of information on different display devices in the driving environment has also been explored.

In terms of information content, clearer and more accurate information delivery can enhance human trust. It has been mentioned that the information provided by the proactive party, the machine, should be highly transparent to the user to allow the user to identify and understand it quickly. Russell et al. mentioned that the characteristics of autonomy that intelligent agents should possess include observation of the environment, action on the environment, and guidance of the activity. Lee suggested that to increase the transparency of automation to the operator, system designers should make the 3Ps (purpose, process, and performance) of the system and its history visible to the operator.

Endsley's SAT proposes three levels, including SA Level 1, perception of elements in the environment, SA Level 2, comprehension of these elements, and SA Level 3, projection of their state in the near future. Based on SAT, Chen et al. proposed the SA-based agent transparency theory to explain what information contributes to transparency. The SAT model argues that as the agent is involved in the execution of the human task, the human needs to be situationally aware of the agent and the environment, which can be achieved through the agent's transparency. SA-based agent transparency theory defines agent transparency as a descriptive quality of an interface where the operator understands the intention, reasoning process, and future plans of an intelligent agent.

The SAT model also consists of three levels. At SAT Level 1 (perception level), the operator should be provided with the agent's goals and perception of the environmental situation. At SAT Level 2 (comprehension level), the operator should be provided with the agent's understanding of the situation and the reasoning process of the action. At SAT Level 3 (projection level), the operator should be provided with the agent's projection of the future outcome. All three levels of the SAT model describe the information that the agent needs to convey to maintain transparent interactions with humans. The operator is, therefore, able to understand the agent's intentions, reasoning process, and predicted outcomes, which leads to better information sharing and a common mental model of communication.

The presence or absence of each level of information in the SAT determines whether the user is able to understand the perception, comprehension, and projection of the current information expressed by robots. The design needs to consider that each level of information can be combined to obtain an appropriate level of transparency that promotes mutual human-robot understanding without increasing the workload. Conducting phase analysis through an SAT model is a way of contributing to transparency between humans and robots. The usability of the intelligent system, the emotional experience of humans, and the level of trust can be measured. Related studies have also applied the SAT model for interface design to conduct explorations, including robot action interfaces and UxV action interfaces.

20.3 Transparency design of proactive interaction

According to SAT, the human cognitive process includes three levels: perception, comprehension, and projection. In the HRI process, the necessity of transparency in the three levels is reflected in the humans and robots. At the same time, anthropomorphism and proactivity require robots to present multiple channels such as voice and visual.

In the design of in-vehicle HRI, it is necessary to consider the appropriateness of the amount of information in different channels and different SAT levels. The impact on several aspects, such as driving safety, usability, and emotion, needs to be examined. The research framework is shown in Fig. 20.1. We apply SAT theory to analyze which SAT level the information is in. Combined with human-robot interface levels, the transparency design assumptions for the in-vehicle robot were carried out based on the proactive interaction scenario of the in-vehicle robot. The design assumptions are then evaluated through experiments, using a driving simulator.

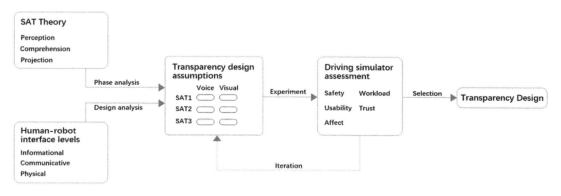

FIGURE 20.1

Research framework.

20.3.1 Human-robot interface levels

To promote appropriate transparency, Lyons argues that an opportunity to foster transparency between the human and the robot is from the human-robot interface. The human-robot interface includes three levels: informational, communicative, and physical. Each level covers a portion of what transparency design needs to consider. The levels in SAT are distinct from the levels of the human-robot interface. The SAT levels focus on the interaction process, where the human-computer interaction process can be well divided into three stages. Moreover, human-robot interface levels function as guidance on the specific design to determine the existence and amount of specific information.

For specific designs, interface features at the informational level need to be considered to avoid too much information or nonintuitive displays, which may confuse and frustrate the user of the robotic system. Interface features at the communication level need to be considered to avoid the robot's inappropriate responses, which affect user trust and performance. Interface features at the physical level may include the robot's emotional expression, effectively describing the robot's emotional state.

In transparency design assumptions, the amount of information and information intuitiveness should be considered at the informational level. At the communication level, communication smoothness and response timeliness should be considered. At the physical level, emotional expressions should be considered.

20.3.2 Transparency design assumptions

Human-robot transparency design requires identifying certain interaction timing. The identification enables the design of in-vehicle robots to present the driver with the appropriate level of transparency to facilitate the driver's understanding. The in-vehicle robots give appropriate information that conforms with the driver's cognition, thus resulting in transparency of the entire interaction process.

We have explored the possible design combinations and excluded the designs that were clearly unreasonable for each interaction timing in a given scenario. For example, considering communication smoothness and response timeliness, the information can be either too much or too little. Then, the design pattern for each transparency level emerged, forming the transparency design assumption.

Transparency was first analyzed from the perspective of information intuitiveness. From the design exploration we conducted before the experiment, we found that when users were given information at the comprehension and perception levels, they could accurately identify the existence of the projection level. However, when the information at the projection level existed alone, users defaulted to the existence of perception and comprehension levels of information. It showed that participants were unable to cognitively recognize low-level transparency information in the presence of high-level transparency information. When analyzed at the physical level, emotional expression providing information at the perception and comprehension levels may affect participants' emotions. Therefore perception and comprehension levels of information should still be taken into account in the transparency design assumptions.

In the proactive interaction condition of the in-vehicle robot, it has more information that needs to be shared with people, and therefore, it requires a higher level of transparency of expression. At the perception level, considering emotional expression from the physical level, voice cues can increase the robot's anthropomorphism and enhance human emotions in the corresponding scenarios. At the comprehension level, considering the informational level aspect, the robot needs to convey the information it perceives to people, and how it comprehends the information needs to be transparent in the voice channel. At the perception and comprehension levels, information in the visual channel, information intuitiveness, and task fluency are considered, and at the comprehension level, information is chosen to be retained. At the projection level, considering information intuitiveness, the voice channel gives easier understanding compared to the visual channel, so the projection information is designed to be expressed through the voice channel. Therefore the design of the projection level is assumed to include voice channel information. It is expected that the visual channel projection level information can play an auxiliary role to the voice channel information to strengthen the robot's expressiveness. Incorporating communication smoothness into the comprehensive consideration, the presence or absence of visual channel SAT Level 3 information can be examined. The design assumptions are listed in Table 20.1, where the question mark represents its inability to analyze the necessity of its existence from the design perspective and the need for experimental verification.

Table 20.1 Transparency design assumptions in proactive interaction condition of the in-vehicle robot.

	Voice	Visual
SAT1	?	/
SAT2	√	√
SAT3	√	?

Notes: The symbol "√" means the information is needed; the symbol "?" means it cannot be determined whether the information is needed and needs proving; the symbol "/" means the information is not needed.

To summarize the above research approach, the transparency design assumes that enhanced understanding between humans and robots leads to more efficient communication and interaction. The kind of design pattern that can reach the above goal needs to be determined. Our research approach was (1) to use the SAT model to analyze the stages, splitting the entire HRI process into three stages. (2) After the stages were analyzed, we adopted the human-robot interface levels of informational, communicative, and physical guidance. (3) We then further expanded human-robot interface levels into the amount of information, information intuitiveness, communication smoothness, response timeliness, and emotional expression for each stage to analyze the information under voice and visual channels and SAT levels.

20.4 Experiment

20.4.1 Participants

For the experiment, 30 participants (25 males and five females), aged between 22 and 40 years ($M = 28.2$, $SD = 5.83$), were selected. All participants had significant driving experience, with 16 driving 2–3 days per week and 14 driving 4 days per week or more. Participants were recruited through an online screening process, with driving experience ranging from 1 to 10 years ($M = 5.8$, $SD = 2.9$). Among them, 21 had never used an in-vehicle robot before, nine had experienced it once before, and three were not existing in-vehicle robot users. Therefore the participants were all regarded as novice in-vehicle robot users. The independent variable of the experiment was the degree of transparency, and between-group experiments were designed. The age and gender distribution of participants were adjusted according to the number of experimental groups in the task so that the demographic attributes of the participants were as balanced as possible.

20.4.2 Design of experiment

The experiment focuses on the proactive interaction scenarios of in-vehicle robots. After conducting real car research and interviews, we collected the scenarios that were used frequently and were more representative. The final experimental tasks for proactive interaction with the in-vehicle robot were identified as telephone and speeding. The telephone task required the driver and the in-vehicle robot to complete a nondriving task together. The speeding task required the driver and the in-vehicle robot to complete a critical task together. The experiment used a between-subjects design. The aim of the experiment was to explore the relationship between the change in transparency levels, the information quantity inside each level, and the driving behavior. Two tasks with different scenarios also enable a comparative exploration of the optimal information quantity design for different scenarios. In the two tasks, the SAT Level 1 was a perception of speeding or incoming calls. The SAT Level 2 was a comprehension of the situation mentioned above. The SAT Level 3 was a projection of the driver's nearest action, expressed as a suggestion or question.

In the telephone task, the experimenter simulated a phone call from the participant's friend Wang. Then, the robot took the initiative to alert and ask the person whether to answer it. The specific experimental group design is as follows:

1. Group 1 contained SAT Level 2 and SAT Level 3 messages. The robot told the driver, "Wang is calling you," with a pleasant expression. Then, the robot asked the driver, "Would you like to answer?" with a phoning expression.
2. Group 2 contained SAT Level 1, SAT Level 2, and SAT Level 3 messages. SAT Level 1 voice channel message "Ring" was added.

The setup for the telephone task is shown in Fig. 20.2. In total, 12 males and 3 females were assigned to experimental group 1; 13 males and 2 females were assigned to experimental group 2.

In the speeding task, the participant was asked to drive in the left lane at a speed of 30 km/hour and then accelerate to 70 km/hour. Once the speed was above 60 km/hour, the robot took the initiative to remind the driver of the speed limit. The specific experimental group design is as follows:

1. Group 1 contained SAT Level 2 and SAT Level 3 messages. The robot told the driver, "The speed limit ahead is 60 km/hour, you have exceeded the speed limit," with a fearful expression. Then, the robot suggested to the driver, "Slow down please."
2. Group 2 also contained SAT Level 2 and SAT Level 3 messages. Based on having all the information in experimental group 1, a speed limit expression, containing an SAT Level 3 visual channel message was added.
3. Group 3 contained SAT Level 1, SAT Level 2, and SAT Level 3 messages. Based on having all the information in experience group 1, an SAT Level 1 voice channel message "Oops," which expressed robot perception to the driver, was added.

The experimental group set up for the speeding task is shown in Fig. 20.3. In total, eight males and two females were assigned to experimental group 1 and experimental group 2; nine males and one female were assigned to experimental group 3.

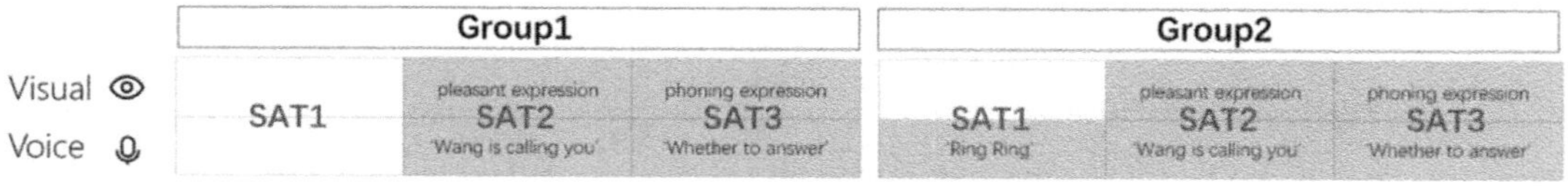

FIGURE 20.2

Telephone task experimental group setup. The gray choices represent information that appeared at a certain level in the experimental group, and the white choices represent that there is no information at that level.

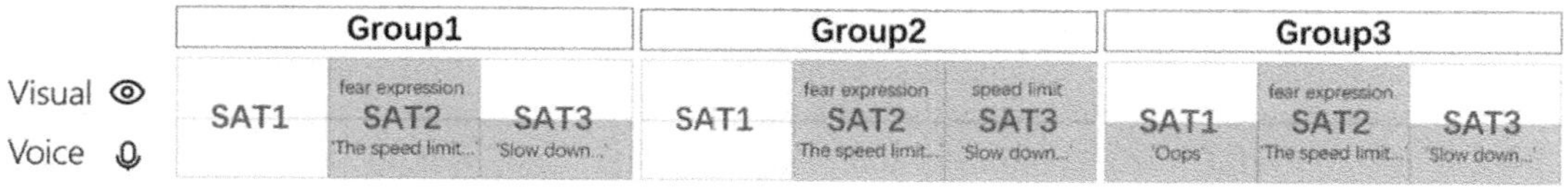

FIGURE 20.3

Speeding task experimental group setup. The gray choices represent information that appeared at a certain level in the experimental group, and the white choices represent that there is no information at that level.

20.4.3 Measurement

The behavior of the in-vehicle robots will draw the attention of the drivers, even if robots perform decisions that conform to the drivers' cognition and give them more helpful information. Appropriate design of HRI strategies based on human cognitive factors can help compensate for human limitations and achieve safety. Therefore it is necessary to examine the multichannel transparency design assumptions of in-vehicle robots to ensure driving safety, improve task execution efficiency, and enhance drivers' trust in the robot.

Harbluk et al. showed that the driver's visual behavior and vehicle control changed when performing tasks with different cognitive requirements. The driver's visual behavior data and vehicle data were collected while participants were performing tasks. The visual behavior data included the total saccade time and total fixation time (seconds), which were extracted from recorded videos. The vehicle data included vehicle speed (km/hour) and driveway offset (dm), which measured participants' driving control in the vertical and horizontal directions, respectively. In terms of subjective data, posttask questionnaires were used to make participants score subjectively on the usability, trust, workload, and affective dimensions using a Likert scale. Usability was measured using the after-scenario questionnaire, which combined the three dimensions: ease of task completion, the time required to complete tasks, and satisfaction with support information to produce an evaluation. The trust scores in the study were based on the model of trust in vehicle automation proposed by Muir, using a posttask trust scale with a comprehensive analysis of the three trust dimensions: predictability, dependability, and faith. It was found that human reliance on automation was influenced by workload. We used the driving activity load index (DALI) scale, which is a scale that concerns multichannel information and includes the effort of attention, visual demand, auditory demand, temporal demand, interference, and situational stress. The DALI scale is more suitable for dynamic driving conditions. The SAM scale designed by Bradley et al. was used to measure the emotional state of the person in terms of pleasure, arousal, and dominance. In the comparison of the analysis of the results between specific experimental groups, priority was given to safety, followed by usability, trust, and workload ratings. Emotional ratings were also taken into account as a secondary evaluation.

20.4.4 Apparatus and materials

The experimental environment was built based on the driving simulator system independently developed by the Car Interaction Design Lab of Tongji University. The simulator was used as the main equipment of the experiment. The scene was developed using Unity software to simulate the real driving environment. The scene used in the experiment was a two-lane straight road with a large number of oncoming cars in the opposite lane. The simulator was equipped with monitoring equipment that automatically collected vehicle data during each simulation. The robot was fixed in a suitable position (Fig. 20.4). The location of the robot was determined by three factors: First, previous work by Williams and others shows that in-vehicle robots were fixed in a position above the center screen. Second, we conducted a real vehicle study on Nomi, the in-vehicle robot of Nio, measuring the relative position of the in-vehicle robot to the center of the steering wheel in three-dimensional space. Third, we placed the robot on our simulator and conducted a small test on the prefixed robot to adjust its position to make it closer to the real driving environment (Fig. 20.5).

FIGURE 20.4

Robot for experiment.

FIGURE 20.5

Experiment environment.

The vehicle robot had three degrees of freedom and was controlled by servos that can raise and lower its head and rotate toward the driver. The robot's face screen displayed features and colorful auxiliary graphics to convey expressions, which were in-depth explored and designed in our previous research. The expression designs are shown in Fig. 20.6, in which the color and brush strokes were adjusted due to the confidentiality of company cooperation. The in-vehicle robot was accompanied by an interactive simulation program to control the robot's movements and expressions. The interactive simulation program can record the robot's behavioral data. Since the program requires the control of the experimenter, factors such as the experimenter's reaction time may pose a threat

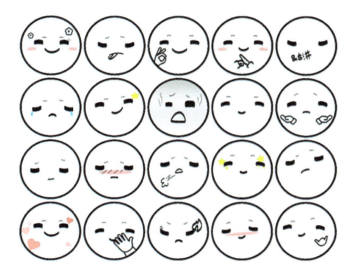

FIGURE 20.6

Expression design set used on robot's face screen.

to the validity of the experiment. The participant's basic information form and scale were used to collect subjective data. Two cameras were used to record the visual behavior of the participants. After the videos were recorded, the user's saccades of the robot were manually checked frame by frame. The number of saccades and fixation time were recorded with a minimum frequency unit of 0.042 seconds (1/24 seconds/frame). Similar simulator construction and data acquisition methods have also appeared in other contemporary studies.

20.4.5 Procedure

The experimental consent forms were signed before the experiment, and the participant was registered for basic information, including driving frequency and in-vehicle robot understanding and experience. Before conducting the experiment, the participant was introduced to the purpose of the experiment and the main tasks. The participant started by driving in the simulator for about five minutes to learn and become familiar with the simulator. Then, the participant interacted with the in-vehicle robot in a simulation. After the participant was familiar with the simulator and the in-vehicle robot, the experiment started, and the experimenter began data recordings.

The tester described the driving task requirements and subtask requirements to the participant. After the participant confirmed, the tester issued the command to start the task. The participant completed the telephone task first and the speeding task second. In each task, participants were assigned to one of the groups. In the telephone task, the participant was asked to keep driving in the left lane at a speed of 30 km/hour. After the participant kept driving stable, the experimenter simulated a phone call from the participant's friend Wang. Then, the robot took the initiative to alert and ask the person whether to answer it. In the speeding task, the participant was asked to drive in the left lane at a speed of 30 km/hour and then accelerate to 70 km/hour. Once the speed reached 70 km/hour, the robot took the initiative to remind the speed limit. After each task was completed, the participant completed the subjective questionnaires and scales. Then, the tester conducted the interview regarding the task. Each task lasted approximately 5–10 minutes. After completing the tasks, participants were interviewed by the tester regarding their overall experience of the in-vehicle robot.

20.5 Results

t-Tests were conducted in the telephone task and one-way analysis of variance (ANOVA) with posthoc pairwise comparison was conducted in the speeding task. A summary of the experimental results in separate tasks can be seen in the following sections.

20.5.1 Telephone task results

As shown in Fig. 20.7, the mean value of the standard deviation of driveway offset for experimental group 1 was 0.18, and the mean value of the standard deviation of driveway offset for experimental group 2 was 0.77. There was a significant difference between the two experimental groups ($p = .00 < .01$).

As shown in Fig. 20.8, the mean saccade time was 3 seconds for experimental group 1 and 5 seconds for experimental group 2. There was a significant difference between the two experimental groups ($p = .02 < .05$). The mean fixation time was 1.671 seconds for experimental group 1 and 3.893 seconds for experimental group 2. There was a highly significant difference between the two experimental groups ($p = .00 < .01$). Participants in experimental group 1 had significantly fewer saccade time and fixation time than participants in experimental group 2.

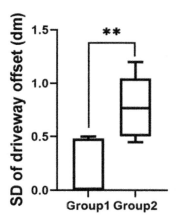

FIGURE 20.7

Standard deviation of driveway offset in telephone task. **Indicates a significant difference ($p < .01$).

20.5 Results   293

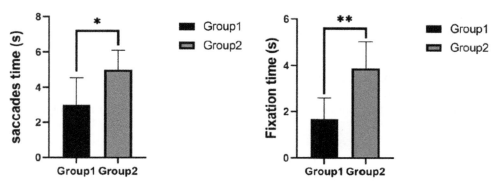

FIGURE 20.8

Saccade time and fixation time in telephone task. *Indicates a significant difference ($p < .05$). **Indicates a significant difference ($p < .01$).

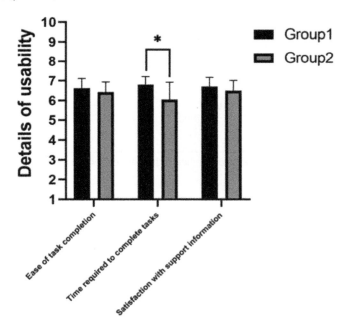

FIGURE 20.9

Usability in telephone task. *indicates a significant difference ($p < .05$).

As shown in Fig. 20.9, it can be concluded from the usability scores that the experimental group 2 participants rated the usability of the robot lower ($p = .049$). In terms of details, the time required to complete tasks was significantly different between experimental groups ($p = .01$), while ease of task completion ($p = .374$) and satisfaction with support information ($p = .262$) were not significantly different.

As shown in Fig. 20.10, it was concluded from the PAD scores that experimental group 2 enabled participants to obtain more positive emotions. The participants were more pleasant ($p = .001$) and aroused ($p = .001$) than experimental group 1. There was no significant difference in dominance ($p = .202$).

As shown in Fig. 20.11, it can be concluded from the workload scores that in experimental group 2, participants had a lower workload than in experimental group 1 ($p = .002$). Judging from the DALI detail scores, the main reasons were lower interference ($p = .003$) and lower situational stress ($p = .002$).

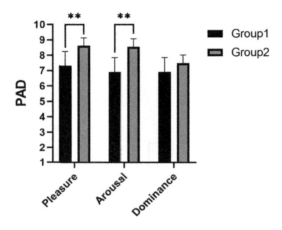

FIGURE 20.10

Emotion score of telephone task. **Indicates a significant difference ($p < .01$).

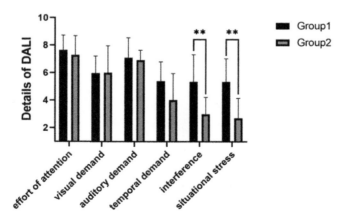

FIGURE 20.11

Workload in telephone task. **Indicates a significant difference ($p < .01$).

20.5.2 Speeding task results

As shown in Fig. 20.12, the mean value of the standard deviation of driveway offset was 0.95 for experimental group 1, 1.54 for experimental group 2, and 1.55 for experimental group 3. The results of the one-way ANOVA performed on the data from the three groups indicated that there was a significant difference in the standard deviation of driveway offset among the three experimental groups ($F(2, 32) = 6.906$, $p = .00 < .01$); posthoc tests revealed that the standard deviation of driveway offset was significantly lower in experimental group 1 ($M = 0.97$, $SD = 0.51$) than in experimental group 2 ($M = 1.60$, $SD = 0.58$) and experimental group 3 ($M = 1.55$, $SD = 0.28$), with no significant differences between experimental group 2 and group 3.

As shown in Fig. 20.13, the mean saccade time was 3.7 seconds for experimental group 1, 1.3 seconds for experimental group 2, and 1.4 seconds for experimental group 3. One-way ANOVA results of the data from the three groups showed that there was a significant difference in the saccade time in the three experimental groups, ($F(2, 18) = 9.479$, $p = .00 < .01$); posthoc tests revealed that the saccade time in experimental group 1 ($M = 3.67$, $SD = 1.73$) was significantly higher than that in experimental group 2 ($M = 1.25$, $SD = 0.50$) and experimental group 3 ($M = 1.38$, $SD = 0.52$), with no significant difference between experimental group 2 and experimental group 3. The mean fixation time of experimental group 1 was 2.389 seconds, the mean fixation time of experimental group 2 was 0.540 seconds, and the mean fixation time of experimental group 3 was 0.834 seconds. A one-way ANOVA of the data from the three groups showed that there was a significant difference in the fixation time of the three experimental groups, ($F(2, 20) = 7.245$, $p = .00 < .01$); posthoc tests revealed that the fixation time of experimental group 1 ($M = 2.39$, $SD = 1.41$) was significantly higher than that of experimental group 2 ($M = 0.54$, $SD = 0.30$) and experimental group 3 ($M = 0.83$, $SD = 0.50$), and there was no significant difference between experimental group 2 and experimental group 3.

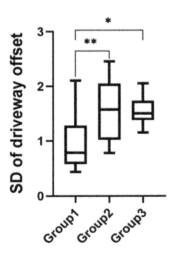

FIGURE 20.12

Standard deviation of driveway offset in speeding task. *Indicates a significant difference ($p < .05$). **indicates a significant difference ($p < .01$).

296 Chapter 20 Design of proactive interaction

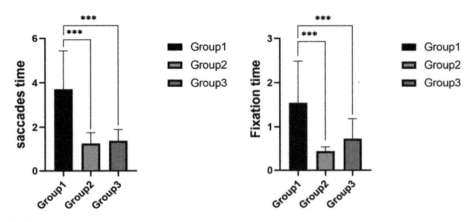

FIGURE 20.13

Saccade time and fixation time in speeding task. ***Indicates a significant difference ($p < .001$).

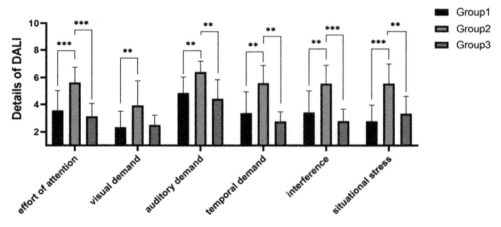

FIGURE 20.14

Workload in speeding task. **Indicates a significant difference ($p < .01$). ***Indicates a significant difference ($p < .001$).

The results of the behavior data showed that the participants in experimental group 1 were significantly higher than those in experimental group 2 and experimental group 3 in terms of saccade time and fixation time.

As shown in Fig. 20.14, significant differences were shown between experimental groups in terms of workload score means ($F\ (2,\ 55) = 10.408$, $p = .000 < .001$) and also in terms of each detailed dimension. Posthoc test analysis yielded that the workload of experimental group 2 was significantly higher than that of experimental group 1 ($p = .000 < .001$) and experimental group 3 ($p = .002 < .01$), while experimental group 1 and experimental group 3 did not show significant differences in both workload score means and workload score details.

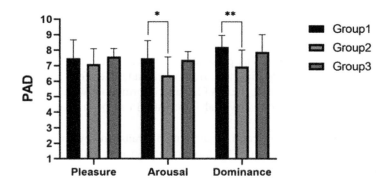

FIGURE 20.15

Emotion score in speeding task. *Indicates a significant difference ($p < .05$). **Indicates a significant difference ($p < .01$).

As shown in Fig. 20.15, it can be concluded from the PAD scores that there were significant differences among the three experimental groups in the arousal ($F\,(2, 38) = 3.430$, $p = .043 < .05$) and dominance ($F\,(2, 42) = 5.945$, $p = .005 < .01$) dimensions. Posthoc test analysis yielded significantly higher arousal ($p = .036 < .05$) and significantly higher dominance ($p = .004 < .01$) in experimental group 1 than in experimental group 2.

20.6 Discussion

In the telephone task, the design with SAT perception Level 1, the voice channel information (hereafter referred to as the ringing group) in group 2, enabled participants to obtain more positive emotions and reduced the workload of the participants. This was probably because the added information enhanced the anthropomorphism of the robot. However, participants had significantly worse horizontal control of the vehicle, and saccade time and fixation time increased significantly. This suggests that the participants were overly distracted by the robot and had reduced concentration on the driving task, which may cause danger. As for the subjective data, participants rated the usability of the ringing group lower, and significant differences were found mainly in the aspect of time required to complete tasks. This suggests that task disfluency due to lengthy speech triggered participants' lower usability ratings of the robot. In addition, the increased information at the expense of time spent did not improve usability ratings in terms of ease of task completion or satisfaction with support information. Overall, the existence of SAT1 perception level provides higher transparency. However, it does not suggest better cooperation between humans and robots.

In the speeding task, the design with the SAT2 comprehension level presents voice and visual channel information, while the SAT3 projection level shows voice channel information has better information transparency. Regarding safety, group 1 had the best data performance in terms of vehicle horizontal control. Adding the SAT3 projection level visual channel information or SAT1 level voice channel information would make the driver's horizontal control of the vehicle worse.

From the subjective data, having SAT3 level visual channel information resulted in a significantly higher workload and lower arousal and dominance. The speed limit expression at the projection level caused participants' greater workload and did not help in driving, indicating that it was an unnecessary distraction for the driver. The presence or absence of the SAT1 level with voice channel information ("oops") did not significantly affect workload. Therefore in the speeding task, the design of the in-vehicle robot can omit the SAT1 stage information, which is not very helpful for the task, and the complicated SAT3 level visual information can be similarly excluded to increase safety and reduce the workload.

In summary, we analyzed the appropriate amount of transparency information for in-vehicle robots in a proactive interaction scenario. The experimental groups with SAT1 performed poorly in both scenarios. To improve the effectiveness of message delivery and help drivers concentrate, designers should carefully consider increasing voice channel information at the SAT1 perception level. The design that performs better in telephone tasks has SAT2 comprehension and SAT3 projection levels of both voice and visual channel information, which is different from the better design for speeding tasks. In the speeding task, the SAT3 projection level and visual information make the workload of participants significantly higher, and driving safety is compromised. Analyzing the task characteristics, we can conclude that the telephone task belongs to function, while the speeding task belongs to critical scenarios. The driver's concentration level on driving tasks in the two scenarios is different; therefore the driver needs different robot transparency.

The advantage of higher transparency for the in-vehicle robot is that it can help drivers judge the situation, reduce drivers' need to reprocess the information provided by the robot and increase the drivers' decision-making confidence. However, higher transparency also has its own disadvantages. In the case of insufficient attention and mental resources, too much information is not only not fully received by the human but also affects the driver's driving performance, especially when implicit information such as robot expressions that require extra mental resources from the driver appears. These results suggest that the "highly transparent" assumption made by previous work needs careful consideration. In critical scenarios, it is recommended to design more direct voice information without adding extra workload. In noncritical driving scenarios, designers can consider giving robots a more transparent response to drivers while maintaining task fluency.

The experiments were based on transparency design assumptions, and they yielded preliminary results for the exploration of suitable transparency. We are able to draw enlightening conclusions from different measurements, and the purposes of the experiments are essentially achieved. However, due to the experimental limitations, it was not possible to exhaust the possible design assumptions, and there may be more appropriate transparency designs for specific scenarios.

The relationship between saccade time and driving safety was also noted in both scenarios. In the no-critical scenario, an increase in saccade time and fixation time on the robot undermines driving safety. In the critical scenario, on the contrary, an increase in the saccade time and fixation time of the robot improves driving control. The reason for this result derives from the difference in the type of tasks that the driver and the in-vehicle robot achieve together in the two scenarios. Taking workload into comprehensive analysis, in the case of joint driving tasks, such as the speeding task, the robot's information alerts can share the workload for the driver and obtain better driving performance. However, when the robot is completing nondriving tasks, such as picking up the phone, the increase in saccade time and fixation time indicates that the driver is more involved in nondriving tasks and does not concentrate on driving tasks, so the driver tends to have poor control of the vehicle. Further analysis revealed a tendency for reduced self-reported workload when drivers were more involved in nondriving tasks. This may be due to the higher workload of the driving task compared with the nondriving task, which is also able to corroborate with the multiresource prediction of dual-task interference in the multiresource theory proposed by Wickens.

20.7 Conclusions

In this paper, we migrate SAT theory into an in-vehicle robot design method and conduct a proactive interaction design of in-vehicle robots based on transparency design assumptions. The results of the experiment show that the driver's driving control, behavior, and self-reporting results produce differences under different in-vehicle robot transparency. Overall, anthropomorphic information on the perception level, though it brings a better emotional experience to the driver, can cause poorer driving performance. In the critical scenario, because of the need to ensure timely and effective information, the driver can accept lower information transparency from the in-vehicle robot than in the noncritical scenario, retaining only key information. The conclusions can be distilled into design guidance: in the in-vehicle scenario, information transparency design assumptions can be made for both voice and visual channels based on SAT theory to help designers arrive at interaction solutions with more appropriate transparency. The guidance can also provide a methodology to solve the transparency challenge that lies in future research on human-centered shared control. Following the guidance, this paper verifies that in the proactive interaction scenario of in-vehicle robots, noncritical scenarios require comprehension and projection level information, while projection level information can be reduced in critical scenarios.

References

[1] A.B. Moniz, B.-J. Krings, Robots working with humans or humansworking with robots? Searching for social dimensions in new human−robot interaction in industry, Societies 6 (3) (2016) 23.
[2] A.A. Mokhtarzadeh, Z.J. Yangqing, Human-robot interaction and self-driving cars safety integration of dispositif networks, in: Proceedings of the IEEE International Conference on Intelligence and Safety for Robotics (ISR), Shenyang, China, 24−27 August 2018, pp. 494−499.
[3] K. Cheverst, H.E. Byun, D. Fitton, C. Sas, C. Kray, N. Villar, et al., Exploring issues of user model transparency and proactive behaviour in an office environment control system, User Model. User-Adapt. Interact. 15 (2005) 235−273.

Further reading

R. Bader, O. Siegmund, W. Woerndl, A study on user acceptance of proactive in-vehicle recommender systems, in: Proceedings of the 3rd International Conference on Automotive User Interfaces and Interactive Vehicular Applications, Salzburg, Austria, 30 November−2 December 2011, pp. 47−54.
D. Billsus, D.M. Hilbert, D. Maynes-Aminzade, Improving proactive information systems, in: Proceedings of the 10th International Conference on Intelligent User Interfaces, San Diego, CA, USA, 10−13 January 2005, pp. 159−166.

M.M. Bradley, P.J. Lang, Measuring emotion: the self-assessment manikin and the semantic differential, J. Behav. Ther. Exp. Psychiatry 25 (1994) 49–59.

M. Buss, D. Carton, B. Gonsior, K. Kuehnlenz, C. Landsiedel, N. Mitsou, et al., Towards proactive human–robot interaction in human environments, in: Proceedings of the 2nd International Conference on Cognitive Infocommunications (CogInfoCom), Budapest, Hungary, 7–9 July 2011, pp. 1–6.

D.N. Cassenti, T.D. Kelley, R.E. Yagoda, E. Avery, Improvements in robot navigation through operator speech preferences, Paladyn J. Behav. Robot. 3 (2012) 102–111.

V. Charissis, J. Falah, R. Lagoo, S.F.M. Alfalah, S. Khan, S. Wang, et al., Employing emerging technologies to develop and evaluate in-vehicle intelligent systems for driver support: infotainment AR HUD case study, Appl. Sci. 11 (2021) 1397.

J.Y. Chen, K. Procci, M. Boyce, J. Wright, A. Garcia, M. Barnes, Situation awareness-based agent transparency; ARL-TR-6905; U.S. Army Research Laboratory: Aberdeen, MD, USA, 2014.

J.Y.C. Chen, M.J. Barnes, Human–agent teaming for multirobot control: a review of human factors issues, IEEE Trans. Hum.-Mach. Syst. 44 (2014) 13–29.

H.M. Cuevas, S.M. Fiore, B.S. Caldwell, L. Strater, Augmenting team cognition in human-automation teams performing in complex operational environments, Aviat. Space Environ. Med. 78 (Suppl. 1) (2007) B63–B70.

M. Endsley, Toward a theory of situation awareness in dynamic systems, Hum. Factors J. Hum. Factors Ergon. Soc. 37 (1995) 32–64.

J.L. Harbluk, Y.I. Noy, M. Eizenman, The impact of cognitive distraction on driver visual behaviour and vehicle control, in Proceedings of the 81st Annual Meeting of the Transportation Research Board, Washington, DC, USA, 8–12 January 2002.

D.R. Large, K. Harrington, G. Burnett, J. Luton, P. Thomas, P. Bennett, To please in a pod: employing an anthropomorphic agent-interlocutor to enhance trust and user experience in an autonomous, self-driving vehicle, in: Proceedings of the 11th International Conference on Automotive User Interfaces and Interactive Vehicular Applications, Utrecht, The Netherlands, 22–25 September 2019, pp. 49–59.

J. Lee, N. Moray, Trust, control strategies and allocation of function in human-machine systems, Ergonomics 35 (1992) 1243–1270.

R. LewisJames, Psychometric evaluation of an after-scenario questionnaire for computer usability studies, ACM Sigchi. Bull. 23 (1991) 78–81.

M.P. Linegang, H.A. Stoner, M.J. Patterson, B.D. Seppelt, J.D. Hoffman, Z.B. Crittendon, et al., Human-automation collaboration in dynamic mission planning: A challenge requiring an ecological approach, in: Proceedings of the Human Factors and Ergonomics Society Annual Meeting, Volume 50, San Francisco, CA, USA, 1 October 2006, pp. 2482–2486.

J. Lyons, Being transparent about transparency: A model for human-robot interaction in Proceedings of the AAAI Spring Symposium: Trust and Autonomous Systems, Standford, CA, USA, 25–27 March 2013.

D. Manzey, J. Reichenbach, L. Onnasch, Human performance consequences of automated decision aids. J. Cogn. Eng. Decis. Mak. 2012, 6, 57–87.

M. Mast, M. Burmester, S. Fatikow, L. Festl, K. Krüger, G. Kronreif, Conceptual HRI design, usability testing, HRI interaction patterns, and HRI Principles for remotely controlled service robots, DeliverableD2.2, European Commission, Bruxelles, Belgium, 2011.

N. Merat, B. Seppelt, T. Louw, J. Engström, J.D. Lee, E. Johansson, et al., The "Out-of-the-Loop" concept in automated driving: proposed definition, measures and implications, Cogn. Technol. Work. 21 (2018) 87–98.

J.E. Mercado, M.A. Rupp, J.Y.C. Chen, M.J. Barnes, D. Barber, K. Procci, Intelligent agent transparency in human–agent teaming for multi-UxV management, Hum. Factors J. Hum. Factors Ergon. Soc. 58 (2016) 401–415.

G.A. Miller, The magical number seven plus or minus two: some limits on our capacity for processing information, Psychol. Rev. 63 (1956) 81−97.
B.M. Muir, Trust in automation. I: Theoretical issues in the study of trust and human intervention in automated systems, Ergonomics 37 (1994) 1905−1922.
K. Park, Y. Im, Ergonomic guidelines of head-up display user interface during semi-automated driving, Electronics 9 (2020) 611.
A. Pauzie, A method to assess the driver mental workload: the driving activity load index (DALI), IET Intell. Transp. Syst. 2 (2008) 315−322.
P.-L.P. Rau, Y. Li, J. Liu, Effects of a social robot's autonomy and group orientation on human decision-making, Adv. Hum. Comput. Interact. 2013 (2013) 263721.
S.J. Russell, P. Norvig, Artificial Intelligence: A Modern Approach, 3rd ed., Tsinghua University Press, Beijing, China, 2011.
A.R. Selkowitz, S.G. Lakhmani, C.N. Larios, J.Y. Chen, Agent transparency and the autonomous squad member, in: Proceedings of the Human Factors and Ergonomics Society Annual Meeting 60, Washington, DC, USA, 19−23 September 2016, pp. 1319−1323.
A. Szajna, R. Stryjski, W. Woźniak, N. Chamier-Gliszczyński, M. Kostrzewski, Assessment of augmented reality in manual wiring production process with use of mobile AR glasses, Sensors 20 (2020) 4755.
K. Van Dongen, P.-P. van Maanen, A framework for explaining reliance on decision aids, Int. J. Hum. Comput. Stud. 71 (2013) 410−424.
J.P. Vasconez, D. Carvajal, F.A. Cheein, On the design of a human-robot interaction strategy for commercial vehicle driving based on human cognitive parameters, Adv. Mech. Eng. 11 (2019). 1687814019862715.
J. Wang, Y. Liu, T. Yue, C. Wang, J. Mao, Y. Wang, et al., Robot transparency and anthropomorphic attribute effects on human−robot interactions, Sensors 21 (2021) 5722.
J. Wang, Y. Wang, Y. Liu, T. Yue, C. Wang, W. Yang, et al., Experimental study on abstract expression of human-robot emotional communication, Symmetry 13 (2021) 1693.
A. Waytz, J. Heafner, N. Epley, The mind in the machine: anthropomorphism increases trust in an autonomous vehicle, J. Exp. Soc. Psychol. 52 (2014) 113−117.
C. Wickens, Multiple resources and mental workload, Hum. Factors J. Hum. Factors Ergon. Soc. 50 (2008) 449−455.
K. Williams, J.A. Flores, J. Peters, affective robot influence on driver adherence to safety, cognitive load reduction and sociability, in Proceedings of the 6th International Conference on Automotive User Interfaces and Interactive Vehicular Applications, Seattle, WA, USA, 17−19 September 2014.
Y. Xing, C. Huang, C. Lv, Driver-automation collaboration for autoated vehicles: a review of human-centered shared control, in: Proceedings of the 2020 IEEE Intelligent Vehicles Symposium (IV), Las Vegas, NV, USA, 19 October−13 November 2020, pp. 1964−1971.

PART 5

Automated vehicles' external human interfaces design and evaluation

CHAPTER 21

Interaction design for environment information surrounding vehicles in parking scenarios

Jianmin Wang[1], Jingyi Pei[1] and Wei Cui[2]

[1]Car Interaction Design Lab, College of Arts and Media, Tongji University, Shanghai, P.R. China
[2]College of Design and Innovation, Tongji University, Shanghai, P.R. China

Chapter Outline

21.1 Introduction and research methodology 306
    21.1.1 Introduction 306
    21.1.2 Research methodology 306
21.2 Design space of environment information surrounding vehicles 307
    21.2.1 Future trends of environment information surrounding vehicles ... 307
    21.2.2 Case analysis 308
21.3 User research of environment information demand in parking scenarios ... 310
    21.3.1 Questionnaire survey on users' demand for information in parking scenarios 311
    21.3.2 In-depth interview on users' demand for information in parking scenarios 311
    21.3.3 Field research 312
21.4 Design practice of environment information surrounding vehicles in parking scenarios 313
    21.4.1 User activity analysis 313
    21.4.2 Scenario analysis 314
    21.4.3 Display scheme design 315
21.5 Usability test of design 319
    21.5.1 Basic information of test 319
    21.5.2 Comparison of text scheme and graphic scheme 320
    21.5.3 Comparison of display positions of environment information 322
21.6 Results and recommendations 323
    21.6.1 Results 323
    21.6.2 Recommendation 323
Further reading 323

21.1 Introduction and research methodology

21.1.1 Introduction

The primary objective of this article is to investigate the interaction design for environment information surrounding vehicles in future urban areas, including the contents and forms of information provided by the surrounding environment for the vehicles. As information display in spatial environments is becoming more and more common in urban areas today, this research attempts to find an innovative, electronic way that utilizes lighting and projection to display environment information and to meet the needs of users who drive. To make the investigation more representative, we chose the parking scenarios as the specific research object and carried out the design practice.

21.1.2 Research methodology

21.1.2.1 Case analysis

At the beginning of the research, the case analysis is conducted to figure out the most cutting-edge interaction design in the field of environment information surrounding cars. For each case, the interactive display mode and the content of the information that appears during each step of the interaction process are analyzed.

21.1.2.2 User research

The user research is conducted after the literature review and the case analysis. To figure out users' demand for environment information when they are in the process of parking, we conducted both online questionnaires and offline in-depth interviews to collect users' pain points during their daily parking experience. We summarized the results of the research and made the personas of the three primary types of users.

21.1.2.3 Field research

To have a clearer understanding of the environment conditions of the parking scenarios, the field visit is conducted in urban spaces and road areas to restore the interaction process of the parking activity. Researchers observed and collected the environment information carriers and contents in each space of the parking lot, and analyzed the space of design optimization.

21.1.2.4 Scenario analysis

Researchers painted out the storyline and the storyboards of the future parking process after the user research and field research had been finished. The scenario analysis was conducted to further clarify the design requirements in detail.

21.1.2.5 Physical prototype

The physical prototype was made not only to demonstrate the feasibility of the design but also to find out the possible application effect of the design in a real urban environment.

21.1.2.6 Usability test
The usability test of the design is carried out to compare the usability of the text and graphic modes of the design and designs displayed in different positions in the environment. During the usability test, online questionnaires and offline in-depth interviews were carried out among 86 participants in total.

21.2 Design space of environment information surrounding vehicles
21.2.1 Future trends of environment information surrounding vehicles
21.2.1.1 Expanded display of environment information
The emergence of stereo traffic to meet the increasing demand for urban traffic volume also means that the current traffic information used to guide land vehicles in urban space will not be able to meet the future needs of urban travelers anymore. The freedom of space and location of information display will be increased, while the area covered by information will be expanded. The freedom in the information display area and height is also greater, and the overall design space in the information display position will be expanded. In the design process of the environment information surrounding the vehicles, it is necessary to consider utilizing a more diversified carrier. The inherent interfaces that already exist in urban areas are one decent choice.

21.2.1.2 More personalized information content
The main reason for the personalization of the information content is the expansion of the types of information carriers and the enrichment of the supporting services in all aspects of travel. The content of the information is closely related to the form, energy, performance, and operating conditions of the carriers. In the driverless age, users' demand for environment information surrounding vehicles will change, the user no longer needs to know the specific traffic signal, but only the driving status, driving intention, and distance information of the carrier. In addition, unmanned vehicles as private travel appliances will also complete the transformation from a simple means of transportation to a more comfortable resting space and a further development in the direction of a "mobile living space." The vehicle may gradually provide users with various services such as scenery viewing, accommodation, and recreation. At the same time, the types of information that users need to receive from the environment will also expand from simple traffic information to various travel service-related information, such as a more detailed display of the surrounding entertainment venues and attractions.

To summarize, in the future city, the environment information surrounding vehicles will contain speed and distance, information of the areas we pass through, and other diversified types of information. The volume and diversity of the information will greatly increase. Moreover, the information needs to be able to meet the increasingly complex personalized needs of users stemming from differences in carrier types and the expansion of services.

21.2.1.3 More attractive display
To make sure that information is adaptable to more diverse external environmental conditions and can fulfill user requirements, a greater amount of information needs to be accommodated within a single display unit. Therefore an increasing number of vehicles equipped with environment

information display units can enhance the display of variable information, timeliness, and personalization of information. In the movie *Ghost in the Shell* (2017), all traffic information displayed in the entire urban environment is based on electronic carriers. The comprehensive electronification of displays empowers all information in the urban space to be variable at all times, making the city more efficient and intelligent in its operation.

In terms of technology, new forms and materials of screens continue to emerge, with foldable screens and flexible screens gradually gaining popularity. The development of screen interaction technology and mixed reality technology has also given rise to products such as interactive workstations and interactive façades. With the display performance of the screen and increasing adaptability to various environments, these products can be applied to urban spaces in the future as a new information display carrier.

Projection has also gradually become a common way of displaying information in urban environments. At present, the mainstream projection method is light projection, as well as holographic projection and semiholographic projection. For example, in the imagination of the future city depicted in the movie *Ghost in the Shell* (2017), holographic information can be seen scrolled around the body of the vehicles, providing the vehicle status more clearly (Fig. 21.1). Moreover, these holographic displays can also extend vertically to match the height of buildings, facilitating the delivery of information at eye level for pedestrians and drivers alike.

21.2.2 Case analysis

This section analyzes the design space of various types of information in the urban environment regarding traffic guidance through case studies. A series of typical cases of information design in the environment information surrounding the vehicles are selected, and their design concepts, design ideas, and design performance are analyzed in detail.

21.2.2.1 Starling crossing

The Starling Crossing, a joint project between UK urban design firm Umbrellium and Direct Line Insurance, is a forward-thinking design attempt to improve the experience of intersection traffic in

FIGURE 21.1

Environment information design in *Ghost in the Shell*.

cities. The main body of the project is a 22-m-long prototype of an interactive road, which shows the prototype of an ideal smart road in the city of the future. The road surface consists of a multivehicle, nonslip LED screen that displays a changeable high-definition pattern on its surface, allowing the prototype road to provide real-time feedback on pedestrian and vehicle behavior. The pattern, layout, configuration, size, and orientation of the various types of signs displayed on the road surface can be adjusted by computer control, and the pattern can be maintained with high brightness and clarity during daylight hours.

The biggest highlight of the project is the smart pavement retrofitting approach to propose targeted solutions to some of the typical sidewalk traffic safety problems, which were selected based on research conducted by the Transport Research Lab in the United Kingdom on the main causes of sidewalk safety problems. The main solutions proposed by the project are organized in Table 21.1.

The design performance of the pavement display solution, based on dynamic signs composed of colorful simple graphics, and the pavement display solution under the corresponding serial number scenes in Table 21.1, is shown in Fig. 21.2. Each sign has a certain trigger mechanism. Take the

Table 21.1 Smart road solutions proposed by starling crossing.

Problematic scenarios	Smart road-based solutions
Crowd accumulation and trampling accidents	The sidewalk width can be scaled at any time to accommodate different numbers of subjects.
Pedestrian jaywalking and other emergencies	The sidewalk will quickly display an emergency status alert message for pedestrian intrusion.
Pedestrians in the blind spot of vehicles	The sidewalk will provide appropriate alert and guidance signals for pedestrians.
Pedestrians not looking down at the road	Use dynamic road and pavement patterns to attract the attention of pedestrians who are preoccupied with their cell phones.
Unclear lane demarcation	Use dynamic signage to vary the function of the lane.

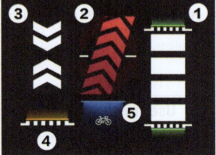

FIGURE 21.2

Graph design of the starling crossing.

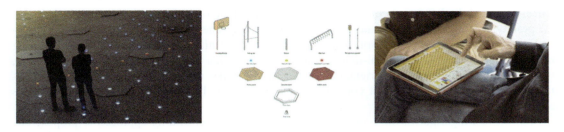

FIGURE 21.3

Application effect of dynamic street.

smart crosswalk as an example, when the user stands behind the sensing area formed by the dotted lines on both sides of the road, the crosswalk will grow in front of the user, and when the user completely passes the road, the crosswalk will retract and disappear.

21.2.2.2 Dynamic Street

Crafted by the international design and innovation firm Carlo Ratti Association in collaboration with Toronto-based Sidewalk Labs, the Dynamic Street project introduces a modular prototype of a reconfigurable paving system. The overall installation consists of 232 special hexagonal tiles, each measuring 4 feet in diameter. Installed on a 2500-square-foot surface, it simulates a 36-foot-wide street. Some of the tiles are equipped with a plug-and-play system that allows for easy installation of various vertical structures such as bollards and utility poles.

Thanks to its modular design, each part of the roadway can be easily moved or replaced. In addition, the system uses recessed lighting to distinguish different traffic zones, allowing for "the creation of an additional lane during rush hour that will 'turn' into a pedestrian plaza on nights when traffic is low" (Fig. 21.3).

In the future, when unmanned vehicles begin to travel on roads, the requirement for the number and configuration of the lanes they occupy and their adjacent areas will change. Flexible and dynamic road systems, such as the one showcased in this project can better meet the changing needs of driverless cars for real-time information about the vehicle's surroundings. With its highly adaptable structure, these roads can enhance safety and accessibility for pedestrians and vehicles, providing an alternative to the conventional vehicle perimeter information systems, consisting of traffic signs, street markings, and so on.

21.3 User research of environment information demand in parking scenarios

To understand the specific needs of users for the environment information in the parking scenarios, this article conducted user research through questionnaires and in-depth interviews. From the interaction process of the parking scenarios and the content of the information in the scene, the user's needs and pain points were uncovered, and the design enhancement points in the parking scene were further summarized.

21.3.1 Questionnaire survey on users' demand for information in parking scenarios

21.3.1.1 Questionnaire setting

The survey was carried out online. A total of 75 questionnaires were collected, and all of them were confirmed to be valid. All the survey users have driver's licenses, and more than 90% of them have 3 years or longer driving experience, which can make the survey results more representative.

The questionnaire includes 15 questions, which are divided into three parts. Since the object of this research is limited to the information displayed in the environment surrounding the vehicle, the questionnaire was set up to investigate the user's satisfaction with several types of information involved in the parking scene, such as road signs, independent road signs, and road markings. After that, the questionnaire investigated the specific content and frequency of users' poor experiences in the parking process. Finally, users' overall satisfaction with environment information was investigated. The 5-point Likert scale was used in evaluating the findings.

21.3.1.2 Result analysis

In the user satisfaction evaluation of independent road traffic signs, more than half of the users chose "satisfied" and "very satisfied," and another 38.1% chose "average." In the evaluation of road signs, most users chose "satisfied" and "very satisfied," and 28.57% chose "average," and the same for road markings; in the evaluation of parking assistance information, most users chose "average," while 9.52% chose "unsatisfied." The users only scored for the environment information assistance in the parking process at 3.57. In addition, when users were asked whether information about the vehicle's surroundings could meet their needs without the use of mobile or other platform applications, the number of users who chose "no" increased. When asked whether they usually need to use mobile applications to better complete the travel process, 90.48% of the users chose "yes" and "very much." This shows that, for users, the clarity of parking assistance information in the environment surrounding the vehicle is not yet comparable to the applications provided by mobile or other platforms, which underscores the necessity of design enhancements to the environment information surrounding vehicles.

In addition, the research found that users have mainly encountered the same problems in the parking process. One of the most important problems is "can't find a parking space when parking," with 90.48% of users reporting that they had encountered this problem. Then there is the problem of information delay; for example, users are not informed of changes in parking space availability in real time. The confusing design of the guidance causes a negative impact on the user experience as well.

21.3.2 In-depth interview on users' demand for information in parking scenarios

21.3.2.1 Content of interview

After receiving the outcome of the questionnaire survey, researchers conducted an in-depth interview regarding the components that users frequently mentioned to be problematic. A total of six users were invited to this interview, two of whom were male and four females, all of whom had driving experience and daily driving needs. The participants belonged to different occupation and age groups, with differences in their frequency of vehicle use and driving experience. The questions provided to the users are listed in Table 21.2.

Table 21.2 Content of in-depth interviews.

Section	Questions
Contextual questions	Your age, occupation, driving experience, and main purpose of driving.
Pain point and new demand discovery	Please describe specifically a parking situation you have encountered that did not go well.
	What kind of location do you prefer to park in? What kind of parking space would you specify and the reasons for your choices?
	Your demands on the guidance in the parking lot.
	What information do you think is necessary to interact with other vehicles during parking?
	What features would you like to see added to the existing parking garage?
Attitude toward future environment information	What is your vision of "intelligent environmental interaction"?
	Do you think intelligent environmental interaction can facilitate your travel if it is applied to parking assistance information?
	Do you think this type of bold design in terms of visual presentation (color, design, dynamics, etc.) is necessary for parking assistance information?

21.3.2.2 Result analysis

The in-depth interview has led to the following knowledge: First, the top four frequent activities users do are finding an available slot, paying, "finding the exit (elevator)" and finding a parked vehicle. Second, the top-mentioned problems users are faced with are too narrow a parking space, not sure about the amount of the parking fee, and uncertainty about the availability of spare parking spaces. In addition to the general problems, we also learned about some details of the problems that users encounter in different parts of the parking process, and an analysis of the users' demands is summarized based on the pain points discovered (Fig. 21.4).

In general, users' demand for functional information during parking is much greater than their demand for emotional information. Most users indicated "clear instructions" as the priority. Users tend to prefer information with the special characteristics of uniformity, standardization, and accuracy. The less impact the information has on driving safety, the more receptive the users are to its visual presentation flexibility.

21.3.3 Field research

Based on the parking scenarios with the high frequency of negative user experience summarized by user research, this article also conducted field research and observation studies. In the process of field research, the author personally experienced and recorded the environment information contacted during the whole process of parking from the driver's perspective, organized and summarized the interactive content and interaction process among people, vehicles, and environment in the actual scenario, and summarized the service process and information architecture of parking service. The 10 parking lots researchers have visited include roadside, underground, and stereo parking lots. The chosen parking lots are in different functional areas and with different volumes. Researchers have summarized the specific facilities and carriers of information in parking scenarios from the field research, and the results can be found in Table 21.3.

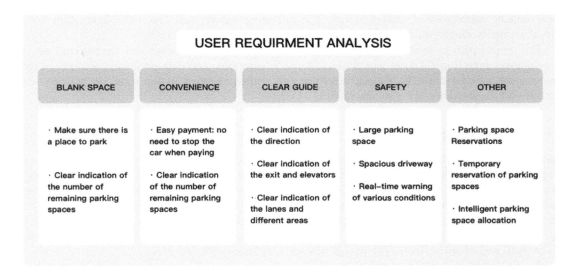

FIGURE 21.4

Summary of users' demands for information in the parking scenario.

Table 21.3 Environment information in the parking scenario.

Position	Environment information		
	Text information	Graphic information	Add-on facilities
Wall (vertical surface)	Directions and exitsCurrent region indicationPosition of the elevator	Directions and exitsGraphical traffic signs	Reflective mirrorEssential infrastructures
Road surface	Directions and exitsCurrent region indication	Passable directions of the lanesPedestrian areaDirections and exitsCurrent region indicationNo parking area indication	Deceleration zoneOccupancy pile
Carport	Parking space number	Parking space availability indication	Charging pile
Ceiling	Directions and exitsCurrent region indicationParking space availability indication	Directions and exitsCurrent region indication	Lights

21.4 Design practice of environment information surrounding vehicles in parking scenarios

21.4.1 User activity analysis

This section carries out a conceptual design based on the previous research. Given that the current urban environment information is mainly based on the visual elements, taking into account the actual conditions, the design of this chapter takes the visual elements (including icons, text, and dynamic effects) in the scenarios as the main design objects, and the output content is in a series of environment information display schemes. The following section will depict the interaction flow, visual elements, design points, and interaction methods of the display schemes.

The typical path of user activity during the parking process is analyzed and summarized into a user journey map (Fig. 21.5) in which users' activity, emotions, and pain points in each part of the process are pointed out. The special conditions that might happen in the roadside parking and garage parking scenarios are also analyzed.

21.4.2 Scenario analysis

Based on the critical behaviors and pain points of users in the parking scenario and the information carriers in the environment summarized from the previous analysis, a series of conceptual storyboards are produced (Fig. 21.8), showing the environment information surrounding the vehicles during the whole process of the parking activity. Thereafter, the specific design requirements for each scene are organized and summarized. In terms of specific visual solution presentation, each dynamic interface of this design output will use at least one of the three types of visual elements: icons, text, and motion effects. The design demands are listed in Table 21.4.

As illustrated in the storyboards (Fig. 21.6), in this design, the inherent interfaces in the urban environment are supposed to be used to complete the renovation of the guide information in parking scenarios, and the display schemes will be completed utilizing lighting and electronic displays. For example, in storyboard number 1, the road surface is capable of automatically adapting to the size of the vehicle and displaying parking space lines and parking status information around the vehicle, and the color shifts when the parking condition of the vehicle changes.

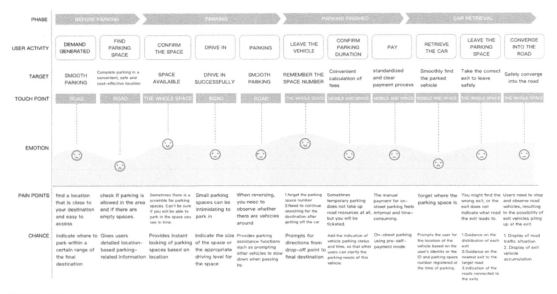

FIGURE 21.5

User journey map of the parking process.

21.4 Design practice of environment information surrounding vehicles

Table 21.4 Design demands of parking scenario.

Steps	Graph	Text	Animation	Specific information demanded
Slot assignment	Yes	Yes	Yes	Assigned slot information display
Finding the slot	Yes	No	Yes	Prohibited area displayAvailable slot display
Parking	Yes	Yes	Yes	Vehicle condition displayRemind the other vehicles
Leaving the slot	Yes	No	Yes	Remind the other vehiclesMoving direction prediction
Other informations	No	Yes	Yes	Parking duration display notes are given by the car owner

Content of design scheme spans Graph, Text, Animation, Specific information demanded.

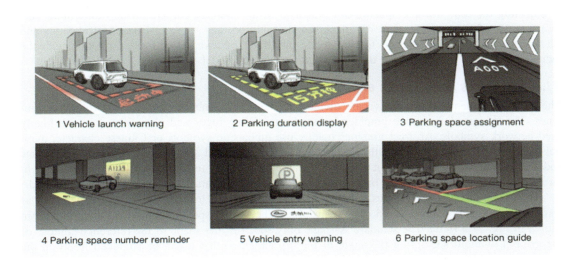

1 Vehicle launch warning
2 Parking duration display
3 Parking space assignment
4 Parking space number reminder
5 Vehicle entry warning
6 Parking space location guide

FIGURE 21.6

Conceptual storyboards of future parking scenario.

21.4.3 Display scheme design

This paper refers to several specifications in the design process, as there is no authoritative and comprehensive dedicated design specification for the field served by this design. The first of which is the nationwide traffic signs and markings design specification in China called Road Traffic Signs and Markings, which has been in use since 2009. This article mainly provides a reference for the morphological design and color selection of the signs and markings in the display scheme. The second aspect is the dynamic design specification, which is the traditional 12 laws of motion of 2D animation proposed by Disney animators Frank Thomas and Ollie Johnston. The third area is the Universal Design principles followed in the design of environmental guidance systems, since the display scheme is about to be presented in urban environments, and the universal design principle is a common principle followed in the design of spatial guidance signage systems. We will follow this principle to design the display scheme for daytime and nighttime, and for text and graphic schemes in the parking scenarios, and consider how the design can be better adapted to the architectural condition of the environment. The fourth aspect is the design specification of graphics. In addition to the signs already specified in Road Traffic Signs and Markings, the Emoji expression graphic library and the Microsoft special symbol library are also referred to.

21.4.3.1 Text design for ground display

The font of the text used is sans serif. Sans serif fonts are more simple and generous and bring more affinity to users' visual experience. When used for ground display, the design displays the text on a solid background with rounded corners. This design reduces the impact of the ground's inherent material and color on the text display compared with using bottomless text and can enhance the clarity and readability of the information to a certain extent. Fig. 21.7 shows the effect of adjusting the font size and spacing of the text to fit the rounded corners of the frame under different needs of the number of words displayed. In addition, according to Road Traffic Signs and Markings, this study is designed to exaggerate the effect of text stretching deformation with the increase in vehicle speed, and the deformed text aspect ratio is 3:1 to 4:1.

21.4.3.2 Graphs and dynamic effects

The graphic elements in this design are divided into abstract symbolic graphics and figurative graphics. The design of figurative graphics is based on existing logos and also refers to some highly used network symbols. Abstract graphics mainly use alphabetic symbols or Chinese character symbols, because abstract graphics without character attributes cannot simulate the form of real objects and have natural disadvantages in terms of ideation, while the use of characters allows users to directly associate the logo content based on the meaning of the characters and is not easily confused.

In the shape design of graphics, when polygons and arcs are involved, standard shapes such as standard orthogonal polygons and circles are used. The design specifications for other shapes are shown in Fig. 21.8 (left). The standard X-shape only specifies the shape of the reference contour of the symbol, without specifying the angle of the nip angle, so that it can be stretched freely in the subsequent practical application according to the needs of the regional scope of placement. The standard guide sign specifies the angle of clamping as well as two different spacings to facilitate the display at different locations according to need. The standard yellow and black isolation strips specify an inclination angle of 45 degrees and equal spacing.

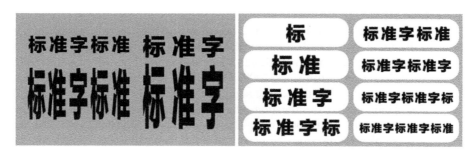

FIGURE 21.7

Text display.

21.4 Design practice of environment information surrounding vehicles

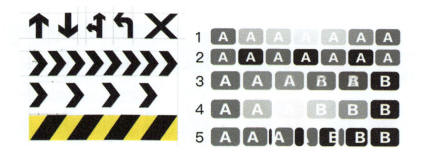

FIGURE 21.8

Graphs (*left*) and the transaction animation (*right*) of the design.

To ensure the accuracy, standardization, and replaceability of the variable display content, the delimitation frame of all individual graphics is designed to be consistent and with the same aspect ratio, such as the first row of graphics in Fig. 21.8 (left). This design ensures that the graphics can be easily replaced by other graphics in their original position when they are in the interface.

Several types of dynamic effects appear frequently in the design as follows: the first and the fourth one in Fig. 21.8 (right) are the blinking dynamic effect, that is, the logo fades in and fades out at fixed intervals on the base image to create a blinking dynamic effect, which can attract users' unintentional attention. The second is the hard handoff effect. The third is the cross-dissolve effect, which refers to the dynamic change of pattern color and transparency. The fifth is the directional flow and jumping dynamic effect, which is used in the directional guidance display scheme in this design.

21.4.3.3 Color scheme

Due to the large number of signs and markings in the parking environment, designing individual colors for each logo is not feasible. It is also difficult for users to remember a large number of colors, which might lead to a decrease in the usability of the signs. Therefore we decided to reduce the number of colors as much as possible. The required colors are divided into different monochromatic colors according to their hues, and each monochromatic color is selected as a standard color from its hue range, which is applied to all designs that require the color of the corresponding hue. For example, only one red color value will appear in this design, which will be used in all display schemes that include the color red. In addition to monochrome, this design also uses gradient colors in part of the visual scheme, which are divided into two kinds: two-tone gradient and monochrome gradient. The specific color scheme is shown in Fig. 21.9.

21.4.3.4 Design output of each scenario

In compliance with the designated design specifications, this study presents the environment information display scheme for each segment of parking. Since most of the graphics in the design are expected to be illuminated on LED electronic screens, they have been displayed on a black background to better match the intended effect of the design. The design is displayed in the nighttime color scheme of the standard color scheme. In addition to the original icon scheme shown in the

318 Chapter 21 Interaction design for environment information

FIGURE 21.9

Color scheme.

table below, a text-only scheme has been designed for some of the icons. The text-only scheme will be shown in the chapter on prototype implementation and compared with the original scheme in the chapter on evaluation. The application effect of some of the scenarios is shown in Fig. 21.10. To show the application effect of the design in the scene more intuitively and clearly, this study is based on the draft of the design to create a concept drawing with the scene modeling as the main effect. The modeling of outdoor scenes was carried out using Cinema 4D R19 software combined with City Rig 1.7 plug-in, which is a plug-in that can facilitate the design and modeling of urban areas by using different sizes of urban buildings, roads, infrastructure, and other model presets. With this plug-in, you can efficiently simulate a basic picture of a smart city with certain details.

21.4.3.5 Physical prototype

Based on the design draft and rendering, researchers used 3D-printed vehicle models, embedded LED strips, and short-focus projection to build the physical prototype. The linear design part of the display scheme was displayed with LED strips, the control of the light color and animation was accomplished by Arduino IDE, and irregular graphic content such as text and complex graphics were displayed by projection. The effect is shown in Fig. 21.11.

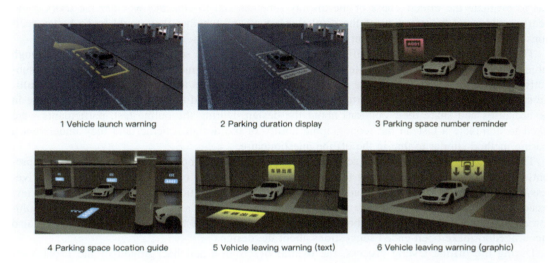

FIGURE 21.10

Application effect of each scenario.

FIGURE 21.11

Physical prototype.

21.5 Usability test of design

21.5.1 Basic information of test

21.5.1.1 Test methods

During the proof-of-concept phase of this project, users were invited to evaluate several different display schemes: floor, wall, icon, and text schemes. One of the main objectives of the usability test was to obtain users' perceptions and feedback on the different visual element-based display schemes and to collect users' opinions and suggestions for their improvement. The second objective was to evaluate the effectiveness of the display schemes and to consider whether the users could clearly understand the meaning of the display schemes and whether they could make better decisions utilizing the display schemes in this design.

The test was conducted both offline and online. The offline segment was conducted through subjective scales and open-ended interviews, while the online segment was conducted through subjective scale scores only. During the offline test, the design drafts and application effects of multiple scenarios were shown to the participants, and they were asked to score each display scheme in the scenarios and share their opinions about the design. During the online test, participants were asked to complete a test on the iconographic ability of the design scheme, and they underwent a scoring method similar to that in the offline questionnaire.

The scale was a 7-point Likert scale, with 1 representing complete disagreement and 7 representing complete agreement, to test users' subjective evaluations of the four indicators: effectiveness, satisfaction, recognizability, and comfort of the display solution. In the open-ended interview session, the inquiries focused on the specific reasons for the user's preference for the design solution and their views and feelings about the interaction process of the design. The users were also encouraged to make open-ended suggestions about the design.

During the offline testing process, the overall description and background of the experiment took about 2 minutes, the scene modeling video took 30 seconds, the scene physical prototype video took 25 seconds, the scale took 5–8 minutes to fill out, and the open-ended interview took 10–15 minutes. The duration of the interview was 10–15 minutes. The total duration was about 20 minutes.

21.5.1.2 Participants

A total of 13 participants were recruited for the offline testing, with ages ranging from 24 to 30, with 4 users aged between 45 and 55, and 3 users aged between 35 and 40. A total of 63 participants were recruited for the online test, 5 of the participants were inexperienced users, while the rest were users with a driver's license and driving experience. Users in the 31 to 50 age group accounted for more than half of the sample, with another 14 users aged between 18 and 25 and 11 users aged 50 or older. The driving experience of the participants was mainly 6 years or more, with 20 of them having more than 10 years of driving experience.

21.5.2 Comparison of text scheme and graphic scheme

The scores of the scales for the comparison of the graphic versus text mode scheme can be seen in Table 21.5. As can be seen from the average of the scores, the overall ratings are above the medium value of 4, indicating that there are no significant shortcomings in the performance of the icon scheme on each indicator. Among them, users have higher average ratings for the two indicators of recognizability and comfort of the display scheme. The users' ratings for the clarity of the solution's ideograms were lower, and the minimum value was below 4, indicating that the icon solution needs to be improved in terms of comprehensibility, and the reasons for this result will be analyzed in detail later in the interview. The lowest rating is for the scheme's usefulness in making the next traffic decision. This result indicates that the display scheme in icon mode has relatively mediocre performance in terms of effectiveness indicators.

Table 21.5 Scores of the text and graphic scheme.

	Offline scores		Online scores	
Evaluation indicators (questions)	Text scheme	Graphic scheme	Text scheme	Graphic scheme
I can clearly understand the meaning of the display scheme.	6.77	4.54	6.22	4.84
The display scheme is confusing.	2.08	4.77	2.71	4.89
The display scheme helps me to make the next traffic decision.	5.92	5.38	6.27	5.16
The display scheme saves me time in completing traffic decisions.	6.15	4.46	6.13	4.9
The display scheme is eye-catching and recognizable.	6.15	4.62	5.98	4.67
The display scheme looks comfortable.	5.31	5.46	5.68	5.16
I am satisfied with the design of the display scheme.	5.77	5.00	6.08	5.29

For the graphic scheme, more than 70% of the users were able to understand the meaning of all the icons in the assessment without any problem. Four users indicated that they needed to think for a second or two before they could understand the icon display scheme in the scenario without textual support. Among the evaluated icons, the two icons that confused the users were the vehicle entry and exit icons. The reason is that the icons only express the driving status of the vehicle, but the expression of the direction of the vehicle is not clear enough. Users were more concerned with visual elements such as the position of the icon display, display area, brightness, and dynamics than with color. Two users suggested that color would affect their judgment of the meaning of the icon only when the meaning of the icon itself is not very clear. For example, for the icon meaning "vehicle is about to leave the garage," one user said, "If the icon is green, I would think the sign means to remind me to leave the garage now, if the icon is yellow, I would think the sign is to remind others that a car is about to leave the garage." In addition, users suggested that whether the icon has a dynamic effect or not would have a great impact on their understanding. Take the parking sign "P" as an example, if the sign has no flashing effect at all, users will quickly determine that it means "a vehicle is parking." But if the sign is flashing, the user will regard it as a warning.

For the text scheme, users scored high (over 6) on the three indicators of comprehensibility, effectiveness, and recognizability of the scheme, while the scores on the indicator of comfort were lower than the aforementioned indicators, indicating that there is still room for improvement in the comfort of the text display scheme. After the interviews, we learned that users tend to prefer text schemes mainly for two reasons: the first is that icons have a higher probability of ambiguity; if the graphic icon is a conventional icon, such as the direction icon, the users will prefer to use the icon, but this kind of icon rarely appears in the parking scenario. Second, users consider that the threshold of icon recognition is high, even if they have learned the icons, they still need to recite them many times to remember them, and they can forget them easily. Text, on the other hand, is very straightforward and does not require any time for the brain to react.

However, users who preferred the graphic scheme suggested that icons are more international than text, have a higher concentration of information, and can be decoded more quickly. In the case of defacement and obscuration, icons are more easily restored by users than text. This may also explain why all but one of the users tested in the interviews indicated that they would prefer the icon scheme if they already had prior knowledge about the meaning of icons. This is why 80% of users agree that the best display scheme is one that combines graphs with text.

21.5.3 Comparison of display positions of environment information

The scores of the scales for the comparison of the ground versus wall scheme can be seen in Table 21.6. From the mean scores, it can be seen that of the wall scheme, the users scored high on the identifiability indicator, indicating that they can identify the wall display scheme clearly and understand the meaning of the wall scheme. However, the mean score for the question "The solution is distracting" is not lower than the median value of 4, indicating that although users recognize the effectiveness of the wall solution, they cannot guarantee that the wall scheme will not distract them during driving. This suggests that while the wall scheme is effective in attracting users' attention, it may also be more likely to distract them from driving due to unintentional attention.

For the ground scheme, the two indicators that scored the highest were satisfaction and effectiveness. Comparing the scores, it can be seen that the wall scheme scores higher than the floor scheme in all indicators, which reflects the overall preference of users for the wall scheme over the floor scheme.

In the follow-up interviews, the researchers found that users expressed three distinct preferences. The first group of users preferred information displayed on the ground because they felt it was easier to look at the ground than left or right while driving, and that the "head-turning" action would distract them from the road ahead. The second reason is that users believe that information on the wall can be easily obscured by other vehicles; The second group of users expressed a preference for information located at the top of their field of view for several reasons: First, there is no possibility of the information at the top being obscured by vehicles, guaranteeing the clarity of the display. The second reason is that they are already used to the existing parking garage design and prefer content displayed above their heads that can be seen at eye level. This is also the reason why this group of users did not choose the side wall scheme. The third group of users expressed a

Table 21.6 Scores of the wall and ground schemes.

Evaluation indicators (questions)	Scores	
	Ground scheme	Wall scheme
I can quickly notice what is being displayed.	4.62	5.00
I can clearly understand the meaning of the display scheme.	4.54	5.46
The display scheme helps me to make the next traffic decision.	4.54	5.62
The display scheme saves me time in completing traffic decisions.	4.77	5.23
The display scheme is distracting to me.	3.54	4.00
The display scheme meets my needs.	4.90	5.23

preference for information on walls, mainly because they believe that wall information enters the human eye with less deformation than ground information, so that the information can be displayed more clearly. The viewing angle of the ground solution flattens with increasing distance, decreasing readability. Finally, one user believes that both the wall and the ground should display content because the garage space is narrow and confined and that there are too many subjects in the space for the displayed content to be easily ignored if not adequate.

21.6 Results and recommendations

21.6.1 Results

In this article, researchers propose a series of environment information display schemes that help users park more efficiently. The design utilizes lighting and projection to display information in the environment surrounding the vehicles. During the investigation and design, the methods of case analysis, user research, field studies, and physical prototypes were used. The feasibility of this design was proved by the physical prototype and the usability of different display schemes was evaluated through online and offline usability tests. The result of the usability tests showed that users preferred environment information that is presented in text form and displayed on vertical interfaces in parking scenarios.

21.6.2 Recommendation

Due to time and resource limitations, the design of this article is only a creative initial exploration of a visual display scheme. There is still a lot of work to be done to design a better scheme for displaying environment information surrounding the vehicles to better meet the needs of the users. For example, environmental effects such as light and weather can be considered more carefully in future research and design. The design practices on the auditory and tactile channels can also be expanded.

Further reading

S. Adams, N. Morioka, T.L. Stone, Logo Design Workbook: A Hands-On Guide to Creating Logos, Rockport Publishers, Massachusetts, 2006, p. 65.

V. Biener, Mixed reality interaction for mobile knowledge work, in: 2022 IEEE Conference on Virtual Reality and 3D User Interfaces Abstracts and Workshops (VRW), 2022, pp. 928–929. Available from: https://doi.org/10.1109/VRW55335.2022.00315.

Carlo Ratti Associati. The dynamic street, 2018. [Online]. <https://carloratti.com/project/the-dynamic-street> (accessed 10.12.21).

G. Das, H.J. Wiener, I. Kareklas, To emoji or not to emoji? Examining the influence of emoji on consumer reactions to advertising, J. Bus. Res. 96 (2019) 147–156.

Expert Witness, Case study – Death by careless driving, June 25, 2018. [Online]. <https://expertwitness.trl.co.uk/case-study-death-by-careless-driving> (accessed 10.10.21).

Future Timeline, Futuristic pedestrian crossing unveiled in London, October 11, 2017. [Online]. <https://www.futuretimeline.net/blog/2017/10/11.htm> (accessed 21.10.21).

L. Jisheng, I.T. Phillips, V. Chalana, R. Haralick, A methodology for special symbol recognitions, Proc. 15th Int. Conf. Pattern Recogn. ICPR-2000 4 (2000) 11–14. Available from: https://doi.org/10.1109/ICPR.2000.902854.

O. Johnston, F. Thomas, The Illusion of Life: Disney Animation, Disney Editions, New York, 1995.

M.Y. Lee, Real time motion graphics produce study of methods that use LED device, J. Korea Soc. Digital Ind. Inf. Manag. 7 1 (2011) 63–74.

S.A. Mouloua, F. James, M. Mustapha, P.A. Hancock, Trend analysis of unmanned aerial vehicles (UAVs) research published in the HFES proceedings, Proceedings of the Human factors and ergonomics society annual meeting, Vol. 62, SAGE Publications, Sage CA: Los Angeles, CA, 2018, pp. 1067–1071.

J. Perlich, D. Whitt, Millennial Mythmaking: Essays on the power of science fiction and fantasy literature, films and games, MCFarland, North Carolina, 2010, p. 167.

Standardization Administration of China(SAC), Road traffic signs and markings, Standards press of China, Beijing, 2009.

M.F. Story, Principles of universal design, Universal Design Handbook, 2001, p. 43.

M. Takaoka, Type Setting, Citic Press, Beijing, 2016.

X.G. Troncoso, S.L. Macknik, S. Martinez-Conde, Corner salience varies linearly with corner angle during flicker-augmented contrast: a general principle of corner perception based on Vasarely's artworks, Spat. Vis. 22 (2009) 211–224.

Umbrellium, Starling crossing, 2017. [Online]. <https://umbrellium.co.uk/projects/starling-crossing> (accessed 22.10.21).

A. Wall, Programming the urban surface, in: J. Corner (Ed.), Recovering Landscape, Essays in Contemporary Landscape Architecture, Princeton Architectural Press, New York, 1999, pp. 233–249.

H.L. Xu, T.X. Zhang, B.M. Tang, T. Liu, Design of road traffic safety color system based on color perception, J. Chongqing Jiaotong Univ. (Nat. Sci.) 1 (2009) 105.

Unmanned vehicle external interaction design based on an automation acceptance model

22

Bihan Zhang and Fang You

Car Interaction Design Lab, College of Arts and Media, Tongji University, Shanghai, P.R. China

Chapter Outline

22.1 Introduction 325
22.2 Automation acceptance model 326
22.3 Scenario-based pedestrian behavior research 327
    22.3.1 Design background 327
    22.3.2 In-depth interview and user research on pedestrian-unmanned vehicle interaction 327
    22.3.3 Behavioral characteristics and demand analysis of pedestrian-unmanned vehicles 328
22.4 Vehicle diplomacy mutual strategy based on automation acceptance model 329
    22.4.1 Construction of vehicle diplomacy mutual strategy model 329
    22.4.2 Anthropomorphic design promotes trust and emotion 330
    22.4.3 Classification and visualization of information outside the vehicle 331
    22.4.4 Multimodal design of external vehicle interaction 332
    22.4.5 Interactive prototype design based on vehicle diplomacy mutual strategy 332
22.5 Design test evaluation 334
22.6 Conclusion 334
Reference 334
Further reading 335

22.1 Introduction

With the continuous development of technology, existing low-level assisted driving systems will gradually be replaced by high-level assisted driving and even fully autonomous driving. The technological revolution in the transportation environment is likely to be accompanied by a revolution in social interaction. Traditional implicit communication methods between drivers and pedestrians, such as communication through eye contact and gestures, will no longer be applicable. In an environment where manually driven, semiautomated, and fully automated vehicles operate simultaneously, advanced human—machine interaction interfaces are needed to inform participants about the current status of vehicles and their impending actions. Therefore, how to enhance people's acceptance of automation technology remains an unresolved issue.

22.2 Automation acceptance model

The automation acceptance model (AAM) incorrectly integrates the technology acceptance model (TAM) with the use of automation, and no reference source has been found. The AAM is designed to describe changes in user acceptance and influencing factors in automated systems. The TAM model, proposed by Davis in 1989 [1], focuses on the acceptance of information systems. Perceived usefulness (PU) and perceived ease of use (PEOU) are the key variables in the model, as shown in Fig. 22.1. PU refers to the subjective impact users believe the system has on improving their job performance. PEOU refers to the extent to which users believe the system is easy to use, representing their evaluation of the system's difficulty or ease of use during the interaction. PEOU can indirectly influence and determine behavioral attitudes by adjusting PU, or it can directly impact user behavioral attitudes.

The AAM introduces compatibility based on the TAM structure. Compatibility refers to the degree of alignment between users, technology, task performance, and the context. More specifically, compatibility is an indicator that measures the consistency between technology and user values, past experiences, and needs. Another influencing factor added to the AAM is trust, which is also considered a direct determinant of behavioral intention. Improper calibration of trust can affect people's acceptance of new technology. If there is excessive trust in the system, meaning high automation but low reliability, it can lead to user misuse or abandonment. The design of new technology should be trustworthy, technically feasible, aligned with tasks, and easily understandable.

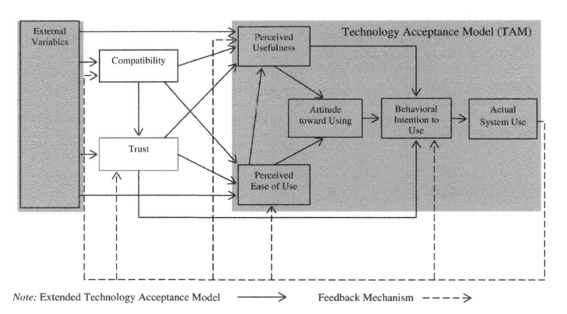

Note: Extended Technology Acceptance Model ⟶   Feedback Mechanism ---->

FIGURE 22.1

Automation acceptance model.

22.3 Scenario-based pedestrian behavior research

22.3.1 Design background

With the continuous growth and widespread adoption of technology, the relationship between humans and technology has become a vital component of everyday life. Automation systems are gradually reshaping the roles of humans in society. The performance of automation systems depends on the relationship established between individuals and the automated systems. This article first simulates common interactions between pedestrians and unmanned logistics vehicles using the methodology inspired by the "Wizard of Oz" approach (Fig. 22.2). In this phase, we prototype unmanned vehicles and record pedestrian reactions by installing GoPro cameras on the vehicle body.

22.3.2 In-depth interview and user research on pedestrian-unmanned vehicle interaction

The interviews and surveys are primarily approached from three perspectives, which include (1) Basic information, which involves the level of understanding of autonomous vehicles; (2) Driving issues, such as what information individuals prioritize for safety assessment and which behaviors they perceive as threatening to their safety or are unacceptable; and (3) Cognitive issues, which include factors that may influence the level of acceptance of autonomous vehicles, and so on. Due to the limited understanding of users regarding Level 4 autonomous driving cabins, directly obtaining specific requirements from users is challenging. Therefore, in conducting user research, the focus is mainly on comparing normal driving scenarios with autonomous driving scenarios. Additionally, users are asked to watch simulated driving videos in specific scenarios to investigate their direct needs during the interaction process, as well as the pain points and expectations they encounter.

FIGURE 22.2

The Wizard of Oz method simulating the interaction between a vehicle and pedestrians.

The quantitative and qualitative data acquisition and analysis were conducted based on an online survey and interviews. Participants meeting the criteria were recruited online, and a closed park area was considered as a simulated scenario. The final sample size was 156, with the age mainly concentrating between 18 and 30 years, with 4 samples aged 35 and above (M = 22.3, SD = 6.25). Upon this foundation, users were grouped into categories. Those who were more familiar with autonomous vehicles and had practical experience were defined as expert users. Users who had heard about autonomous vehicles but only at a conceptual level were defined as regular users. Users who had never encountered or heard about autonomous vehicles were defined as novice users. From these groups, two expert users, two regular users, and one novice user were selected for in-depth interviews (Fig. 22.3).

22.3.3 Behavioral characteristics and demand analysis of pedestrian-unmanned vehicles

Based on the previous on-site inspections using the Wizard of Oz method, questionnaire surveys, and in-depth user interviews, relevant information about user scenarios as road participants encountering autonomous vehicles on the road was gathered. Utilizing user journey mapping, the information was organized and summarized, as shown in Table 22.1.

The behavior of pedestrians and driverless cars driving on the road is divided into four stages, each stage includes several subbehaviors. In the first stage, both the pedestrians and the unmanned vehicle coexist on the road, with minimal interaction due to the considerable distance between them. The second stage involves the mutual identification between the unmanned vehicle and the user. The user needs to discern the driving information of the unmanned vehicle by the vehicle's shape and other driving information. Simultaneously, the unmanned vehicle also needs to inform the user that it has noticed the pedestrian, ensuring the safety of the pedestrian as a participant on the road. In the third stage of information transmission and interaction, users need to make behavioral decisions by receiving information from unmanned vehicles. Unclear information will aggravate the insecurity of pedestrians and increase the probability of accidents.

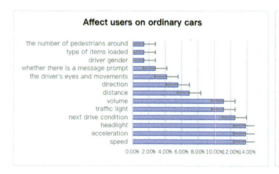 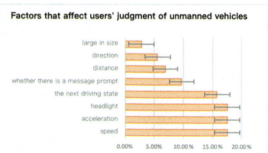

FIGURE 22.3

Interview and questionnaire quantitative research results.

Table 22.1 Pedestrian behavior characteristics and needs analysis.

Stage	Scenario	Decision maker	Key issue	Information requirement
Driving stage	Determine whether it is an unmanned vehicle and the driving state of the vehicle	user	The class of vehicle cannot be determined in time	Driverless vehicle driving data: speed, acceleration and deceleration, distance
Prompt attention stage	Alert to incoming vehicles and indicate that pedestrians have been noticed	Driverless car	There is no guarantee that you will be safe as a road participant; It is impossible to accurately judge the status of the unmanned car	Driverless car driving intention, identified prompts, attention prompts; Driverless vehicle driving intention, right-of-way information
Interaction stage	Choose the next step according to the driving state of the unmanned vehicle and give way/continue to move forward	user	Without control of the road situation, it is impossible to accurately make the next behavior judgment	Driverless vehicle driving data: speed, acceleration and deceleration, distance
Decision stage	Pedestrian waiting/ driverless car waiting	User/driverless car	The problem of power distribution of roads	Emotional communication

22.4 Vehicle diplomacy mutual strategy based on automation acceptance model

22.4.1 Construction of vehicle diplomacy mutual strategy model

Based on the above pedestrian behavior analysis, we can attribute the poor user experience of user-unmanned vehicle interaction to the following four aspects, and build a vehicle diplomacy mutual design strategy model based on these four aspects, as shown in Fig. 22.4.

22.4.1.1 Trust

Trust is the most significant factor affecting user adoption of autonomous vehicles. For pedestrians, as well as passengers in autonomous vehicles, vulnerable groups are particularly at risk and more susceptible to harm. Pedestrians cannot be familiar with every autonomous vehicle, making it crucial to implement standardized strategies for external vehicle interactions and ensure clear and transparent information transmission. This is essential for building a core relationship among humans, vehicles, and the road.

22.4.1.2 Information recognizability

This aspect evaluates several factors: (1) Whether color can enable pedestrians to recognize the size from a distance, (2) Whether color can enable pedestrians to understand the transmitted

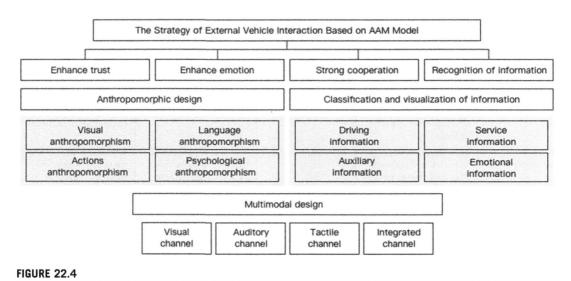

FIGURE 22.4

A strategy model of external vehicle interaction based on automation acceptance model.

Chinese information at a distance and under strong lighting conditions, (3) Whether symbolic language is intelligible to pedestrians, (4) Whether the speed of information rolling is in line with the cognitive ability of pedestrians, and (5) Whether there is a multimodal display mode.

22.4.1.3 Collaboration

This aspect focuses on three key elements: (1) What the car is doing: the driverless car needs to tell the user its current state, enabling the user to make informed judgments; (2) What the car will do next: the driverless car needs to tell the user its next behavior, reassuring the user of its presence; and (3) What pedestrians can do: providing assistance to pedestrians who may be confused on the road.

22.4.1.4 Right of way and emotionalization

To make the relationship between humans and vehicles more harmonious, invoking feelings of fondness and mutual understanding, a natural and multimodal interaction approach is employed. This goes beyond pure text and involves multiple channels, allowing information to be conveyed more clearly. Autonomous vehicles driving in the area can also serve as part of the cityscape through external screens, providing convenient services and enhancing the urban environment.

22.4.2 Anthropomorphic design promotes trust and emotion

Trust is defined as "the general expectation of one person of a written, promised, oral, or written statement by another person or group." One approach to enhancing human trust and making automated technology easy to interact with people is to implement anthropomorphic features and characteristics of vehicles, which can provide a sense of natural interaction. In design, we can divide anthropomorphism into several different dimensions, namely visual anthropomorphism, linguistic anthropomorphism, action anthropomorphism, and psychological anthropomorphism.

22.4.2.1 Visual anthropomorphism
Visual anthropomorphism involves using visual cues to imbue vehicles with human−like characteristics. For example, headlights of vehicles can be designed to resemble eyes, and light strips around the front cabin can be arranged to create an animal-like outline. This represents the fundamental level of anthropomorphic design. At the visual level, using the expression screens can further personify the car, enhancing its human−like qualities.

22.4.2.2 Language anthropomorphism
At the linguistic level, we consider that when interacting with people, the selection of interaction strategies is more natural, such as displaying the character of the system through different speech techniques, adding some necessary modal words and pause words to the voice interaction mode, increasing the sense of reality, making the voice role match the system role, and so on.

22.4.2.3 Action anthropomorphism
Action anthropomorphism involves designing and shaping the movement trajectory of the automated system according to the actions of humans or animals. For example, combined with visual anthropomorphism, the "ears" can be wiggled to express emotions during interaction with humans. The design of headlights can incorporate variations in intensity, color, and size of the light beam to simulate changes in human eyes, and express emotions.

22.4.2.4 Psychological anthropomorphism
Psychological anthropomorphism represents the pinnacle of anthropomorphic design, representing the peak stage we aim for after crossing the "uncanny valley" effect. We can envision the system as if it were human, modeling the automated system to align more closely with human thoughts at the psychological level. This involves comprehensive planning for the behavior, tone, appearance, voice, and other aspects that the system should exhibit under this imagined human role. When combined with the system's learning capabilities, this approach enables the system to achieve anthropomorphism from the inside out.

22.4.3 Classification and visualization of information outside the vehicle
Based on the interaction between the unmanned vehicle and the pedestrian, we can divide the area around the unmanned vehicle into three parts: the front interaction area, the rear interaction area, and the side interaction area. In the front interaction area, the driving status information is mainly displayed when the unmanned vehicle and pedestrian are driving opposite or crossing. The rear interaction area is mainly for the situation of driving in the same direction. In addition to displaying the driving status information, it also provides emotional interaction such as saying

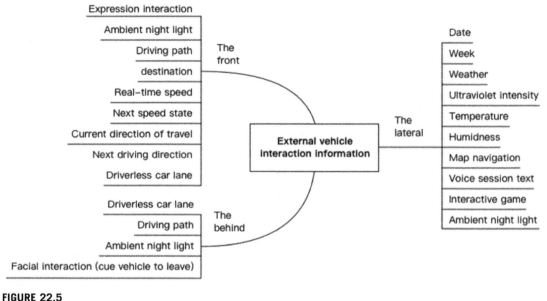

FIGURE 22.5

External vehicle interaction information architecture.

hello, waving goodbye, and expressing gratitude to pedestrians for waiting. The side interaction area mainly displays convenience services and entertainment functions unrelated to driving information (Fig. 22.5).

22.4.4 Multimodal design of external vehicle interaction

To fully simulate the interaction between people, multimodal interaction is considered to integrate multiple senses and conduct human–computer interaction through text, speech, vision, action, environment, and other ways. It is primarily designed to cater to visual, auditory, and tactile channels, as well as multichannel integration, which is in line with users' habits and fulfills their expectations (Table 22.2).

22.4.5 Interactive prototype design based on vehicle diplomacy mutual strategy

The outside LED display on the car body indicates the driving state of the vehicle through different expressions. Compared with the text display, the graphic method has a small cognitive load on the user, which can be understood by pedestrians of different ages. For example, the excited expression is used to accelerate and the leisurely expression of drinking coffee is used to slow down. In addition, considering the great influence of the physical environment of vehicle interaction, such as lighting conditions, two sets of schemes are designed for day and night, and the color matching with the highest contrast and clarity under the two lighting conditions is selected as the final scheme (Fig. 22.6).

Table 22.2 Multimodal design of external vehicle interaction.

Multimodal interaction	Design scheme
Visual channel	The driverless car simulates the interaction between an ordinary car driver and a pedestrian by reading the user's realization and responding with a neutral expression.
Auditory channel	Communicating through simple natural language. The voice interaction of the driverless car conforms to the role model in the user's mind, adopting characteristics that may be playful or endearing.
Tactile channel	Expressing gratitude through gestures such as tipping off and simple touch interactions, such as tapping. Implement multichannel synthesis to align with the character modeling image of the unmanned school bus, incorporating expressions, natural language communication, and gesture-based communication.

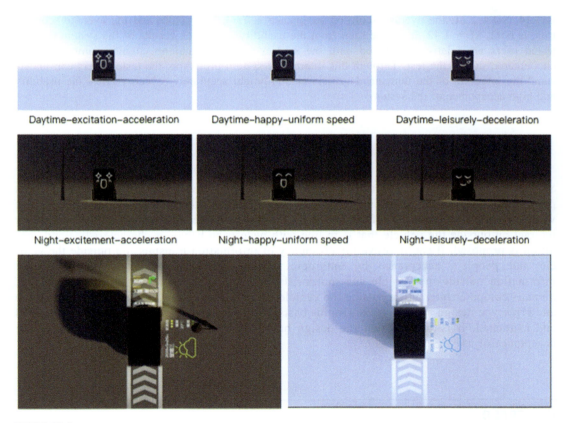

FIGURE 22.6

High-fidelity prototype of car exterior interaction.

22.5 Design test evaluation

Testing and evaluation with user participation in the conceptual design phase can help improve the design of the human−machine interface. Closed and semiclosed parks were the first landing scene of driverless cars. We set the test scene on campus and recruited college students to conduct the usability test of human−computer interaction on the exterior screen of the car. A total of six subjects were recruited, three female and three male. The test evaluation began with the introduction of the experimental background and participation criteria. After confirming the consent statement, the participants were introduced to the autonomous vehicle, which can travel at speeds up to 25 km/hour, with a size of $1.5 \times 1.0 \times 0.4$ m, and is allowed to travel on all road types.

In this test and evaluation, the user test method is adopted to collect real user usage data, including PU, PEOU, acceptance, and trust to measure the usability of the interaction strategy prototype. The scores were assessed using a Likert scale, with 0 indicating strongly disagree and 9 indicating strongly agree. The average score of PU is 6.62, $SD = 0.93$; the mean score of PEOU is 7.14, $SD = 1.21$; the average score of acceptance is 7.74, $SD = 0.56$; and the average score of trust is 6.92, $SD = 1.42$. It can be seen from the usability evaluation results of the system that the tested users are relatively satisfied with the system prototype, and no cognitive impairment is encountered during the test. Through this interactive system prototype, the tested users can understand the intention of the unmanned vehicle smoothly and accurately, which verifies the usability and high acceptance of the system.

22.6 Conclusion

Starting from the common interaction scenes between unmanned vehicles and pedestrians, this paper studies the design requirements of a human−machine interface on the exterior screen of vehicles. Based on the AAM of automatic acceptance, a strategy model of human−computer interaction outside the vehicle that is in line with pedestrian behavior characteristics and cognition is proposed. It can improve the efficiency of information transmission outside the vehicle and people's acceptance by anthropic design, classification, and visualization of information outside the vehicle and multimodal interaction. Combined with the strategy of human−computer interaction outside the vehicle, the vehicle diplomacy mutual design scheme and prototype design are proposed. On this basis, the preliminary prototype design is validated and evaluated. It is verified that the design based on this strategy can indeed improve the acceptance of users.

Reference

[1] F.D. Davis, Technology acceptance model: TAM, in: M.N. Al-Suqri, A.S. Al-Aufi (Eds.), Information Seeking Behavior and Technology Adoption, 1989, pp. 205−219.

Further reading

M. Ghazizadeh, J.D. Lee, Extending the technology acceptance model to assess automation, Cogn. Tech. Work. 14 (2012) 39−49.

T. Hao, Z. Dan-hua, Z. Jiang-hong, Research on automotive human machine interface design based on complex interaction context, Packaging Eng. 18 (2012) 26−30.

J.M. Kraus, F. Nothdurft, P. Hock, et al., Human after all: effects of mere presence and social interaction of a humanoid robot as a co-driver in automated driving, Automotiveui, 2016.

Y.-B. Lu, H.-M. Xu, A comparison study of TAM and its theory basis, Sci. & Technol. Prog. Policy 000 (001) (2012) 2−6.

D. Moore, R. Currano, G.E. Strack, D. Sirkin, the case for implicit external human-machine interfaces for autonomous vehicles, in: Proceedings of the 11th International Conference on Automotive User Interfaces and Interactive Vehicular Applications (Utrecht, Netherlands) (AutomotiveUI'19), Association for Computing Machinery, New York, NY, USA, 2019, pp. 295−307. https://doi.org/10.1145/3342197.3345320.

B.M. Muir, N. Moray, Trust in automation: part II. Experimental studies of trust and human intervention in a process control simulation, Ergonomics 39 (3) (1996) 429−460.

R. Parasuraman, V. Riley, Humans and automation: use, misuse, disuse, abuse, Hum. Factors 39 (2) (1997) 230−253.

X.-W. Zhong, J.-F. Wang, C.-H. Liu, An analysis of the causes of uncanny valley effect based on emotional fuzzy computation, Packaging Eng. 039 (014) (2018) 190−196.

Designing communication strategies of autonomous vehicles with pedestrians: an intercultural study

23

Mirjam Lanzer[1], Franziska Babel[1], Fei Yan[1], Bihan Zhang[2], Fang You[2], Jianmin Wang[2] and Martin Baumann[1]

[1]Department of Human Factors, Institute of Psychology and Education; Faculty of Engineering, Computer Science and Psychology, Ulm University, Germany [2]Car Interaction Design Lab, College of Arts and Media, Tongji University, Shanghai, P.R. China

Chapter Outline

- 23.1 Introduction 337
- 23.2 Related works 338
- 23.3 Method 339
  - 23.3.1 Sample 339
  - 23.3.2 Procedure 340
  - 23.3.3 Materials 341
  - 23.3.4 Questionnaires 342
- 23.4 Results 343
  - 23.4.1 Manipulation check 343
  - 23.4.2 Compliance 343
  - 23.4.3 Acceptance 345
  - 23.4.4 Trust in automation 346
- 23.5 Discussion 347
  - 23.5.1 Practical implications 348
  - 23.5.2 Limitations 348
- 23.6 Conclusion 349
- Further reading 349

23.1 Introduction

Developments in automated driving are transitioning from highway maneuvers to navigating urban areas. In cities, autonomous vehicles (AVs) have the potential to decrease accident rates involving vulnerable road users (VRUs) like pedestrians, for instance, by providing pedestrian protection systems. However, this requires that AVs are able to interact safely and efficiently with pedestrians. Since the majority of interactions between vehicles and pedestrians occur in either crossing scenarios or in shared spaces, such as parking lots, these situations have been focused on in prior research. In many situations, implicit communications, such as the movement pattern of the vehicle, can be used to provide sufficient information for the pedestrian to make a decision whether to cross or not. However, in complex or ambiguous situations, explicit communication between the AV and pedestrians might be needed. This could be the case when there are less clear traffic regulations (e.g., in shared spaces), or when the vehicle is moving slowly or has already come to a standstill. In such situations, external human-machine interfaces (eHMI) can help to guide pedestrians' decisions and facilitate the interaction. While there is a quickly growing body of research on eHMIs concerning modalities, visualizations, colors, and message content, very few studies up to this point have focused on communication strategies that the AV should adopt when interacting with pedestrians (e.g., warning displays with color codes).

Thus we conducted a user study with the following objectives: (1) to compare the influence of different communication strategies (varying degrees of politeness) of an AV on pedestrians' behavior, acceptance, and trust, (2) to test these strategies in two common interaction scenarios (crosswalk and street), and (3) to compare these strategies for two countries, Germany and China, with different (traffic) cultures.

23.2 Related works

There is an increasing number of studies on eHMIs and their effects on pedestrians. When eHMIs were employed to communicate the AV's intention to yield in crossing scenarios, pedestrians reported felt more comfortable and safer. So far, different eHMI approaches have mainly used two modalities: (1) visual and (2) auditory, or a combination of both. Visual approaches include displays that are installed on the external surface of a vehicle, light bands that transfer information through light color and motion, and projections onto the street in front of the pedestrian. For auditory approaches, the AV either uses verbal or nonverbal signals to communicate with the pedestrian. When eHMIs have a visual component, they often show symbols (e.g., eyes, smile, icons) or text. Although text has certain limitations to visually impaired pedestrians, illiterate pedestrians, or pedestrians who do not speak the language that is used, they are also perceived as the most clear. Besides different possible visualizations of eHMIs, the message that is conveyed is also being investigated. Messages include displaying the status of the AV (e.g., automated mode), the intent of the AV (e.g., braking), or instructions for pedestrians (e.g., walk).

In a study by Ackermann et al., a projection-based instruction was preferred. Building upon this, text-based eHMIs containing instructions for pedestrian-AV communication have focused on short, precise requests such as "Don't Walk" or "Go ahead." However, the specific communication strategies considering politeness that are represented by these requests (e.g., command vs polite invitation) have not been systematically compared yet in traffic. Under the assumption that an AV is similar to a nonhumanoid robot, findings regarding human-robot communication can be transferred to pedestrian-AV communication. Here, research has mostly focused on conveying a robot's request in a polite manner to achieve user compliance, trust, and acceptance. However, a polite request has shown to be sometimes inefficient in acquiring user compliance. Hence, a more dominant approach, that is, commands, has also been tested. Although a dominant form of communication (e.g., a command) can be perceived as controlling, it is also a precise and brief form of communication. Contrarily, polite wording is more acceptable but it is harder to interpret due to its subjunctive nature and embellishments and potentially less time-efficient due to the lengthy wording. Commands can be effective to achieve compliance but do sometimes lead to negative effects and non-compliance (i.e., reactance). Preferences for indirect, polite, or direct, dominant communication strategies are hereby grounded in culture.

Politeness is one of the most culture-specific conventions for behavior, varying significantly between Western and Asian cultures. Whereas direct, efficient communication in Germany is preferred, indirect and nonface-threatening approaches are desired in China. Consequently, when developing communication strategies for AVs, culture-specific expectations regarding politeness need to be considered. Consequently, systematically comparing direct and indirect communication strategies for pedestrian-AV communication could provide hints for an acceptable and efficient eHMI message design.

23.3 Method

To evaluate different communication strategies regarding politeness, we designed and conducted a study with a $2 \times 2 \times 3$ mixed design. The country served as the between-subjects factor with two levels: Germany and China. As within-subjects factors, each participant experienced six conditions in a 2×3 design (scenario with two levels: crosswalk versus street and communication strategy with three levels: polite versus neutral versus dominant). This exploratory study was guided by the following research question:

> **RQ:** What impact do the variables communication strategy of an autonomous vehicle and scenario have on pedestrians in terms of (1) compliance, (2) acceptance, and (3) trust in Germany and China?

23.3.1 Sample

The German sample comprised $N = 34$ participants (19 females, 15 males). All German participants were living in Germany at the time of the survey. They were on average $M = 30.68$ ($SD = 11.63$) years old, ranging from 20 to 60 years. The sample consisted of students (44%), employees (41%), and other job descriptions (15%). Participants stated that they had prior experience with a robot (27%) or with an AV (27%). Participant recruitment was conducted through flyers, social media, and various mailing lists. Participants were requested to be at least 18 years old, a German native speaker, and complete the online questionnaire using a tablet or computer.

The Chinese sample comprised $N = 56$ participants (44 females, nine males, three unknown). All Chinese participants were living in China at the time of the survey. They were on average $M = 22.31$ ($SD = 6.24$) years old, ranging from 18 to 30 years. The sample consisted of students (89%), employees (7%), and other job descriptions (4%). Participants stated that they had prior experience with a robot (20%) or with an AV (38%). Participant recruitment was conducted through an online questionnaire system and on social media. Participants were requested to be at least 18 years old, a Chinese native speaker, and complete the online questionnaire using a tablet or computer.

23.3.2 Procedure

The online survey started with a brief introduction, participation criteria, and data protection regulations. After confirming the declaration of consent, the autonomous delivery vehicle was introduced to the participants (Fig. 23.1).

It was explained that the autonomous delivery vehicle could drive up to 25 km/hour, had a size of $0.5 \times 0.5 \times 0.5$ m, and was allowed to drive on all road types. Then, it was randomized whether the participants saw the crosswalk or the street scenario first. Each scenario was shortly described to the participants alongside a picture (Fig. 23.2).

The introductions to the scenarios were given as follows: "Imagine that you are walking from one building to another. You are in a hurry because you have an appointment in a few minutes." (1) Crosswalk scenario: "To reach your destination you have to cross the road at a zebra crossing. You check if the road is clear. While doing so, you see an autonomous delivery vehicle approaching the zebra crossing." (2) Street scenario: "As many passers-by are on the sidewalk, you decide to walk in the street. As there is rarely much traffic at this point, the road can usually also be used as a pedestrian.

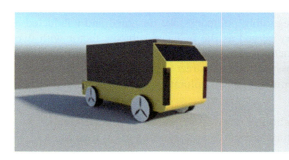 

FIGURE 23.1

Front and back view of the autonomous delivery vehicle.

(A) Crosswalk. (B) Street.

FIGURE 23.2

Pictures used for scenario descriptions.

After some time you notice that an autonomous delivery vehicle approaches from the front." Within each scenario, the three different strategies were randomized again. For each strategy in each scenario, a video was shown to the participants. Participants had the opportunity to replay the video once more. After each video, participants filled in the questionnaires described below. In the end, demographic questions were asked and a manipulation check was performed. On average, the study lasted 30 minutes for the German and 40 minutes for the Chinese sample. Participants were compensated with course credit or a chance to win one out of 10 Amazon gift certificates (German sample only).

23.3.3 Materials

The combination of scenario and communication strategy resulted in six interaction messages. The polite, neutral, and dominant self-developed interaction messages are presented in Table 23.1. The messages were developed based on research regarding politeness markers. For the polite message, the politeness strategy "please" in combination with the counterfactual modal (*Could you...*) was applied. For the neutral message, only "please" was used as a politeness marker. For the dominant message, no politeness markers were used and it was formulated as a direct command. Interaction messages were first developed in German, then translated into English, and finally to Chinese. The Chinese messages were then reviewed alongside the German messages by an independent researcher who speaks both German and Chinese. Every interaction message was presented in the form of a 20-second-long video. The videos were produced using CINEMA 4D software, which built a scene model and used animation and camera tracking technology to make the autonomous delivery vehicle move in the scene. In both scenarios, a short overview of the situation was given in a birds-eye view in the beginning. Then the camera switches to the pedestrian's perspective and the participant experiences the scenario from a first-person perspective. A minimalist design was chosen so that participants could fully focus on the presented strategies.

In the crosswalk scenario, the autonomous delivery vehicle drives in the street toward a crosswalk. The pedestrian representing the participant walks toward the crosswalk with the intention to cross. Then the autonomous delivery vehicle stops at the crosswalk and projects one out of three text messages onto the street according to the respective communication strategy (Fig. 23.3A). In the street scenario, the autonomous delivery vehicle drives in the street and the pedestrian representing the participant walks in the street. Both are moving toward each other until the autonomous delivery vehicle projects its message on the ground (Fig. 23.3B).

Table 23.1 Interaction messages of the autonomous delivery vehicle used in the online experiment.

		Crosswalk	Street
Communication strategy	Polite	(Chinese: 请您稍等下, 好吗?)修改为: (English: Please, hold on)	(Chinese: 请您让开, 好吗?)修改为: (English: Would you please step aside ?)
	Neutral	(Chinese: 请您稍等.)修改为: (English: Please, wait)	(Chinese: 请您让开)修改为: (English: Please step aside)
	Dominant	(Chinese: 稍等一下!)修改为: (English: Wait !)	(Chinese: 让开)修改为: (English: Go away!)

(A) Crosswalk, polite strategy, Chinese. ("Could you please wait?") (B) Street, dominant strategy, German. ("Get out of the way!")

FIGURE 23.3

Strategies projected onto the street.

23.3.4 Questionnaires

After each video, we measured:

> **Acceptance:** Van der Laan acceptance scale (e.g., useless-useful) with two additional items each for politeness (e.g., impolite-polite) and respect (e.g., disrespectful-respectful).
> **Affective state:** Intensity and valence dimensions of the self-assessment manikin (SAM) (participants choose pictures that suit their affective state).
> **Compliance:** Self-developed single items adapted to the respective scenario (see below).
> **Fear:** Fear of robots scale adapted from (e.g., I would be afraid of the system).
> **Power distribution:** Power distribution scale adapted from (e.g., Who had the most control of what happened in the situation?).
> **Trust in automation:** Three-item version of the trust in automation scale (e.g., I trust the system).

Only the results for compliance, acceptance (without the additional items for politeness and respect), and trust are discussed in this paper. The compliance item was used to assess how participants would behave in this situation. They were given the following options for the crosswalk scenario:
(1) *I stop immediately and wait for the vehicle to pass.*
(2) *After some hesitation, I wait for the vehicle to pass.*
(3) *After some hesitation, I cross the street, so that the vehicle has to stop.*
(4) *I cross the street immediately, so that the vehicle has to stop.*
For the street scenario, they were given the following options:
(1) *I go out of the way immediately.*
(2) *After some hesitation, I go out of the way.*
(3) *After some hesitation, I do not go out of the way.*
(4) *I do not go out of the way.*
For both scenarios, participants had a fifth option of "I would behave as follows": followed by a text field to write down their answer. Additionally, participants had the chance to write down any other remarks on how they experienced the behavior of the AV after each video. In the end, they were asked to rate each strategy: "How polite would you rate this sentence?" and "How dominant would you rate this sentence?" on a 7-point Likert scale. This served as a manipulation check to make sure that participants experienced the strategies as intended.

23.4 Results

In this section, descriptive and inferential statistics are presented. For all metric variables, factorial repeated measures analysis of variance (ANOVA) were conducted for the German and Chinese samples separately since assumptions for three-way mixed ANOVA were not met according to Levene tests (homogeneity of error variances) and Box tests (homogeneity of covariances). Shapiro-Wilk tests showed that the manipulation check and some of the acceptance items were not normally distributed. Nevertheless, ANOVAs were conducted since they are relatively robust against not normally distributed variables and the nonparametric Friedman test revealed the same result patterns in all cases. When there were violations of sphericity detected with the Mauchly test, Greenhouse-Geisser corrections were applied. For ordinal data, Friedman ANOVAs were conducted. For all posthoc tests, Bonferroni corrections were used.

23.4.1 Manipulation check

A manipulation check was conducted to assess whether each communication strategy was perceived as polite, dominant, or neutral as intended (Fig. 23.4). For ratings of politeness, repeated measures ANOVA showed statistically significant differences for the crosswalk scenario in the German ($F\ (2, 66) = 99.13, p < .001$) and the Chinese sample ($F\ (2, 110) = 70.58, p < .001$), as well as for the street scenario in the German ($F\ (2, 66) = 102.38, p < .001$) and the Chinese sample ($F\ (2, 110) = 69.32, p < .001$). Hence for both samples in two scenarios, the polite strategy was perceived as the most polite and the dominant strategy as the least polite. The neutral strategy was in between. The same was found for the dimension of dominance. Dominance ratings were also statistically significant for both the crosswalk scenario in the German ($F\ (2, 66) = 5.32, p = .007$) and the Chinese sample ($F\ (2, 110) = 126.33, p = <.001$) and the street scenario in the German ($F\ (2, 66) = 12.69, p = <.001$) and Chinese ($F\ (2, 110) = 61.06, p = <.001$) sample. In all scenarios and both samples, the dominant strategy was rated to be the most dominant and the polite strategy as least dominant with the neutral strategy in between. All posthoc comparisons between strategies were significant except for the dominance rating of the neutral versus dominant strategy in the crosswalk scenario in the German sample.

23.4.2 Compliance

Descriptive results can be seen in Fig. 23.5. There were differences between the scenarios. In the German sample, fewer people complied in the crosswalk scenario compared with the street scenario while in the Chinese sample, fewer people complied in the street scenario compared with the crosswalk scenario. In both samples and for both scenarios, participants complied more often when the AV's strategy was polite compared with neutral or dominant. However, Friedman-ANOVAs indicated no significant differences between strategies in the German sample for the crosswalk ($\chi^2\ (2) = 1.47, p = .481$) and street scenario ($\chi^2\ (2) = 1.65, p = .437$). Contrary to this, Friedman-ANOVAs in the Chinese sample were significant both for the crosswalk ($\chi^2\ (2) = 7.67, p = .022$) and the street scenario ($\chi^2\ (2) = 20.15, p < .001$). Wilcoxon posthoc tests in the Chinese sample revealed that for the crosswalk scenario, the polite versus dominant strategy significantly differed ($z = -2.25, p = .025$). For the street scenario, the polite versus dominant ($z = -3.61, p < .001$) and the neutral versus dominant ($z = -3.02, p = .003$) strategy significantly differed.

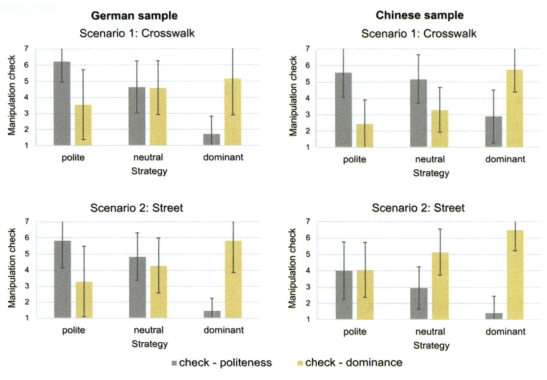

FIGURE 23.4

Means of politeness and dominance ratings for the three communication strategies for crosswalk and street scenario for the German ($N = 34$) and Chinese ($N = 56$) sample. Means are based on a 7-point Likert scale. The error bars represent the standard deviations.

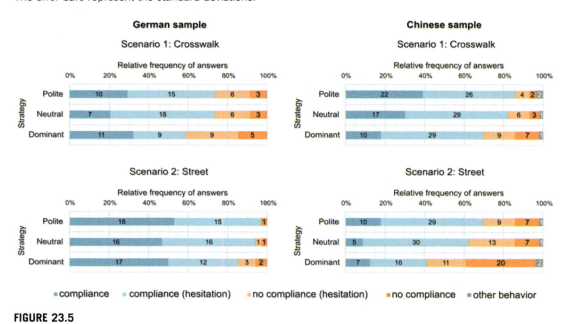

FIGURE 23.5

Absolute and relative compliance frequencies according to *Strategy* (polite strategy increases compliance) and *Scenario* for the German ($N = 34$) and Chinese ($N = 56$) sample.

23.4.3 Acceptance

The two subscales of acceptance, usefulness, and satisfaction, were analyzed separately. The polite strategy was rated most useful and satisfactory followed by the neutral and the dominant strategy in both scenarios (Fig. 23.6).

There was a significant main effect of strategy on usefulness ($F (2, 66) = 23.17$, $p < .001$) and satisfaction ($F (2, 66) = 42.44$, $p < .001$) in the German sample showing that the polite strategy received higher ratings on usefulness and satisfaction. There was no significant main effect of the scenario and no significant interaction between strategy and scenario for either usefulness or satisfaction. Posthoc tests revealed that in the German sample, all differences between strategies were significant ($p < .05$) for usefulness and satisfaction except for the polite versus neutral comparison in usefulness.

In the Chinese sample, there was also a significant main effect of strategy on usefulness ($F (2, 110) = 56.55$, $p < .001$) and satisfaction ($F (2, 110) = 37.58$, $p < .001$). Additionally, there was a

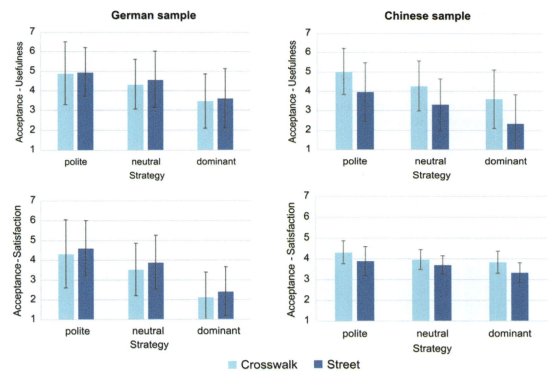

FIGURE 23.6

Main effects on Acceptance (usefulness and satisfaction) of *Strategy* (polite strategy increases acceptance) and *Scenario* in the German ($N = 34$) and Chinese ($N = 56$) sample. Means are based on a 7-point Likert scale. The error bars represent the standard deviations.

significant main effect of the scenario on usefulness ($F(2, 110) = 66.14$, $p < .001$) and satisfaction ($F(2, 110) = 55.87$, $p < .001$). For the crosswalk scenario, usefulness and satisfaction ratings were higher than for the street scenario. Interaction effects were not significant. Posthoc tests showed that in the Chinese sample, all differences between strategies were significant ($p < .001$) for usefulness and satisfaction.

23.4.4 Trust in automation

For both samples in two scenarios, trust was highest for the polite strategy, followed by the neutral strategy, and lowest for the dominant strategy (Fig. 23.7). Comparing scenarios, trust was rated higher for the street scenario in the German sample and the crosswalk scenario in the Chinese sample.

For the German sample, a repeated measures ANOVA showed a significant main effect for strategy [$F(2, 66) = 26.96$, $p < .001$; cf. Fig. 23.7 (*left*)]. There was no significant main effect for the scenario and no significant interaction in the German sample. Posthoc tests showed that the polite strategy received significantly higher trust ratings compared with the neutral ($p = .014$) and dominant ($p < .001$) strategies and that the neutral strategy had higher ratings than the dominant strategy ($p < .001$). In the Chinese sample, there was a significant main effect of strategy [$F(1.81, 110) = 48.95$, $p < .001$; cf. Fig. 23.7 (*right*)], a significant main effect of scenario ($F(2, 110) = 36.88$, $p < .001$), as well as a significant interaction ($F(1.70, 110) = 4.05$, $p = .026$). Posthoc tests showed that all comparisons between strategies and scenarios were significant ($p < .01$) except for the polite versus neutral strategy in the crosswalk scenario and between the polite strategy in the crosswalk versus street scenario.

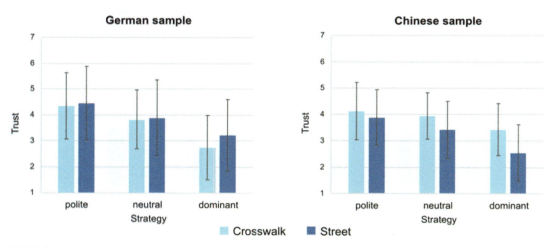

FIGURE 23.7

Main effects on Trust score of *Strategy* (polite strategy increases trust) and *Scenario* in the German ($N = 34$) and Chinese ($N = 56$) sample. Means are based on a 7-point Likert scale. The error bars represent the standard deviations.

23.5 Discussion

This exploratory study examined whether different communication strategies regarding the politeness of an autonomous delivery vehicle influence pedestrians' compliance with its requests as well as acceptance and trust toward the autonomous delivery vehicle in two different scenarios, at a crosswalk and in the street. This was explored in an online study with a German and a Chinese sample. The communication strategy was either polite, neutral, or dominant and was projected as a text message onto the street. A manipulation check showed that participants in Germany and China perceived the polite, neutral, and dominant communication sentences as intended. After the autonomous delivery vehicle made its request to the pedestrian to either wait in the crosswalk scenario or go out of the way in the street scenario, compliance with these requests was assessed. In the German sample, there was no significant effect of the communication strategy on participants' compliance. In the Chinese sample, however, more participants complied with the request when it was presented in a polite manner. In the German sample, the crosswalk scenario with its accompanying traffic rules possibly had a greater influence than the communication strategy. Some participants ($N = 5$) stated that they did not comply with the AV's request to wait at the crosswalk because "at a zebra crossing ALL vehicles must stop, no exceptions." On the other hand, in the road scenario, some participants ($N = 3$) indicated that they were giving way to the automated vehicle because "the vehicle is right that I should not be on the street." Both explanations of the described behaviors are congruent with German road traffic regulations, and the majority of drivers also adhere to those regulations (e.g., yielding rates of up to 73% at crosswalks in an observational study in Germany). Therefore the chosen scenarios were possibly too closely linked to learned behavior for German participants. Moreover, in the German sample, compliance was higher in the street scenario, than in the crosswalk scenario, while in the Chinese sample, it was vice versa. Again, the already described rule-compliant behavior might explain this effect in the German sample. In the Chinese sample, however, the higher compliance rate in the crosswalk scenario could be associated with a generally low rate of vehicles stopping at crosswalks, even though Chinese drivers are also required by law to yield to pedestrians at crosswalks. Chinese pedestrians may be accustomed to giving priority to vehicles at zebra crossings and therefore gave priority to the AV more often in this study. In addition, it was found that their understanding and judgments of the interaction strategy regarding politeness are not only based on the punctuation and salutation, but also related to the semantic understanding, personal habits, and mood while crossing the road. For example, in the street scenario, some users ($N = 5$) reported that the words "Hao Ma?" (meaning "all right?" in English) and "Qing" (meaning "please" in English), when used in the polite interaction strategy might cause ambiguities in tone and sound provocative, which made them feel uncomfortable, and therefore they did not give way to the autonomous delivery vehicle.

Besides compliance, it was examined whether the communication strategy or the scenario affected pedestrians' acceptance and trust toward the autonomous delivery vehicle. The results show that the strategy changed the acceptance of the vehicle. In both samples, the lowest acceptance was observed for the dominant communication strategy and the highest acceptance for the polite communication strategy. The autonomous delivery vehicle was rated significantly more useful and satisfactory when its request was communicated in a polite way instead of in a dominant way. This is in line with previous studies regarding politeness. In the German sample, there were no differences in acceptance ratings between the different scenarios. In the Chinese sample, however, the autonomous delivery vehicle's request had higher acceptance ratings at the crosswalk than in the street. This is consistent with the compliance finding for the Chinese sample, where more participants complied in the crosswalk scenario than in the street scenario. The results for trust follow the pattern. The autonomous delivery vehicle was trusted more when it communicated its request politely instead of dominantly. This applies to both the German and the Chinese sample. Politeness has indeed been described as an antecedent of trust in human-robot interaction. Again, in the Chinese sample trust ratings were higher for the crosswalk than the street scenario. Additionally, there was a significant interaction effect in the Chinese sample. In the polite condition, trust ratings were equally high for the crosswalk and the street scenario. When the strategy became more dominant, however, trust ratings between the scenarios diverged more, with participants trusting the AV more in the crosswalk scenario than in the street scenario. So when the vehicle communicated politely, it did not differ in which situation the request took place concerning pedestrians' trust toward the vehicle.

23.5.1 Practical implications

These findings can be used to design communication strategies between automated vehicles and pedestrians and to develop eHMIs that enhance the acceptance and trust toward automated vehicles. The polite communication strategy has been shown to have the most positive impacts on compliance, acceptance, and trust. Even though space might be limited when displaying messages to pedestrians, the communication strategy should be as polite as possible. Furthermore, there were cultural differences between German and Chinese participants. While in both groups the polite strategy achieved better results, there were differences between scenarios. While in the Chinese sample, there was higher compliance in the crosswalk scenario, the opposite was true for the German sample. In the street scenario, when the same polite strategy was applied in the German sample, almost all of the people complied with the autonomous delivery vehicle's request to move out of the way while in the Chinese sample, a substantial amount of people did not. Therefore the communication between automated vehicles and pedestrians should be looked into in more cross-cultural studies. Systems that are designed and tested in one country might not obtain the same effects in another country, even if they are used in similar situations.

23.5.2 Limitations

There are some limitations to this study. First, this study only measured intended behavior, which does not necessarily reflect real behavior in a direct interaction. Surprise or habituation effects can play a major role in real interactions with the automated delivery vehicle. However, there are indications that results from online studies are comparable to results from other data collection methods. Second, the type of eHMI in this study was a text message projected onto the street. This was chosen to ensure that the polite or dominant strategy tested in this study came across very clearly. However, text might not be ideal in all traffic situations (e.g., takes long to read) and is not inclusive for all pedestrians (e.g., visually impaired). Therefore differences in communication strategies should be systematically tested in other eHMI modalities as well. Furthermore, more implicit communication channels should be considered since pedestrians use the vehicle's movement and approaching strategy to decide whether to cross or not. Third, in the presented scenarios, one AV interacted with one pedestrian. In real traffic, however, more complex scenarios with multiple AVs and multiple pedestrians can occur. The issue of scalability should therefore be considered in future studies. Fourth, the results in this study refer to an AV that is a small delivery vehicle. With a full-sized vehicle, these results might differ since a conflict or collision between the AV and pedestrians would have more severe consequences for the pedestrian. Fifth, samples were not equal in size and differed in participant characteristics like age or occupation. Even though an age difference of 10 years was shown to not have an influence on politeness perception, as was the case in this study, future studies should aim for a balanced sample and take sample characteristics into account.

23.6 Conclusion

This experimental study presents the first systematic evaluation of different communication strategies regarding politeness in pedestrian-AV interaction. In addition, the strategies were tested in two common interaction scenarios, and the results were compared between participants from Germany and China. While the polite strategy led to more acceptance and trust toward the AV among Germans and Chinese, only the Chinese complied more often with a polite request from the AV. This again shows the necessity of intercultural studies: that eHMI concepts cannot simply be transferred from one country to another. Overall, a polite communication strategy seems to be a relevant factor for the development of eHMIs that should be taken into account. This contributes to the investigation of effective and accepted communication strategies between automated vehicles and pedestrians.

Further reading

C. Ackermann, M. Beggiato, S. Schubert, J.F. Krems, An experimental study to investigate design and assessment criteria: what is important for communication between pedestrians and automated vehicles? Applied Ergonomics 75 (2019) 272−282. Available from: https://doi.org/10.1016/j.apergo.2018.11.002. 2019.

A. Rasouli, J.K. Tsotsos, Autonomous vehicles that interact with pedestrians: a survey of theory and practice, IEEE Transactions on Intelligent Transportation Systems 21 (3) (2019) 900−918. Available from: https://doi.org/10.1109/TITS.2019.2901817. 2019.

P. Bazilinskyy, D. Dodou, J. de Winter, Survey on eHMI concepts: the effect of text, color, and perspective, Transp. Res. Part F Traffic Psychol. Behav. 67 (2019) (2019) 175−194. Available from: https://doi.org/10.1016/j.trf.2019.10.013.

M.M. Bradley, P.J. Lang, Measuring emotion: the self-assessment manikin and the semantic differential, Journal of Behavior Therapy and Experimental Psychiatry 25 (1) (1994) 49−59. Available from: https://doi.org/10.1016/0005-7916(94)90063-9. 1994.

G. Briggs, M. Scheutz, The pragmatic social robot: toward socially-sensitive utterance generation in human-robot interactions, in: 2016 AAAI Fall Symposium Series Artificial Intelligent Human-Robot Interact. Tech. Rep. FS-16-01, Association for the Advancement of Artificial Intelligence (AAAI), Paola Alto, California, USA, 2016, pp. 12−15. <https://www.aaai.org/ocs/index.php/FSS/FSS16/paper/viewFile/14138/13657>.

M.-P. Böckle, A.P. Brenden, M. Klingegård, A. Habibovic, M. Bout, SAV2P: exploring the impact of an interface for shared automated vehicles on pedestrians' experience, in: Proceedings of the 9th International Conference on Automotive User Interfaces and Interactive Vehicular Applications Adjunct (Oldenburg, Germany) (AutomotiveUI '17), Association for Computing Machinery, New York, NY, USA, 2017, pp. 136–140. <https://doi.org/10.1145/3131726.3131765>.

C.-M. Chang, K. Toda, T. Igarashi, M. Miyata, Y. Kobayashi, A video-based study comparing communication modalities between an autonomous car and a pedestrian, in: Adjunct Proceedings of the 10th International Conference on Automotive User Interfaces and Interactive Vehicular Applications (Toronto, ON, Canada) (AutomotiveUI '18), Association for Computing Machinery, New York, NY, USA, 2018, pp. 104–109. <https://doi.org/10.1145/3239092.3265950>.

C.-M. Chang, K. Toda, D. Sakamoto, T. Igarashi, Eyes on a car: an interface design for communication between an autonomous car and a pedestrian, in: Proceedings of the 9th International Conference on Automotive User Interfaces and Interactive Vehicular Applications (Oldenburg, Germany) (AutomotiveUI '17), Association for Computing Machinery, New York, NY, USA, 2017, pp. 65–73. <https://doi.org/10.1145/3122986.3122989>.

M. Colley, M. Walch, J. Gugenheimer, A. Askari, E. Rukzio, Towards inclusive external communication of autonomous vehicles for pedestrians with vision impairments, in: Proceedings of the 2020 CHI Conference on Human Factors in Computing Systems (Honolulu, HI, USA) (CHI '20), Association for Computing Machinery, New York, NY, USA, 2020, pp. 1–14. <https://doi.org/10.1145/3313831.3376472>.

M. Colley, M. Walch, J. Gugenheimer, E. Rukzio, Including people with impairments from the start: external communication of autonomous vehicles, in: Proceedings of the 11th International Conference on Automotive User Interfaces and Interactive Vehicular Applications: Adjunct Proceedings (Utrecht, Netherlands) (AutomotiveUI '19), Association for Computing Machinery, New York, NY, USA, 2019, pp. 307–314. <https://doi.org/10.1145/3349263.3351521>.

M. Colley, M. Walch, E. Rukzio, Unveiling the lack of scalability in research on external communication of autonomous vehicles, in: Extended Abstracts of the 2020 CHI Conference on Human Factors in Computing Systems (Honolulu, HI, USA) (CHI EA '20), Association for Computing Machinery, New York, NY, USA, 2020, pp. 1–9. <https://doi.org/10.1145/3334480.3382865>.

C. Danescu-Niculescu-Mizil, M. Sudhof, J. Dan, J. Leskovec, C. Potts, A computational approach to politeness with application to social factors, in: ACL 2013 - 51st Annual Meeting Association Computer Linguistics Proceedings Conference, Vol. 1, 2013, pp. 250–259. <https://doi.org/10.48550/arXiv:1306.6078>.

S. Deb, L.J. Strawderman, D.W. Carruth, Investigating pedestrian suggestions for external features on fully autonomous vehicles: a virtual reality experiment, Transp. Res. Part F Traffic Psychol. Behav. 59 (2018) (2018) 135–149. Available from: https://doi.org/10.1016/j.trf.2018.08.016.

D. Dey, A. Habibovic, B. Pfleging, M. Martens, J. Terken, Color and animation preferences for a light band EHMI in interactions between automated vehicles and pedestrians, in: Proceedings of the 2020 CHI Conference on Human Factors in Computing Systems (Honolulu, HI, USA) (CHI '20), Association for Computing Machinery, New York, NY, USA, 2020, pp. 1–13. <https://doi.org/10.1145/3313831.3376325>.

D. Dey, J. Terken, Pedestrian interaction with vehicles: roles of explicit and implicit communication, in: Proceedings of the 9th International Conference on Automotive User Interfaces and Interactive Vehicular Applications (Oldenburg, Germany) (AutomotiveUI '17), Association for Computing Machinery, New York, NY, USA, 2017, pp. 109–113. <https://doi.org/10.1145/3122986.3123009>.

S.M. Faas, L.-A. Mathis, M. Baumann, External HMI for self-driving vehicles: which information shall be displayed? Transp. Res. Part F Traffic Psychol. Behav. 68 (2020) (2020) 171–186. Available from: https://doi.org/10.1016/j.trf.2019.12.009.

F.H. Gerpott, D. Balliet, S. Columbus, C. Molho, R.E. de Vries, How do people think about interdependence? A multidimensional model of subjective outcome interdependence, Journal of Personality and Social Psychology 115 (4) (2018) 716. Available from: https://doi.org/10.1037/pspp0000166. 2018.

G.V. Glass, P.D. Peckham, J.R. Sanders, Consequences of failure to meet assumptions underlying the fixed effects analyses of variance and covariance, Review of Educational Research 42 (3) (1972) 237−2881972. Available from: https://doi.org/10.3102/00346543042003237.

A. Habibovic, V.M. Lundgren, J. Andersson, M. Klingegård, T. Lagström, A. Sirkka, J. Fagerlönn, C. Edgren, R. Fredriksson, S. Krupenia., Communicating intent of automated vehicles to pedestrians, Frontiers in Psychology 9 (2018) 1336. Available from: https://doi.org/10.3389/fpsyg.2018.01336. 2018.

A.-C. Hensch, I. Neumann, M. Beggiato, J. Halama, J.F. Krems, Effects of a light-based communication approach as an external HMI for automated vehicles—a wizard-of-oz study, Transactions on Transport Sciences 10 (2) (2019) 18−32. Available from: https://doi.org/10.5507/tots.2019.012. 2019.

P. Hock, F. Babel, K. Muehl, E. Rukzio, M. Baumann, Online experiments as a supplement of automated driving simulator studies: a methodological insight, in: Proceedings of the 11th International Conference on Automotive User Interfaces and Interactive Vehicular Applications: Adjunct Proceedings (Utrecht, Netherlands) (AutomotiveUI '19), Association for Computing Machinery, New York, NY, USA, 2019, pp. 282−286. <https://doi.org/10.1145/3349263.3351334>.

Y. Huang, Politeness principle in cross-culture communication, English Language Teaching 1 (1) (2008) 96−1012008. Available from: https://files.eric.ed.gov/fulltext/EJ1082589.pdf.

J.-Y. Jian, A.M. Bisantz, C.G. Drury, Foundations for an empirically determined scale of trust in automated systems, International Journal of Cognitive Ergonomics 4 (1) (2000) 53−71. Available from: https://doi.org/10.1207/S15327566IJCE0401_04. 2000.

J.D. Van Der Laan, A. Heino, D. De Waard, A simple procedure for the assessment of acceptance of advanced transport telematics, Transp. Res. Part C Emerg. Technol. 5 (1) (1997) 1−10. Available from: https://doi.org/10.1016/S0968-090X(96)00025-3. 1997.

N. Lee, J. Kim, E. Kim, O. Kwon, The influence of politeness behavior on user compliance with social robots in a healthcare service setting, International Journal of Social Robotics 9 (5) (2017) 727−743. Available from: https://doi.org/10.1007/s12369-017-0420-0. 2017.

Y.M. Lee, R. Madigan, J. Garcia, A. Tomlinson, A. Solernou, R. Romano, G. Markkula, N. Merat, J. Uttley, Understanding the messages conveyed by automated vehicles, in: Proceedings of the 11th International Conference on Automotive User Interfaces and Interactive Vehicular Applications (Utrecht, Netherlands) (AutomotiveUI '19), Association for Computing Machinery, New York, NY, USA, 2019, pp. 134−143. <https://doi.org/10.1145/3342197.3344546>.

Y.M. Lee, R. Madigan, O. Giles, L. Garach-Morcillo, G. Markkula, C. Fox, F. Camara, M. Rothmueller, S.A. Vendelbo-Larsen, P.H. Rasmussen, Road users rarely use explicit communication when interacting in today's traffic: implications for automated vehicles, Cognition, Technology & Work (2020) 1−142020. Available from: https://doi.org/10.1007/s10111-020-00635-y.

Y. Li, M. Dikmen, T.G. Hussein, Y. Wang, C. Burns, To cross or not to cross: urgency-based external warning displays on autonomous vehicles to improve pedestrian crossing safety, in: Proceedings of the 10th International Conference on Automotive User Interfaces and Interactive Vehicular Applications (Toronto, ON, Canada) (AutomotiveUI '18), Association for Computing Machinery, New York, NY, USA, 2018, pp. 188−197. <https://doi.org/10.1145/3239060.3239082>.

Y. Li, M. Dikmen, T.G. Hussein, Y. Wang, C. Burns, To cross or not to cross: urgency-based external warning displays on autonomous vehicles to improve pedestrian crossing safety, in: Proceedings of the 10th International Conference on Automotive User Interfaces and Interactive Vehicular Applications (Toronto, ON, Canada) (AutomotiveUI '18), Association for Computing Machinery, New York, NY, USA, 2018, pp. 188−197. <https://doi.org/10.1145/3239060.3239082>.

D.F. Llorca, V. Milanes, I.P. Alonso, M. Gavilan, I.G. Daza, J. Perez, M.Á. Sotelo, Autonomous pedestrian collision avoidance using a fuzzy steering controller, IEEE Transactions on Intelligent Transportation Systems 12 (2) (2011) 390−4012011. Available from: https://doi.org/10.1109/TITS.2010.2091272.

A. Löcken, C. Golling, A. Riener, How should automated vehicles interact with pedestrians? A comparative analysis of interaction concepts in virtual reality, in: Proceedings of the 11th International Conference on Automotive User Interfaces and Interactive Vehicular Applications (Utrecht, Netherlands) (AutomotiveUI '19), Association for Computing Machinery, New York, NY, USA, 2019, pp. 262–274. <https://doi.org/10.1145/3342197.3344544>.

K. Mahadevan, S. Somanath, E. Sharlin, Communicating awareness and intent in autonomous vehicle-pedestrian interaction, in: Proceedings of the 2018 CHI Conference on Human Factors in Computing Systems (Montreal QC, Canada) (CHI '18), Association for Computing Machinery, New York, NY, USA, 2018, pp. 1–12. <https://doi.org/10.1145/3173574.3174003>.

N. Mavridis, A review of verbal and non-verbal human–robot interactive communication, Robotics and Autonomous Systems 63 (2015) (2015) 22–35. Available from: https://doi.org/10.1016/j.robot.2014.09.031.

C.H. Miller, L.T. Lane, L.M. Deatrick, A.M. Young, K.A. Potts, Psychological reactance and promotional health messages: the effects of controlling language, lexical concreteness, and the restoration of freedom, Hum. Commun. Res. 33 (2) (2007) 219–240. Available from: https://doi.org/10.1111/j.1468-2958.2007.00297.x. 2007.

D. Moore, R. Currano, G. Ella Strack, D. Sirkin, The case for implicit external human-machine interfaces for autonomous vehicles, in: Proceedings of the 11th International Conference on Automotive User Interfaces and Interactive Vehicular Applications (Utrecht, Netherlands) (AutomotiveUI '19), Association for Computing Machinery, New York, NY, USA, 2019, pp. 295–307. <https://doi.org/10.1145/3342197.3345320>.

T.T. Nguyen, K. Holländer, M. Hoggenmueller, C. Parker, M. Tomitsch, Designing for projection-based communication between autonomous vehicles and pedestrians, in: Proceedings of the 11th International Conference on Automotive User Interfaces and Interactive Vehicular Applications (Utrecht, Netherlands) (AutomotiveUI '19), Association for Computing Machinery, New York, NY, USA, 2019, pp. 284–294. <https://doi.org/10.1145/3342197.3344543>.

K.E. Oleson, D.R. Billings, V. Kocsis, J.Y.C. Chen, P.A. Hancock, Antecedents of trust in human-robot collaborations, in: 2011 IEEE International Multi-Disciplinary Conference on Cognitive Methods in Situation Awareness and Decision Support (CogSIMA), IEEE, 2011, pp. 175–178. <https://doi.org/10.1109/COGSIMA.2011.5753439>.

M.A. J. Roubroeks, J.R. C. Ham, C.J. H. Midde, The dominant robot: threatening robots cause psychological reactance, especially when they have incongruent goals, in: International Conference Persuas. Technology, Springer Heidelberg, Berlin, Heidelberg, Germany, 2010, pp. 174–184. <https://doi.org/10.1007/978-3-642-13226-1_18>.

M. Salem, G. Lakatos, F. Amirabdollahian, K. Dautenhahn, Would you trust a (faulty) robot?, in: Proceedings of the Tenth Annual ACM/IEEE International Conference on Human-Robot Interaction – HRI '15. ACM, New York, NY, USA, 2015, pp. 141–148. <https://doi.org/10.1145/2696454.2696497>.

M. Salem, M. Ziadee, M. Sakr, Effects of politeness and interaction context on perception and experience of HRI, in: International Conference on Social Robotics, Springer, Springer International, Switzerland, 2013, pp. 531–541. <https://doi.org/10.1007/978-3-319-02675-6_53>.

V. Srinivasa, L. Takayama, Help Me please: robot politeness strategies for soliciting help from people, in: Proceedings of the 2016 CHI Conference on Human Factors Computer System - CHI '16 (San Jose, California, USA).Association for Computing Machinery, New York, NY, USA, 2016, pp. 4945–4955. <https://doi.org/10.1145/2858036.2858217>.

M. Strait, P. Briggs, M. Scheutz, Gender, more so than age, modulates positive perceptions of language-based human-robot interactions, in: 4th International Symposium on New Frontiers in Human-Robot Interactions, Canterbury, UK, 2015, pp. 54–61. <https://hrilab.tufts.edu/publications/straitetal15aisb.pdf>.

A. Weiss, R. Bernhaupt, M. Tscheligi, E. Yoshida, Addressing user experience and societal impact in a user study with a humanoid robot, in: Adaptive and Emergent Behaviour and Complex Systems – Proceedings of the 23rd Convention of the Society for the Study of Artificial Intelligence and Simulation of Behaviour, AISB 2009. AISB, 2009, pp. 150–157. <http://citeseerx.ist.psu.edu/viewdoc/download?doi = 10.1.1.160.765&rep = rep1&type = pdf>.

X. Zhuang, C. Wu, Pedestrian gestures increase driver yielding at uncontrolled mid-block road crossings, Accident Analysis & Prevention 70 (2014) (2014) 235–244. Available from: https://doi.org/10.1016/j.aap.2013.12.015.

PART 6

Intelligent cockpit HMI evaluation

CHAPTER 24

Augmented reality cognitive interface in enhancing human vehicle collaborative driving safety: a design perspective

Fang You[1], Yuwei Liang[1], Qianwen Fu[2] and Jun Zhang[2,3]

[1]Car Interaction Design Lab, College of Arts and Media, Tongji University, Shanghai, P.R. China [2]College of Design and Innovation, Tongji University, Shanghai, P.R. China [3]School of Design, The Hong Kong Polytechnic University, Hong Kong, P.R. China

Chapter Outline

24.1 Introduction 358
24.2 Context and theoretical framework 359
  24.2.1 Human-vehicle team and cooperation mechanism 359
  24.2.2 Cognitive interface and augmented reality information elements 362
24.3 Design method for augmented reality human-machine interface 364
  24.3.1 Collaborative contextual analysis 365
  24.3.2 Cognitive information visualization 368
  24.3.3 Component adjustment and overlay 370
24.4 Empirical evaluation 371
  24.4.1 Hypotheses 372
  24.4.2 Experimental design 372
  24.4.3 Apparatus and materials 374
  24.4.4 Procedure 375
  24.4.5 Measures and analysis 375
24.5 Results 377
  24.5.1 Subjective scales 377
  24.5.2 Fixation metric 378
  24.5.3 Fixation heatmap 380
  24.5.4 Interviews 380
24.6 Discussion 381
  24.6.1 Visualization components and quantities 381
  24.6.2 Attention and cognitive demands 382
  24.6.3 Training and mental models 383
  24.6.4 Limitations 384

| 24.7 Conclusions | 384 |
| References | 384 |

24.1 Introduction

In highly automated vehicles (AVs), the vehicle can act as an independent agent with some autonomous authority to collect data and make decisions [1]. The human driver, on the other hand, needs to assume a supervisory role within the control loop, constantly aware of changes in the dynamic environment and vehicle state, to respond to emergencies or complex situations [2]. A key aspect is that the cognitive gap between human and automation agents in perceiving hazardous situations can result in conflicts in decision-making and behavior. For instance, the information conveyed by the AV may confuse the driver's perception, leading to misinterpretation of task intentions and ultimately resulting in unharmonious interventions. Therefore, designing intuitive and efficient human-machine interfaces (HMIs) is essential to support information communication and establish a flexible and enjoyable collaboration basis for both.

The augmented reality-human machine interface (AR-HMI) is considered an attractive method for supporting humans in better understanding decision-making processes by visualizing information shared with AVs [3–5]. Compared with traditional screen interfaces, a significant advantage of AR is its ability to present visual feedback that is tightly coupled with the physical space within the driver's line of sight, thereby eliminating the competition for visual resources between proxy interfaces and the external environment during primary driving tasks. Therefore, the four main applications of AR-HMI interaction between human drivers and AVs include: (1) enhancing the attributes and dynamic information of environment-related elements [6,7] (e.g., collision warning for pedestrians or obstacles); (2) conveying intention or system status [8–11] (e.g., displaying automated trajectories and anthropomorphized system capability features); (3) projecting invisible future states [12–14] (e.g., predicting the future trajectories of other vehicles); and (4) visualizing recommendations and actions [15–17] (e.g., AR route guidance and handover requests for shared control). Previous research has recognized that cleverly designed AR-HMIs have a positive effect on enhancing situational awareness and improving collaboration. However, specific human factors related to the design of AR information to support collaborative driving have not yet been explored.

This article provides important insights into AR information design for human-vehicle collaboration, which can improve both the safety and user experience of future AVs. This work develops a method for designing augmented reality cognitive interfaces for autonomous driving. It aims to identify novel collaborative interface visualizations and provide a common language and inspiration for design space. The main issues discussed in this paper are as follows:

RQ1: How do driver and AV collaboration activities unfold, and what stages are involved in this collaboration?

RQ2: How do differences in AR interface information properties affect driver cognition of task environments and team-shared information?

RQ3: What is the impact of varying degrees of AR cognitive information visualization on human-vehicle collaborative driving performance? Is this impact contingent upon the contextual complexity of road traffic?

To address RQ1, we introduce a comprehensive human-machine collaboration ladder framework that explains the interaction mechanisms and the associated design requirements. To answer RQ2, we classify visual information elements that can be applied to AR interface design based on cognitive levels. This novel categorization method helps designers and researchers identify the cognitive characteristics of information elements. Next, we create a design space that captures the functional elements in human-vehicle collaborative driving and supports modular AR-HMI information design. To better understand the context and manner in which AR interfaces support collaborative driving, we apply the proposed interface design method to design exemplary information elements in typical dangerous scenarios. We then conduct a study ($N = 24$) on a driving simulator that includes eye-tracking equipment, a simulated driving scene built with Unity, and videos with overlaid augmented information. To address RQ3, we controlled the behavior of other traffic participants in the scene design to manipulate the complexity of the traffic environment. We investigated the impact of information component types and scene complexity on drivers' driving performance and collaborative perception, using them as independent variables in our study. We use three subjective scales (workload, trust, and usability) and key eye-tracking indicators and conduct structured interviews for evaluation. The results show that using visual components in multiple scenarios can improve workload, trust, usability, and attention performance.

24.2 Context and theoretical framework

24.2.1 Human-vehicle team and cooperation mechanism

The complementary interaction between humans and intelligent agents (IAs) has long been a topic of interest in the field of human-computer interaction [18–21]. In autonomous driving, integrating intelligence into the context while maintaining human self-efficacy is a significant challenge [22]. Many studies have shown that applying cooperative methods is essential for safety, design, and future research. Examples include driving control and switching [23,24] and collaborative task allocation [25,26]. Traditional human-machine collaboration research focuses more on the arrangement of resources, tasks, and activities rather than the form of relationships, such as task sharing and decision transmission [24,27]. These works deconstruct cooperative interactions within human-vehicle teams from a task perspective, emphasizing how members adjust their interactions in task environments according to task expectations.

Although there is some consensus in the conceptualization and application of collaboration, the key factors within the human-vehicle team have not yet been fully understood. Mapping human social collaboration to the human-automation setting is the mainstream approach for exploring collaboration mechanisms. Based on the shared cooperative activity (SCA) framework, Cali [28] identified seven core characteristics of human and automation collaboration: behavioral norms, task authorization, autonomy and control, understandability, common ground, offering help, and requesting help. Crouser and Chang [29] summarized previous research on human-machine collaboration for problem-solving and merged five characteristics of human and automation collaboration: (1) reaching a consensus on goals; (2) planning, task allocation, and coordination; (3) shared context; (4) communication; and (5) adaptation and learning. Klien et al. [30] emphasized that joint activities in human and automation collaboration should involve four basic commitments: team agreement (the basic contract of team members), mutual predictability in action, mutual guidance, and maintenance of common ground. Lee et al. [31] identified a 3C organizational framework to provide different views of drivers and automation on tasks and relationships across time scales. In these perspectives, we particularly focus on the temporal dependencies among collaborative features, which collectively constitute the process of collaboration from construction to maintenance. Collaboration starts with construction (building behavioral rules, team protocols, etc.) and progresses into the ongoing maintenance phase (understanding, communication, assistance, etc.). It is meaningful for our research, indicating that human-vehicle team collaboration can be modeled as a continuous engagement process with a dual perspective of system and partner. The former can share task information, trigger decisions, and guide behavioral efficacy, while the latter can use the communication strategies of social participants.

Trust is a crucial factor in fostering collaboration between drivers and AVs, and a stable collaborative relationship relies on mechanisms of mutual trust maintenance [32,33]. The widely used definition of trust can be described as follows: "Trust is a relationship between at least two agents, where one agent expects the other to help achieve their goals in situations characterized by uncertainty and vulnerability" [34]. In the context of traditional human-machine teams, trust specifically refers to the human perspective on trust in automated functions [33,35]), as automation agents within a team lack the intrinsic understanding of trust in the human sense. In this paper, our definition of trust is inspired by certain relational trust frameworks [36,37], with a particular focus on the reciprocal influence between interacting entities. Trust represents a facilitative relationship from a bidirectional perspective within the human-vehicle team, encompassing both "drivers' belief that the system will act in accordance with their expectations, willing to accept vulnerabilities in anticipation of positive outcomes" and the "system's belief in and responsive calibration trust process toward the driver, aimed at achieving coordinated team goals." To foster trust, providing transparent and predictable information within the human-vehicle team is necessary. On one hand, human drivers set an appropriate level of trust by understanding automation's capabilities, limitations, learning, and reasoning processes [38,39]. On the other hand, automation adjusts its role in the lifecycle of collaborative relationships based on evolving needs and trust [40]. For instance, drivers' requests for explanations might stem from appropriate trust but a lack of knowledge (system as a mentor) or insufficient trust (system as a conversational partner).

Based on the above understanding, we introduce a human-machine collaborative ladder framework (Fig. 24.1) to guide the design of HMI interfaces, which comprehensively explains the interactive mechanism and accompanying design requirements behind collaborations.

24.2.1.1 Collaborative ladder

An efficient collaborative solution between humans and intelligent systems is divided into three progressive stages: mutual explanation, mutual support, and mutual trust. The first two stages correspond to observability and detectability [41,42]. For example, the intelligent system explains the current task and environmental information to maintain transparency. The third stage tends to encompass higher-level social and emotional dynamics beyond communication [43], such as automation's commitment to completing tasks and achieving cooperation. We assumed a comprehensive collaborative process in which, during the collaborative explanation phase, humans can perceive the current status of automation and how it processes and plans the current task (such as intent and action). Moving into the support phase, roles within the team gain a reliable understanding of the explanatory content, demonstrating adaptive behavior. Finally, entering the trust

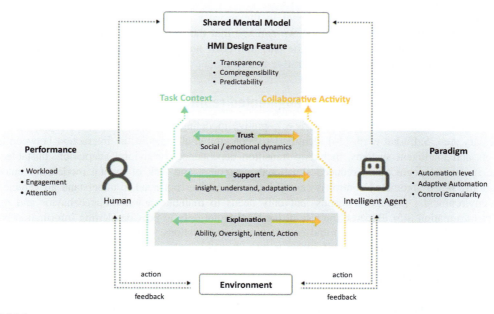

FIGURE 24.1

Human-machine collaborative ladder framework. Describes a ladder-like progression of levels in the process of collaboration between humans and intelligent agents, containing content cues and perspective components that can inspire collaborative interface analysis, making it practically meaningful.

development phase, they align their respective intentions and jointly construct a shared mental model.

24.2.1.2 Content clues

Task context and collaborative activity are two unique but related types of information exchange. For task context, the driver and automation need to understand the task goals, involved procedures, and expected possible results. For collaborative activities, attention should be paid to the team structure, roles, and characteristics of the joint activities.

24.2.1.3 Perspective components

We classify the collaborative factors mentioned in the previous literature based on the objects involved, namely humans, IAs, and HMIs. We create components from the three perspectives to examine the overall collaborative efficiency. In the human performance component, workload, engagement, and attention are included. In the IA schema component, automation level, adaptive automation, and control granularity are included. In the HMI feature component, transparency, comprehensibility, and predictability are included. We show some examples in the experimental methods section.

Overall, humans and intelligent systems participate in environmental actions and feedback together, forming a shared mental model in the process of stepping on the collaborative ladder. Through this collaborative framework, we can see that the mutual dependence of humans and IAs on tasks and relationships constitutes the part that HMI needs to design, which responds to RQ1's questions about human-vehicle collaborative activities.

24.2.2 Cognitive interface and augmented reality information elements

In a human-centered collaborative interface, cognitive modeling of the driver and intelligent system is a necessary step. Saffarian et al. [44] identified six potential challenges in the interaction between human drivers and intelligent systems, including transition dependence, behavioral adaptation, unstable workload, functional degradation, reduced situational awareness, and lack of mental models for automated functions. When automated control is engaged, the driver does not need to understand all vehicle status and environmental perception results, but only needs to maintain autonomous behavior and planned safety information [45]. Because the role of the HMI is not only to provide alerting information but also to ensure sufficient attention level and correct mutual expectations, Carsten and Martens [33] proposed six key elements for in-vehicle HMI design to provide a comprehensive understanding of intelligent system functions and states: ensuring correct trust calibration, eliciting appropriate attention and intervention, minimizing automation surprises, providing comfort and usability for human users, facilitating effective communication between humans and systems, and ensuring user control and override capabilities. Most HMI designers hope to add transparent information such as vehicle relative position and road markings to display that "the automation system is in detection state." Although such design helps to establish cooperative trust, drivers still need to reason and understand other undetected content, which leads to negative effects of attentional distraction and increased cognitive load.

To address RQ2, we investigated how to translate cognitive modeling into a design space that supports AR interface information. Kunze et al. [46] listed 11 variables from the perspective of visual features, including position, size, shape, value, orientation, hue, texture, arrangement, saturation, clarity, opacity, and resolution. Tönnis and Klinker [47] defined six principles for AR information presentation: continuity (continuous or discrete), information presentation (2D or 3D), spatial registration (tactile simulation or unregistered), reference frame (world-fixed or screen-fixed), and reference type (direct or indirect). Wiegand et al. [48] demonstrated a design space for 3D AR applications in vehicles, including categories such as user, context, visualization, interaction, and technology. While they captured a small portion of the content included in our classification, their framework lacks a clear organization to specifically and deeply apply it.

Previous work inspired us to introduce cognitive analysis methods into HMI information design [49] to extract HMI information that meets cognitive needs. In the concept of the decision ladder, human information processing involves different levels of cognitive processing, with more cognitive resources required at higher levels. We first listed three cognitive levels of perception, understanding, and planning based on widely recognized information processing stages in cognitive psychology. Rasmussen [50] proposed three corresponding types of information for skill behavior, rule behavior, and knowledge behavior: signal, symbol, and icon. Based on his explanation, we proposed an AR information type classification that is consistent with user needs and cognitive processes (Fig. 24.2). We then derived a set of interface information design dimensions that match the characteristics of each level to help designers consider and analyze various options and aspects of cognitive interface information.

24.2 Context and theoretical framework

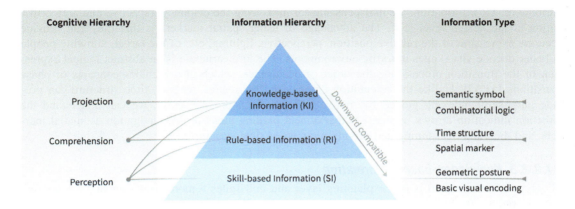

FIGURE 24.2

Multilayer mapping of cognitive levels and interface information. The information at the perception level corresponds to skill-based information (SI), derived from basic features and patterns extracted from visual inputs. This category encompasses basic visual encoding, geometric posture, and other information types closely associated with specific factual details. Information at the comprehension level corresponds to rule-based information (RI), constructed upon the integration of perceptual-level information into rule-based frameworks. This category includes spatial markers, time structures, and other information types that facilitate information organization and contextual understanding. At the projection level, information corresponds to knowledge-based information (KI), engaging with more abstract concepts and metacognition. This category comprises semantic symbols, combinatorial logic, and other consciously engaged intricate reasoning processes.

24.2.2.1 Skill-based information

This type of information corresponds to the perception layer, which is the preattention process that occurs unconsciously. In driving situations, human drivers can react instantly to environmental conditions through direct perception of visual signals. Skill-based information (SI) includes basic visual encoding, such as scale, color, and shape, which are very useful in compensating for the limitations of human vision. These encodings conceptually resemble other visualization taxonomies [51,52]. A common way to indicate physical quantities is through state variables, for example, when a driver sees a red line in a reverse image (indicating close distance and collision), they involuntarily brake to slow down. In addition, geometric posture related to physical motion (position, direction, dynamics, etc.) can sometimes trigger instantaneous perception, such as guiding lines extending into the distance that keep the driver's gaze following (with only direct perception of direction in mind) [53].

24.2.2.2 Rule-based information

It is related to matching patterns with appropriate rules, triggered by familiar cues in the environment, and feedback stored rules from prior learning. We demonstrate two dimensions of information: spatial marking and sequential structure. A spatial marker is an indicator with added spatial effects, reflecting the depth perception in cognitive features. Depth parameters, as the basic attribute of AR information, are indispensable for measuring the display perspective and area distribution at the design level [54]. The advantage of using spatial markers is that they can provide accurate expression of the relative position, perspective, lighting, etc. of the target, allowing people to better decode visual cues in traffic environments. The time structure is an abstract logical expression of the transition process and the order of occurrence, which also includes concepts of phase transitions [51,55]. Unlike the orientation emphasized by spatial markers, time structure can provide purposeful indexing guided by motivation [56]. In a broader sense, we can understand the direction indication arrows, virtual landmarks, and parking areas in AR navigation as spatial markers, and the numerical timing and driving progress as time structures.

24.2.2.3 Knowledge-based information

This type of information is in the planning layer and undergoes a more conscious and deeper reasoning process. In the collaborative driving process, the driver needs to obtain information on the AV's goals, plans, and strategies. Generally, this decision-making and judgmental information is conveyed through semantic symbols and combinatorial logic. We attach a social collaborative perspective to semantic symbols, such as displaying the AV's activity status and history of communication trace or "OK gesture," to reach a consensus with the AV [57]. Therefore, understanding the interaction between collaborative participants and extracting it into semantic symbols plays a crucial role in our design. Combinatorial logic shows multiple logical relationships between multiple objects or sets in space. For instance, AR assembly instructions that incorporate geometric attributes, physical properties, and part constraints have been investigated [58], as well as the inference made by human-vehicle teams based on the relative distances or accelerations of traffic participants in the environment [59], to judge the impact of possible future changes (the front car changes lanes to avoid obstacles, causing interference with the ego car's path). It is worth noting that the complex encoded information at this level should be used with caution or simplified as much as possible, as accurately interpreting it requires support from knowledge or past experience. If the driver cannot effectively process and draw inferences from it, it often leads to the allocation of imbalanced attention resources and distraction in dangerous situations [60].

Overall, the match between information types and cognitive features is downward compatible, which means that higher-level information can sometimes be compatible with lower-level cognition. Consistent with Vicente's views [61], we believe that interface design should transform higher-level information requirements into lower-level cognitive requirements to reduce users' cognitive load while still supporting the three levels. Therefore collaborative interfaces can extract and transform corresponding information based on cognitive levels, taking into account design factors. This addresses the parts of RQ2 related to differences in information attributes.

24.3 Design method for augmented reality human-machine interface

Following our discussion of human and psychological factors and the collaborative interface that arises from them, in this section we will introduce the design method of the AR cognitive interface (Fig. 24.3). This method allows the design factors in the collaborative driving scenario to be decoupled into two different forms of abstraction, and to effectively visualize interface information

24.3 Design method for augmented reality human-machine interface

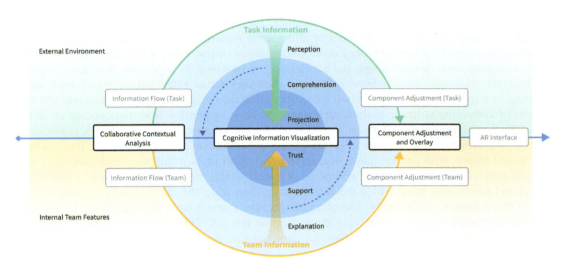

FIGURE 24.3

Augmented reality cognitive interface design method. The design factors within the collaborative driving context are decoupled into two distinct levels of abstraction—task-related and team-related. This is achieved through a layered mapping of visual features, resulting in effective visualization of interface information. The method consists of three steps: (1) Collaborative contextual analysis, aimed at acquiring finer-grained and comprehensive information; (2) Cognitive information visualization, involving the concretization of cognitive functions and outcomes of target activities; and (3) Component adjustment and overlay, wherein visual components are constructed to reflect the cognitive layers. This approach serves to enhance the design of cognitive interfaces for AR applications in the domain of collaborative driving.

through hierarchical mapping of visual features. Specifically, designers and researchers can break down the scenarios in which they want to apply AR cognitive interfaces (e.g., the handling process of the human-vehicle team when encountering lateral hazards), and transform them into cognition-based design components through bidirectional information flow, which can be easily deduced and used together in the interface (e.g., AR warning signs combined with AR region ranges). To rationally and intuitively integrate the design process of 3D space and cognitive interfaces, this method provides (1) Collaborative contextual analysis aimed at acquiring finer-grained and comprehensive information; (2) Cognitive information visualization, involving the concretization of cognitive functions and outcomes of target activities; and (3) Component adjustment and overlay, wherein visual components are constructed to reflect the cognitive layers. We will introduce these steps one by one below, combined with specific scenarios.

24.3.1 Collaborative contextual analysis

Collaborative driving situations are characterized by long-term and constantly changing conditions, which create obstacles for designing systems that can adapt to different stages of functionality. We believe that such obstacles arise from information noise that needs to be filtered, especially when different action and role states are mixed together. Therefore it is necessary to divide the design scheme into stages and disperse the focus of information requirements at the most appropriate time. For example, overtaking can be broken down into multiple stages, such as following the car several times, changing lanes several times, and accelerating to overtake. It is possible that designers may overlook the transition process between stages when they focus solely on analyzing each stage, resulting in a loss of critical information elements during the design process. To overcome the limitations of conventional linear analysis structures, we drew inspiration from the concept of hierarchical finite-state machines to support a more flexible collaborative driving framework construction process. As shown in the diagram (Fig. 24.4), the collaborative context is divided into multiple interrelated stages, with similar stages organized into logical units within the same hierarchy, such as escalation of danger and danger mitigation. For each stage, swift identification and analysis of human and automation states within the collaborative team allow the discernment of potential information requirements. Additionally, interconnecting stages through input lines delineate the information transformation process, simplifying the expansion process. This structured approach enhances the efficiency of identifying and addressing information needs within each stage of the collaborative context.

After determining the contextual framework, we further analyzed the required information flow for each stage, namely the task chain and team chain. The task chain mainly includes information on task objects, object states, and action effects, while the team chain includes information on roles, responsibilities, expectations, and intentions. This approach connects all potentially needed information together and conveys it as completely as possible to enable a unified design for similar features in subsequent stages. In the safety collaborative scenario studied in this paper, the collaborative cognitive information flow within the human-vehicle team is illustrated in Fig. 24.5.

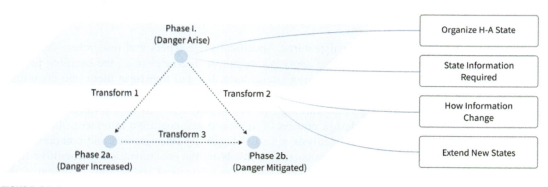

FIGURE 24.4

Collaborative contextual framework. Complex contexts are built more efficiently. The collaborative context is divided into multiple interrelated phases, and similar phases are organized into logical units within the same hierarchy, such as escalation of danger and danger mitigation. For each phase, swift identification and analysis of human and automation states within the collaborative team allow the discernment of potential information requirements. Additionally, interconnecting stages through input lines delineate the information transformation process, simplifying the expansion process. This structured approach enhances the efficiency of identifying and addressing information needs within each stage of the collaborative context.

24.3 Design method for augmented reality human-machine interface

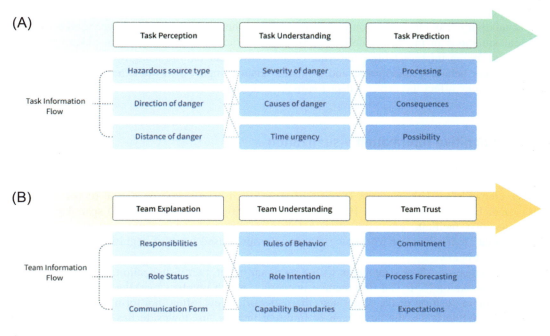

FIGURE 24.5

Example of two information flows (in the scenario when danger occurs). Task information flow and team information flow correspond to the perspectives on the two sides of the collaborative ladder framework. The horizontal axis further expands to list various levels of information: (A) Task information flow: encompasses information relevant to the task object, its current state, and the potential consequences of different actions. (B) Team information flow: includes information about team collaboration activities, such as the roles team members play, their respective responsibilities, the shared expectations, and their intended actions. These two types of information flows together form a comprehensive repository of necessary cognitive information within the context of the collaborative human-vehicle team.

It begins with perceiving the specific danger: What is the hazard? Where is it occurring? How close is the danger? This leads to generating task comprehension: Is the danger high? What is causing the danger? Is time a critical factor? Finally, it culminates in establishing expectations: What action should be taken? What could be the potential outcomes? How likely is the situation to change? On the other hand, the team information flow, as depicted in Fig. 24.5B, initiates with elucidating foundational team information: What are each member's responsibilities? What is the current status of roles? What forms of communication are in place? This then facilitates the development of team comprehension: What are the behavior norms? What are the roles' intentions? Where are the boundaries of capability? Ultimately, it leads to cultivating team trust: What assurances are being provided? What is the predictive process? What are the specific expectations? This structured information flow framework encompasses both task-related and team-related elements, enabling a systematic understanding of the cognitive processes within the collaborative driving context.

Additionally, we also considered whether the information in the collaborative scenario is necessary or auxiliary. Taking danger sources as an example, the necessity of the danger source varies in different situations in the safety collaborative driving studied in this paper. An unknown danger source may be an unpredictable pedestrian or a speeding car exiting a highway, with the former having greater behavior randomness and a smaller change impact than the latter. If the lack of this information may affect the driver's decision-making and behavior, it needs to be provided to avoid interference.

24.3.2 Cognitive information visualization

In this step, we transform the previously extracted information into basic design factors based on cognitive characteristics and propose feasible suggestions. Spatial visual and depth perception of graphic elements are key elements in our design, inspired by research on digital representation of workspaces by intelligent systems [62]. Our core proposition is to transform 3D spatial perception into clues that facilitate communication between drivers and automation systems, providing a shared visual environment or workspace. Three commonly used types of cognitive information will be introduced below.

In the context of AR environments with directional intent, two visualization methods depend on the specific situation [53]. The first method is overlay, which involves marking or directly overlaying the absolute position of the target onto the real-world view, often in a form that follows spatial movements. While it has the advantage of intuitive linkage, its downside lies in blurring the driver's perceptual range [63,64]. Particularly in hazardous scenarios studied in this paper, overlay demands that drivers must search for hazard indicators with uncertain positions across their entire visual field. The second method is conformal, where elements are displayed in a fixed area to show relative orientations while being registered in the vehicle's reference frame. Examples include directional arrows with angular analogies [65] and virtual shadows [66]. This approach associates elements with targets rather than directly appending them, which can to some extent converge attention and thereby simplify detection formats [67]. In our design, we visualize the relative direction of hazardous objects and provide distance information, with the length of the line negatively correlated with spatial distance. We did not categorize hazardous sources for several reasons. First, hazard source types exhibit diversity and uncertainty. When multiple dynamic hazard sources are present in the environment, overly detailed categorization can lead to complexity and confusion. For instance, when pedestrian hazards and vehicle hazards occur simultaneously, presenting a unified "degree of danger" would be easier for drivers to comprehend than enumerating hazard source types. Additionally, hazard sources functioned as confounding variables in our study. The diversity of hazard source information could potentially elicit varying participant responses, such as differences in attentiveness and anxiety levels toward approaching pedestrians and vehicles. Therefore, in our experimental design, we solely focused on the hazard of approaching vehicles, allowing information presentation to be independent of hazard source classification.

For information on status representation, fixed or changing attribute values may be displayed, with some emphasis on the changing process. We suggest using the visual coding information (SI) mentioned earlier to clearly distinguish between various states. Based on the predictive features of process perception, state visualization should be familiar and smoothly transitioned, allowing drivers to easily predict its changing form based on experience and imagination. In our design, we use culturally familiar warning gradient colors to intuitively express the degree of danger when visualizing the hazardous level. Compared with discrete identification, the continuous dynamic change between three colors enables drivers to better predict the trend of danger transformation. In addition, we visualize the behavioral criteria in the collaborative information, representing the activation of safety functions with frequently displayed AR graphics, intuitively expressing the system's responsibilities and response mechanisms. This enhanced pattern perception is necessary for drivers to maintain their expectation of collaborative driving with the vehicle:

For relational information with intrinsic connections, identification should be fully considered. Based on human causal perception characteristics, as the intermediate results of the process exhibit a certain regular trend, the ongoing activity, to some extent, reflects the next stage of the process. When visualizing, we use basic visual coding to simplify the expression of this higher-level information and map the deceleration behavior of the system through graphic scaling. It is assumed that drivers can gain this insight based on trajectory, speed change indications, and other content.

Next, we organized the AR graphic drafts based on (1) Visualization, (2) Information details, (3) Cognition properties, and (4) Classification descriptions. After finalizing the concrete visualizations of the AR graphics, we engaged in discussions with N = 6 experts. Subsequently, we made the necessary updates based on their feedback. Only the updated designs are presented in Table 24.1.

In the design of the first visual, we began by identifying a series of crucial information keywords to be visualized, such as depth, direction, and distance. The collaborative clue category is "task," and the information type category is "direction (SI)." Subsequently, drawing upon cognitive theory, we selected "selective and directional" design references. Our objective was to present both potential positions and distances under limited information conditions, without generating visual clutter after context changes. We introduced an adaptive visualization approach: marked angles indicate the source direction of the hazard, while the length of the marks signifies the depth distance of the hazard. In comparison with the design provided by Müller et al. [53], all experts unanimously agreed that our visualization could accommodate more complex real-world driving scenarios, such as hazards outside the driver's line of sight and simultaneous occurrence of multiple hazards.

Table 24.1 Visualization element list, showcasing the updated augmented reality graphic drafts.

Visualization	Information detail	Cognition properties	Classification description	
			Collaborate clues	Information type
	Indicate depth information, orientation points, direction, and distance	Selectivity and directionality	Task	Direction (SI)
	Identify hazards and consistently demonstrate the level of danger	Process perception	Task	Status (RI)
	As a reference to the space body of your own car, Role status	Global perception	Team	Status (RI)
	Communication of the system's deceleration intentions, code of conduct	Causal perception	Team	Relation (KI)

Moving to the second visual, the dynamic hazard identification process was our focal point. The collaborative clue category is "task," and the information type category is "status (rule-based information, RI)." This design allows drivers to form an inertia-based preparatory expectation and displays intermediate results, informing timely states [68]. We contemplated using the "yellow-orange-red" color progression to map different hazard levels during stages of change since these colors are widely recognized with predefined meanings. While we initially considered additional forms such as size variation, we later discarded this notion after evaluating users' visual experience.

The third visual, serving as a visualization of the vehicle's behavior in spatial reference, simultaneously responds to environmental perception. Its collaborative clue category is "team," and the information type category is "status (RI)." We adopted the design reference of "global perception" and proposed a well-known semicircular arc symbol (symbolic of the scanning radar on vehicles) to display fundamental team information support [33]. The semicircular arc symbol is easy to locate in the driver's forward view and doesn't obstruct the view of other real-world elements in the driving environment. This facilitates easy global perception for the driver and reduces visual overload compared with concrete vehicle graphics.

In the fourth visual, we aimed to convey the behavioral guidelines of the AV (e.g., slowing down when danger approaches). Its collaborative clue category is "team," and the information type category is "relationship (knowledged-based information, KI)." We initially considered projecting a shape area in front that scales with speed to achieve "causal perception." This visualization approach increases certainty and is expected to enhance trust [69]. However, some experts noted that the causal connection in this visualization might not be closely related enough and could demand higher training and learning requirements.

24.3.3 Component adjustment and overlay

The objective of this step is to assemble each subdesign element into a complete AR visualization while minimizing redundancy during the composition process. We seek to identify the relationships between design elements, addressing questions that may arise during the design process, such as: Which subelements can jointly determine the state of a system? Which subelements require global observation across time? Should the composition process involve adjustments to form or direct abandonment?

In the display shown in Fig. 24.6A, we have identified several possible element associations. Based on functional purpose links, we further modified and completed the latest design. First, the danger state information and change process are interrelated, determining the task state, and should therefore be combined into a visual element for display. Second, from a higher level of information, we recognize that the danger state and role state in cooperative situations are also logically related, both serving as descriptive information for the danger detection task. Referring to the role state, we have modified the danger warning graphic to make it more compatible with the role state graphic. In addition, we have integrated behavior guidelines into the visual expression of role state information, using the same style to emphasize the connection between these graphics in terms of means and purpose, thus enabling drivers to form a globally consistent understanding. Finally, we produce the interface prototype. Combined with the first-person view (Fig. 24.6B) and the overhead view (Fig. 24.6C), we offer a clearer presentation of the spatial visual and depth perception characteristics of the graphical elements.

24.4 Empirical evaluation

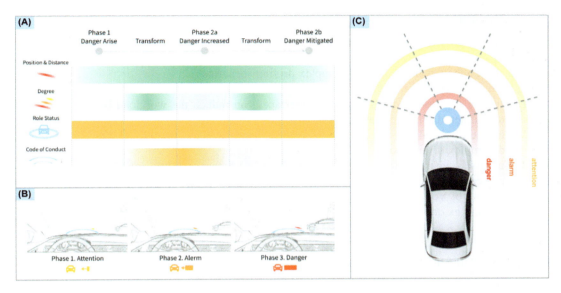

FIGURE 24.6

Component adjustment and overlay. (A) Visualization sequence presented in the contextual phase. (B) Final interface prototype. (C) Overhead Illustration of AR-HMI.

In conclusion, our design methodology offers detailed design steps as guidelines, serving as a design framework and source of inspiration for designers. The examples presented in this section are rooted in safety requirements and are driven by users' cognitive characteristics. Following the proposed design methodology, these examples generate design solutions through design steps, ensuring that the approach remains cognition-driven and collaboratively supportive. By using this approach, we were able to design an AR cognitive interface that enhances the safety of human-vehicle collaboration by providing real-time alerts for potential safety hazards and system feedback. This design effectively avoids multiple graphic clutter and can display the visualization of ADAS functions such as forward collision warning and lane departure assistance in a unified manner. It is compatible with the comprehensive scenarios of front-side merging vehicles, passing pedestrians, and rear-side merging vehicles, helping drivers form a mental model of the dangerous situation in dynamic and complex environments. It can be said that the coordination and simplification strategy provides a clear advantage in designing complex human-vehicle interactions.

24.4 Empirical evaluation

After defining the design method of the AR cognitive interface, we explored whether the use of the AR cognitive interface by drivers under autonomous driving conditions has advantages.

24.4.1 Hypotheses

In accordance with RQ3 (proposed in this paper), we juxtapose the three categories of cognitive information schemes designed in Section 24.3. Our particular interest lies in contrasting the performance of these schemes in simple and complex driving scenarios.

The following hypotheses are proposed:

H1: The use of visualized AR cognitive interfaces can reduce the workload of drivers in both simple and complex driving scenarios.

H2: The use of visualized AR cognitive interfaces can enhance cooperative driving trust in both simple and complex driving scenarios.

H3: The use of visualized AR cognitive interfaces has better usability in both simple and complex driving scenarios.

H4: The use of visualized AR cognitive interfaces can improve the driver's attention in both simple and complex driving scenarios.

24.4.2 Experimental design

We conducted a within-subject repeated measures experiment with two independent variables: the level of cognitive information visualization and the complexity of driving scenarios. Building upon the collaborative ladder framework (Fig. 24.1) and the cognitive information mapping diagram (Fig. 24.2), we differentiated various levels of AR cognitive information visualization and developed three distinct HMI design schemes. Each interface presented variations in terms of collaborative cues and information hierarchy. We provide a clear depiction of our interface design examples (Fig. 24.7), illustrating the correlation between the degree of visualization and corresponding interface design elements. The dependent variables of the experiment encompassed several aspects, including cognitive workload, trust level, interface usability, and attention, among others.

24.4.2.1 Scenarios and tasks

The experimental scenario was set on a clear daytime double-lane urban road with two-way traffic. The AV (referred to as self-car) maintained a speed of 30 km/hour in the right lane, while the surrounding vehicles traveled at speeds ranging from 20 to 40 km/hour. The self-car encountered collision risk events at specific locations during its travel and notified the driver through the AR-HMI. The driving route is depicted in (Fig. 24.8A), showing six intersections with traffic signals. In each drive, events occurred randomly at two of these locations to reduce driver-learning effects. The duration of each drive was approximately 2 minutes.

Fig. 24.8B and C illustrate the specific event contexts, corresponding to different levels of scenario complexity. Scenario complexity is closely associated with the behaviors of surrounding traffic participants. In the simple scenario, there were no relevant traffic participants other than the lateral approaching vehicle and the self-car. In the complex scenario, the self-car needed to pay extra attention to potential conflicts other than the lateral approaching vehicle, such as the oncoming traffic from the side front (even though right-of-way rules apply in reality, conflicts could still arise).

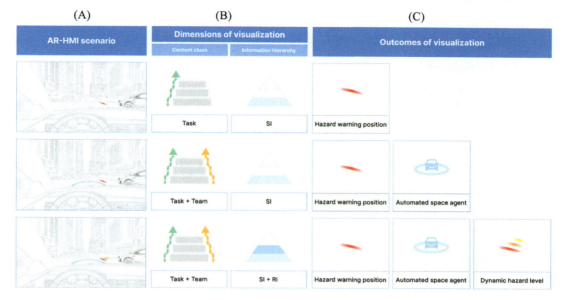

FIGURE 24.7

Augmented reality-human-machine interface (HMI) design schemes. (A) HMI1: Interface provides only task context-related cues with visualization at the skill-based information (SI) level. (B) HMI2: Interface combines task context and collaborative cues with visualization at the SI level. (C) HMI3: Interface integrates task-context and collaborative cues with visualization at multiple levels, employing both SI and rule-based information visualizations.

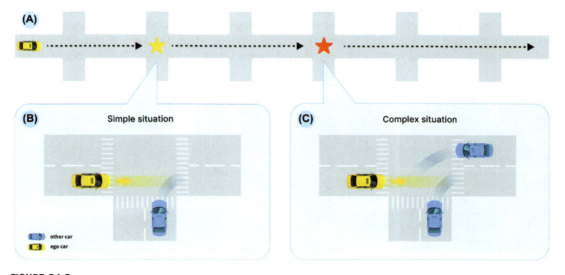

FIGURE 24.8

Scene design. (A) Driving route and event triggering illustration. (B) Simple collision risk scenario with only lateral approaching vehicle conflict. (C) Complex collision risk scenario with conflicts involving multiple directions of traffic participants. (The road layout has been simplified to highlight the event logic.)

24.4.2.2 Participants

The required sample size was calculated using G*Power prior to the experiment, and 19 participants were needed for repeated measures analysis of variance with a power of 0.80 and α level of 0.05. Participants were recruited primarily through various online media platforms. They were required to hold a driver's license for at least 2 years, self-report driving frequency of at least once a month, have normal corrected visual acuity, and have no other visual impairments. A sample of 24 participants was obtained, with a sample rate of correct eye movement recognition above 90%. The age distribution was concentrated in the range of 18–31 years ($M = 23.7$, $SD = 2.8$). The recruitment process and the entire experimental procedure were approved by the Automotive Interaction Laboratory of Tongji University and participants were compensated for their participation.

24.4.3 Apparatus and materials

The driving simulator provided by the Automotive Interaction Lab for the experiment consisted of a steering wheel control system, a main computer equipment, and a three-screen display (Fig. 24.9A). The Tobii Pro wearable eye tracker was used to capture the driver's forward field of view at a frequency of 30 Hz for primary eye movement feature extraction and gaze behavior analysis. The instrument recording accuracy was 0.5 degrees with a spatial resolution of 0.2 degrees and an error of no greater than 0.1 degrees. The Tobii I-VT (attention) filter was used to filter the fixation point data. The testing environment had good lighting conditions and no interference.

We built the experimental environment using the Unity 3D engine and C#, simulating a realistic city traffic driving scene that includes roads, pedestrians, vehicles, traffic systems, and other elements (Fig. 24.9B). The driving environment was a two-way city road with sunny weather conditions, with surrounding vehicle speeds set within the range of 20–40 km/hour. Pre-recorded video segments were used to simulate L3-level automated driving, where the vehicle traveled steadily and safely along a planned route, with several road events set along the way. The number of events

FIGURE 24.9

Experimental equipment and materials. (A) Driving simulator. (B) Unity 3D program. (C) Simulated experimental scenarios with set traffic rules and side vehicle behavior with overlay augmented reality visualization.

was kept consistent across segments (see experimental variables for details), with a total duration of approximately 2 minutes. Finally, an AR-HUD interface prototype was quickly developed using video editing tools and overlaid onto the video scene with matching timing (Fig. 24.9C). The interface prototype was displayed on a three-screen display composed of three 1920×1090 pixel resolution screens.

24.4.4 Procedure

Participants read and signed informed consent forms before completing a pretest questionnaire. They were then invited to sit in the driving simulator, wear eye-tracking equipment, and complete calibration. Participants were instructed to imagine themselves as drivers while watching a first-person video, to simulate the state of real users when using autonomous driving. To familiarize themselves with the experimental environment, participants needed to complete a few minutes of a practical trial. The road setting in this stage was the same as in the formal experiment but without any events. We instructed participants to focus only on the driving scenario and the vehicle's operating state during the formal experiment, as shown in Fig. 24.10. Participants experienced all three HMI conditions in a balanced sequence determined by a Latin square design. Each journey consisted of continuous urban roads with six intersections, with red and yellow dots representing potentially dangerous locations that appeared randomly to reduce the impact of learning effects. After each journey, participants were asked to complete relevant questionnaires. After completing all experiments, we conducted semistructured interviews with the participants to understand their feelings about the interface and specific reasons.

24.4.5 Measures and analysis

To compare participants' subjective evaluations of using different AR interfaces, we used three scales: driving activity load index (DALI), trust, and system usability scale (SUS), to assess self-reported workload, trust, and usability. DALI is more suitable for driving tasks than NASA-TLX [70] and includes six workload dimensions: level of attention, visual demand, auditory demand, temporal demand, interference, and situational stress. The six dimensions are weighted and combined to obtain the subjective workload score. The trust scale covers 12 potential factors that influence trust between humans and automation systems and is commonly used to measure trust in automation. Automation

FIGURE 24.10

Latin square-based route configuration.

trust scale is commonly employed to assess trust in automation, encompassing 12 underlying factors that influence trust between humans and automated systems [71,72]. These factors include individual predisposition, system complexity, task difficulty, level of expertise, system reliability, and dependability. SUS is a widely used posttask subjective assessment questionnaire [73] that can be divided into three dimensions: effectiveness, efficiency, and satisfaction.

In addition, we conducted eye-tracking analysis to investigate participants' cognitive processes, and the obtained data were quantitatively analyzed, as shown in Table 24.2. When specific visual behaviors contribute to better performance and more anticipation of danger, these eye-tracking features can be attributed to more effective visual scanning and better situational awareness structures generated from the information received from the HMI. Considering the distribution of basic AR interface elements, we defined the center area of the front roadway as area of interest (AOI). Attention metrics include time to first fixation (TTFF) and first fixation duration (FFD). Cognitive performance metrics include total fixation duration (TFD), average fixation duration (AFD), and pupil dilation rate (APD). It is noteworthy that an earlier study validated the correlation between drivers' trust in automation and their eye-tracking behavior, indicating that increased trust leads to decreased monitoring frequency [75]. Drawing upon this finding, we integrated it with the conclusions of the automation trust scale to conduct a comprehensive predictive analysis.

Table 24.2 Overview of the analytics framework.

Dependent variables	Measures	Interpretation
Workload (subjective)	Driving activity load index [70]	Includes six dimensions: effort of attention, visual demand, auditory demand, temporal demand, interference, and situational stress.
Trust	Automation trust scale [71]	Twelve potential influencing factors that affect trust between people and automation, such as individual preferences, system complexity; difficulty, level of expertise, system reliability and dependability, and so on.
Useability	System usability scale [73]	Includes three dimensions: effectiveness, efficiency, and satisfaction.
Visual attention	Fixation heatmap [74]	Visualize areas of focus and attraction on materials through shades of color.
	Time to first fixation	Time before sightline first enters AOI, related to visual prominence.
	First fixation duration	Duration of the first fixation on a particular AOI, related to the level of interest.
	Total fixation duration	All durations of fixation on one AOI, related to cognitive difficulties.
Workload (objective)	Average fixation duration	The average duration of a single fixation for one AOI.
	Rate of average pupil diameter	Differences in mean pupil dilation between the stimulation and baseline periods, related to cognitive difficulties and emotional arousal.

Note: AOI is used as an abbreviation for area(s) of interest.

24.5 Results

The analysis considered all subjective data from 24 participants, while excluding some erroneous eye-tracking data, including two participants with an effective sampling rate lower than 80% during the tasks due to persistent identification issues encountered during calibration. First, the results of the analysis sample display the subjective evaluations of workload, system trust, and AR interface usability for each task-related aspect. Second, objective data are presented, including eye-tracking behavior during a designated time window (TOI) that covers an 8-second interval from 3 seconds prior to the occurrence of a hazardous state to 3 seconds after it ended (the HMI interface remains visible for 2 seconds). To simplify the experiment and analysis process, subjective data are only employed for the single-variable analysis of HMI types (Table 24.3), whereas the analysis of eye-tracking physiological data incorporates the dual-variable of HMI types and scene complexity (Table 24.4). Table 24.4 provides the mean and standard deviation of all metrics presented by the HMI elements and scene complexity variables for analysis. In the independent variable rows, "type" refers to the type of AR interface elements, while "complexity" pertains to the task scenario's simple or complex form. The significant differences related to the hypotheses will now be reported.

24.5.1 Subjective scales

24.5.1.1 Subjective workload

The DALI scale was utilized to evaluate the subjective workload of users, where a lower weighted score indicates better performance. Overall, HMI visualization has a significant impact on

Table 24.3 Means (and standard errors) for subjective scales data used for analysis.

Metrics	HMI1	HMI2	HMI3
Subjective workload	65.61 (3.08)	40.00 (3.29)	47.96 (5.49)
Trust	3.69 (0.11)	4.28 (0.12)	4.31 (0.16)
Usability	69.00 (3.04)	75.17 (2.59)	60.07 (5.20)

Table 24.4 Means (and standard errors) for eye-tracking data used for analysis.

	HMI1		HMI2		HMI3	
Metrics	Simple	Complex	Simple	Complex	Simple	Complex
Time to first fixation	1.12 (0.17)	1.59 (0.19)	0.54 (0.11)	0.66 (0.07)	0.75 (0.16)	0.97 (0.29)
First fixation duration	0.44 (0.04)	0.32 (0.08)	0.40 (0.04)	0.71 (0.07)	0.38 (0.03)	0.51 (0.10)
Total fixation duration	3.35 (0.28)	3.12 (0.59)	3.44 (0.38)	3.85 (0.66)	3.23 (0.35)	2.61 (0.72)
Average fixation duration	0.57 (0.05)	0.56 (0.08)	0.65 (0.06)	0.50 (0.06)	0.49 (0.03)	0.58 (0.11)
Average pupil diameter	4.96 (0.11)	4.98 (0.36)	4.94 (0.12)	4.83 (0.18)	5.06 (0.08)	4.25 (0.23)

Note: *Time metrics in seconds.*

subjective workload ($F(2, 48) = 10.19$, $p < 0.001$, Fig. 24.11A). Compared with HMI1 without any visualization, the other two AR cognitive interfaces resulted in a significant reduction in workload ratings.

24.5.1.2 Trust
The trust questionnaire was used to assess users' trust in the autonomous driving system behind the AR interface, with higher scores indicating higher levels of cooperative trust. Overall, HMI visualization had a significant impact on trust levels ($F(2, 45) = 6.34$, $p < 0.05$, Fig. 24.11B). Compared with HMI1 without visualization, the trust scores were significantly higher in the two AR cognitive interfaces.

24.5.1.3 Usability
SUS was employed to assess the user experience of the AR interface during driving, with higher scores indicating better system usability. Overall, the HMI visualization had a significant impact on usability ($F(2, 42) = 4.01$, $p < 0.05$, Fig. 24.11C). Interestingly, between the two AR cognitive interfaces that employed visualization methods, an increased number of multimodal combinations resulted in a significant decrease in usability scores.

24.5.2 Fixation metric
Using a two-factor multivariate analysis of variance method, this study evaluated the effects of HMI type and scene complexity on various eye-tracking behavior data (Fig. 24.12), including TTFF, FFD, TFD, AFD, and APD. Unless otherwise specified, this study used mean ± standard error of the mean to describe the data. Box plots were used to check for outliers, the Shapiro-Wilk test was used to check for normality, and the Levene test was used to test for homogeneity of variances. The results showed that there were no outliers after including data from the 13 participants, and the residuals of the dependent variables were close to a normal distribution ($p > 0.05$) and had homogeneity of variances ($p > 0.05$).

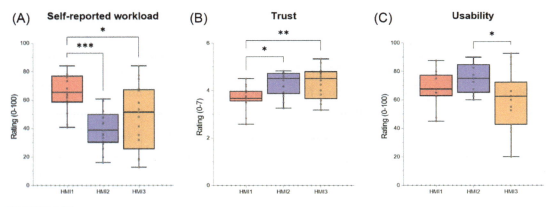

FIGURE 24.11

Self-reported workload scores (0–100), system trust ratings (0–7), and usability scores (0–100) for all HMI tasks from left to right. *Note*: Error bars indicate ± 1 standard error of the mean.

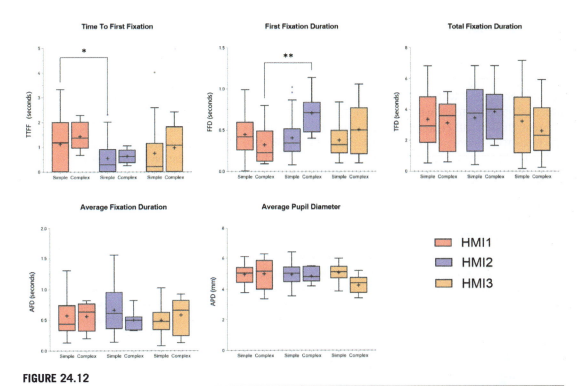

FIGURE 24.12

Comparison of first fixation duration, first fixation duration length, total fixation duration, mean fixation duration, and mean pupil diameter across all human-machine interface tasks in different scenarios.

24.5.2.1 Time to first fixation

Overall, visualization shows a significant difference in the TTFF. In simple scenes, cognitive interfaces lead to lower TTFF. Main effect analysis shows that HMI visualization has a statistically significant effect on TTFF ($F = 4.690$, $p = 0.010$, partial $\eta^2 = 0.080$). A pairwise comparison analysis was conducted to compare the main effect of HMI visualization. In simple scenes, the difference between the HMI1 and HMI2 groups is 0.575 (95%CI: 0.071–1.080, $p = 0.019$), indicating that cognitive interfaces significantly decrease TTFF in simple scenes.

24.5.2.2 Fixation to first displacement

The Friedman analysis of variance indicated a significant interaction between HMI visualization and scene complexity on the (FFD) ($F = 4.975$, $p = 0.008$, partial $\eta^2 = 0.080$). Separate effects analysis showed that different HMI types had different FFD in different complexity scenes: simple scenes ($F = 0.960$, $p = 0.386$, partial $\eta^2 = 0.016$); complex scenes ($F = 4.197$, $p = 0.017$, partial $\eta^2 = 0.068$). Pairwise comparisons of each category's separate effects were performed. In complex scenes, the FFD difference between HMI1 and HMI2 was 0.384 (95%CI: 0.060–0.690, $p = 0.014$), indicating that cognitive interfaces significantly increased FFD in complex scenes.

24.5.3 Fixation heatmap

A comparison of three types of HMI was conducted to examine visual attention. Fig. 24.13 displays the fixation heatmap, with color intensity indicating the duration or frequency of fixations (darker colors representing longer durations/more frequent fixations). Intuitively, HMI1 has more scattered fixations compared with the other two AR cognitive interfaces. This indicates that users had more dispersed gaze patterns when using HMI1. For HMI2 and HMI3, fixations were concentrated on different clusters, all centered on key clues. Overall, the fixation patterns when using AR cognitive interfaces were more systematic, with clear clusters in AOI areas, demonstrating a significant preference for key clues. It is worth noting that the darker blocks of the fixation clusters in HMI1 were larger, indicating a wider visual search area throughout the danger process, resulting in reduced fixation time and percentage of fixation points in the information area.

24.5.4 Interviews

During the interviews, most participants stated that they believed the curved areas in the cognitive interface had a focal effect on their gaze. Additionally, some participants expressed doubts about the usability and usefulness of the latter two cognitive interfaces, as some of the elements were deemed unnecessary (possibly redundant). Furthermore, almost half of the participants reported that HMI1 conveyed more negative emotions, as they were not prepared for the sudden appearance of

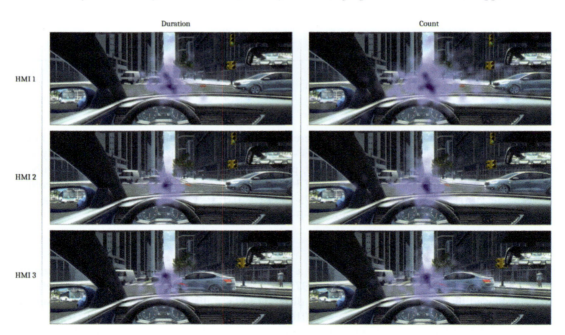

FIGURE 24.13

Gaze hotspot diagram of the three human-machine interfaces.

patterns. In contrast, the explicit rules in the latter two cognitive interfaces provided them with a greater sense of safety. Some participants had difficulty understanding the metaphorical components that contained information about the status of the vehicle, particularly as they did not notice the graphical changes caused by a decrease in speed (which resulted in the shrinking of the graphic). However, after learning about these interesting visualizations, the vast majority of participants found them to be more advantageous than text and icons.

24.6 Discussion

This empirical study revealed the significant role of AR cognitive interfaces in addressing human factor issues in human-vehicle collaboration systems, but in a much more complex way than we initially predicted. Most likely, the various characteristics of the scene and HMI altered the drivers' experience in visual attention and situational understanding, compoundedly affecting the shared mental models of the human-vehicle team.

24.6.1 Visualization components and quantities

We used visual components as a variable in comparing HMI1 and HMI2 to explore the role of necessary cognitive information. Results from our study on the lateral danger warning task indicate that cognitive interfaces that use visual methods significantly improve the trust performance of the human-vehicle team and reduce workload compared with HMI1. This finding supports hypotheses H1 and H2 and is consistent with other research in this field that links interface information to shared cognition [33]. One possible explanation for this is trust calibration. Specifically, cognitive information develops a trust mechanism that satisfies predictability and timeliness in trust calibration. Because this information is based on the collaboration stage, it reflects the functions and status of the AV while displaying task summaries to the driver. Providing context and situation context reduces uncertainty and allows drivers to establish mental expectations for collaboration with the system to take appropriate action in critical moments, compared with emphasizing only sudden dangerous situations. On the other hand, the reduction in workload may be due to the assistance of a higher level of situational awareness. Related to previous research on shared situational awareness [76], AVs take over most of the perception and understanding of the surrounding environment and provide drivers with a small portion of interesting and abstract cognitive information. The complementarity and robustness of team relationships mitigate time and situational pressures, which may help us understand the positive effects of shared situational awareness on driver performance.

Another important finding is that, compared with HMI2, the average workload of the cognitive interface for HMI3 with more visual components has increased, and the average usability score has significantly decreased, supporting H3. This result suggests that changes in the form of interface information elements may have a negative impact on driving performance. We believe that this difference can be attributed to visual complexity. The HMI design in this study specifically emphasizes the visual display complexity of graphics, without delving into the number of information metric categories, and caution must be exercised in interpretation. As most users reflected in the interviews, long-lasting schemes make it more difficult for them to identify key information. Therefore it can be preliminarily hypothesized that drivers prefer concise and straightforward information to meet moderate understanding needs, and additional processing may result in information overload and accompanying stress. For example, diverse information may require drivers to exert more effort in attention and understanding phases, and content related to cognitive load and net benefits will be discussed in the next section. We suggest that designers consider subtle trade-offs between sparse monotony and dense variability when representing information in new HMI designs.

Additionally, the box plot shows that the data for the HMI3 group is relatively scattered, with distributions in both high and low regions. This finding is unexpected and indicates differentiated perceptions among participants regarding the scheme with multiple visual components. Currently, it is unclear why some supportive parts are beneficial, but it may be related to interface transparency. We speculate that information-rich content provides a smooth and predictable automated driving experience, reducing uncertainty and ultimately leading to positive results. Of course, the influence of individual driver styles cannot be ruled out, and calm and careful drivers are more likely to accept gradual information guidance. We suggest that AR information for collaborative interfaces can be designed based on context relevance and time urgency.

24.6.2 Attention and cognitive demands

Previous research has highlighted that visual scene analysis in actual driving environments is a more complex process than previously thought, involving visual and cognitive resources, as well as metatasks for managing attention allocation between the road environment and the HMI [77]. Therefore, it is crucial to consider attentional and cognitive factors when designing HMIs for drivers.

In our study, we observed differences in the first fixation duration and duration of the first fixation between conditions with and without visual components. Specifically, the component scheme significantly reduced the first fixation duration in both types of scenes. Surprisingly, the component scheme only significantly increased the duration of first fixation in complex scenes. These results suggest that cognitive information has a certain advantage in guiding attention and arousing interest, supporting H4. According to attentional set theory [78,79], observers establish an attentional set including features of the target object based on visual demand, making it easier to detect stimuli matching these features. For driving tasks, attentional sets can contain object-, situation-, or behavior-related features of the driving environment, while dynamically adjusting the elements in the set over time. We speculate that the AR scheme using context establishes a mapping relationship between the elements in the attentional set, linking various attentional elements like a network and maximizing the matching range. Therefore as long as the driver observes one of the collaborative features or task object features, they can easily observe the other under its guidance. This event reveals why participants using context-based AR were able to capture the information provided by the HMI more quickly. However, the influence of the visual information area on the results is currently uncertain, and future research needs to strictly control this factor to further verify the results.

As previously stated, visual saliency alone cannot account for all eye movements during active viewing. Therefore we additionally explore the relationship between cognitive load and net gain. The most interesting finding is that when participants use HMI2, the duration of their initial fixation significantly differs between scenes of varying complexity. Specifically, more complex scenes lead to a higher attentional focus on the HMI2 region. According to the visual resource theory, drivers' visual information processing allocates limited resources to different visual tasks, which can create interference and competition between tasks. Generally, human peripheral vision is responsible for simple and rough detection tasks (such as detecting approaching objects and lane deviations), while focal vision supports more refined object localization or classification (such as object spatial location and attributes). The former can be completed in a short or even momentary time, while the latter requires more time and cognitive resources. During driving, visual focus shifts between the road environment and HMI, which is in line with the view of Baldwin et al. [80]. When the complexity of the road environment requires higher analytical demands, drivers rely more on pattern-fixed (operational state known) HMI to reduce cognitive load. This is consistent with Henderson et al.'s finding that cognitive factors are also a major factor in gaze control [81]. Currently, it is unclear why scene differences were not observed in the same type of HMI3, which may be related to the complexity of multiple information changes. Overall, visual processing is a compatible one-way process from low to high precision. If presented information can satisfy the bottom-up feedforward flow, it may more naturally attract the driver's attention and bring better performance for road safety. Therefore we suggest that the design of AR information should balance the fastest perception with a minimal effort of understanding. It should fully extract the user's familiar environmental orientation and relevant conditions. Future research can provide insights into this aspect.

24.6.3 Training and mental models

In addition to understanding discrepancies in scenes, cognitive prompts are necessary for researchers to explain the system when using visual components in cognitive interfaces, despite the fact that each individual was informed of the system's explicit cooperation agreement before the experiment began. According to the researchers' subjective impressions, the significant challenge faced by the subjects is learning and understanding the system's behavioral guidelines. For example, the system metaphorically represents deceleration handling behavior through AR spatial graphics shrinking. Eight subjects acknowledged this feature at the end of the experiment, but only two subjects stated that their understanding matched its original meaning. In fact, the subjective workload ratings of these subjects were better than those of the subjects who did not understand. One possible explanation is the difference in user ability and experience to collaborate with the system. For example, experienced drivers were found to recover situational awareness faster than inexperienced drivers when relinquishing vehicle control [82]. If users gain usage experience by learning, their workload may decrease. To verify this hypothesis, future investigations need to evaluate performance before and after learning.

However, conveying the design features of a system to novice users who do not have a mental model of how to collaborate with the system is a challenge. In situations where the human-machine collaboration system faces dangerous scenarios, asking users to actively decelerate by stepping on the pedal in an attempt to help the system is not a good strategy. A user who lacks a cooperative mindset is likely to become frustrated at this stage or even stop using the system that was initially perceived as having poor usability. Therefore a better understanding of the system's behavior will increase users' willingness to collaborate and enhance the potential advantages of AR cognitive interfaces over conventional interfaces.

24.6.4 Limitations

The sample used in this study mainly consisted of young Asian adults with higher education, and some of the participants had limited driving experience. The extent to which this affects the overall communication of collaboration information is unknown. In addition, the smaller sample size in the complex scenario group limits the confidence interval of some results. With an increase in the number of participants, some trends may reach statistical significance, such as the overall superiority of the AR cognitive interface over the conventional warning in complex scenarios. A more representative sample will be an important consideration in future studies. Furthermore, although as many realistic scenario elements as possible were incorporated into the software, the lack of perceived realism in the driving simulator and video scenario combination remains a major weakness, which impacts the perception of high-risk emergency situations. We believe that future research can be conducted in test facilities with a higher sense of spatial presence to maximize the practical relevance of the results. The method of measuring cognitive demand based on region-based gaze duration is not exhaustive in current research, and different measurement methods may provide another perspective. Despite these limitations, this work provides a preliminary practice of cognitive information design methods, which certainly requires more research.

24.7 Conclusions

In summary, this study presents a novel AR cognitive interface design approach to support cognitive demands in human-vehicle collaborative scenarios. The approach involves three stages: collaborative scenario analysis, cognitive information visualization, and component adjustment and overlay. We decoupled design factors in collaborative driving scenarios and cleverly mapped the visual features of interface information to cognitive levels. Moreover, we conducted exemplary information design through this approach, ensuring that design elements are minimal and well combined. In a subsequent study with $N = 24$ participants, we found that using visualized components of cognitive interfaces could lead to lower cognitive workload and better usability, trust, and attention performance in various scenarios. However, using multiple visualized components could lead to reduced attention performance in complex scenes. To our knowledge, our work is one of the first to specifically investigate cognitive and collaborative information design in AR and further extends research on improving collaborative driving performance and safety through cognitive interfaces.

References

[1] G.H. De Almeida Correia, B. van Arem, Solving the user optimum privately owned automated vehicles assignment problem (UO-POAVAP): a model to explore the impacts of self-driving vehicles on urban mobility, Transport. Res. Part. B: Methodol. 87 (2016) 64–88.
[2] S.E. Shladover, The truth about "self-driving" cars, Sci. Am. 314 (6) (2016) 52–57.
[3] J.L. Gabbard, G.M. Fitch, H. Kim, Behind the glass: driver challenges and opportunities for AR automotive applications, Proc. IEEE 102 (2) (2014) 124–136.
[4] K. Bark, C. Tran, K. Fujimura, V. Ng-Thow-Hing, Personal navi: benefits of an augmented reality navigational aid using a see-thru 3d volumetric hud (2014).

[5] D. Van Krevelen, R. Poelman, A survey of augmented reality technologies, applications and limitations, Int. J. Virtual Real. 9 (2) (2010) 1–20.

[6] H. Kim, J.L. Gabbard, A.M. Anon, T. Misu, Driver behavior and performance with augmented reality pedestrian collision warning: an outdoor user study, IEEE Trans. Vis. Comput. Graph. 24 (4) (2018) 1515–1524. Available from: https://doi.org/10.1109/TVCG.2018.2793680.

[7] F. Schwarz, W. Fastenmeier, Visual advisory warnings about hidden dangers: effects of specific symbols and spatial referencing on necessary and unnecessary warnings, Appl. Ergonom. 72 (2018) 25–36. Available from: https://doi.org/10.1016/j.apergo.2018.05.001.

[8] A. Eriksson, S.M. Petermeijer, M. Zimmermann, J. de Winter, K.J. Bengler, N.A. Stanton, Rolling out the red (and green) carpet: supporting driver decision making in automation-to-manual transitions, Ieee Trans. Human-Mach. Syst. 49 (1) (2019) 20–31. Available from: https://doi.org/10.1109/THMS.2018.2883862.

[9] D. He, D. Kanaan, B. Donmez, In-vehicle displays to support driver anticipation of traffic conflicts in automated vehicles, Accid. Anal. & Prev. 149 (2021) 105842.

[10] C.P. Janssen, S.F. Donker, D.P. Brumby, A.L. Kun, History and future of human-automation interaction, Int. J. Human-Comput. Stud. 131 (2019) 99–107.

[11] H. Guo, L. Song, J. Liu, F. Wang, D. Cao, H. Chen, C. Lv, P.C. Luk, Hazard-evaluation-oriented moving horizon parallel steering control for driver-automation collaboration during automated driving, IEEE CAA J. Autom. Sinica. 5 (6) (2018) 1062–1073. Available from: https://doi.org/10.1109/JAS.2018.7511225.

[12] S. Kim, R. van Egmond, R. Happee, Effects of user interfaces on take-over performance: a review of the empirical evidence, Information 12 (4) (2021) 162.

[13] Y. Wu, M. Abdel-Aty, O. Zheng, Q. Cai, L. Yue, Developing a crash warning system for the bike lane area at intersections with connected vehicle technology, Transport. Res. Rec. 2673 (4) (2019) 47–58. Available from: https://doi.org/10.1177/0361198119840617.

[14] Z. Yang, J.L. Shi, Y. Zhang, D.M. Wang, H.T. Li, C.X. Wu, et al., Head-up display graphic warning system facilitates simulated driving performance, Int. J. Human-Comput. Interact. 35 (9) (2019) 796–803. Available from: https://doi.org/10.1080/10447318.2018.1496970.

[15] S. Langlois, B. Soualmi, Augmented reality versus classical HUD to take over from automated driving: an aid to smooth reactions and to anticipate maneuvers (2016).

[16] L. Lorenz, P. Kerschbaum, J. Schumann, Designing take over scenarios for automated driving: how does augmented reality support the driver to get back into the loop? (2014).

[17] P. Wintersberger, C. Schartmüller, A. Riener, Attentive user interfaces to improve multitasking and take-over performance in automated driving: the auto-net of things, Int. J. Mob. Hum. Comput. Interact. (Ijmhci) 11 (3) (2019) 40–58.

[18] B.J. Grosz, Beyond mice and menus, Proc. Am. Philos. Soc. 149 (4) (2005) 529–543.

[19] B.J. Grosz, S. Kraus, Collaborative plans for complex group action, Artif. Intell. 86 (2) (1996) 269–357. Available from: https://doi.org/10.1016/0004-3702(95)00103-4.

[20] S.L. Brandt, J. Lachter, R. Russell, R.J. Shively, A human-autonomy teaming approach for a flight-following task, Adv. Neuroergonom. Cognit. Eng. (2017) 12–22.

[21] C. Rich, C.L. Sidner, N. Lesh, Applying collaborative discourse theory to human-computer interaction, Ai Mag. (No. 4)(2001) 15.

[22] C. Stephanidis, G. Salvendy, M. Antona, J.Y.C. Chen, J. Dong, V.G. Duffy, et al., Seven HCI grand challenges, Int. J. Human–Comput. Interact. 35 (14) (2019) 1229–1269. Available from: https://doi.org/10.1080/10447318.2019.1619259.

[23] C.P. Janssen, S.T. Iqbal, A.L. Kun, S.F. Donker, Interrupted by my car? Implications of interruption and interleaving research for automated vehicles, Int. J. Human-Comput. Stud. (No.C)(2019) 221–233.

[24] J. Pichen, T. Stoll, M. Baumann, C.M. Assoc, From SAE-levels to cooperative task distribution: an efficient and usable way to deal with system limitations? In: AutomotiveUI '21: 13th International ACM Conference on Automotive User Interfaces and Interactive Vehicular Applications, 2021, pp. 109–115.
[25] F.O. Flemisch, K. Bengler, H. Bubb, H. Winner, R. Bruder, Towards cooperative guidance and control of highly automated vehicles: H-Mode and Conduct-by-Wire, Ergonomics (3)(2014) 343–360.
[26] C. Wang, A framework of the non-critical spontaneous intervention in highly automated driving scenarios, in: Paper presented at the AutomotiveUI '19: Proceedings of the 11th International Conference on Automotive User Interfaces and Interactive Vehicular Applications: Adjunct Proceedings, Utrecht Netherlands, 2019.
[27] S. Geyer, M. Baltzer, B. Franz, S. Hakuli, M. Kauer, M. Kienle, et al., Concept and development of a unified ontology for generating test and use-case catalogues for assisted and automated vehicle guidance, Iet Intell. Transp. Syst. 8 (3) (2014) 183–189. Available from: https://doi.org/10.1049/iet-its.2012.0188.
[28] N. Cila, Designing human-agent collaborations: commitment, responsiveness, and support, in: Paper presented at the CHI '22: Proceedings of the 2022 CHI Conference on Human Factors in Computing Systems, New Orleans LA USA. 2022.
[29] R.J. Crouser, R. Chang, An affordance-based framework for human computation and human-computer collaboration, IEEE Trans. Vis. Comput. Graph. 18 (12) (2012) 2859–2868. Available from: https://doi.org/10.1109/TVCG.2012.195.
[30] G. Klien, D.D. Woods, J.M. Bradshaw, R.R. Hoffman, P.J. Feltovich, Ten challenges for making automation a "team player" in joint human-agent activity, Intell. Syst., IEEE (NO.6)(2004) 91–95.
[31] J. Lee, H. Rheem, J.D. Lee, J.F. Szczerba, O. Tsimhoni, Teaming with your car: redefining the driver–automation relationship in highly automated vehicles, J. Cognit. Eng. Decis. Mak. (2022). Available from: https://doi.org/10.1177/15553434221132636. 432651100.
[32] Y. Xing, C. Lv, D. Cao, P. Hang, Toward human-vehicle collaboration: review and perspectives on human-centered collaborative automated driving, Transport. Res. Part. C: Emerg. Technol. 128 (2021) 103199. Available from: https://doi.org/10.1016/j.trc.2021.103199.
[33] O. Carsten, M.H. Martens, How can humans understand their automated cars? HMI principles, problems and solutions, Cogn. Technol. Work. 21 (1) (2019) 3–20. Available from: https://doi.org/10.1007/s10111-018-0484-0.
[34] A.G. Mirnig, P. Wintersberger, C. Sutter, J. Ziegler, A framework for analyzing and calibrating trust in automated vehicles (2016).
[35] F.M.F. Verberne, J. Ham, C.J.H. Midden, Trusting a virtual driver that looks, acts, and thinks like you, Hum. Factors 57 (5) (2015) 895–909. Available from: https://doi.org/10.1177/0018720815580749.
[36] E.J. De Visser, R. Pak, T.H. Shaw, From 'automation' to 'autonomy': the importance of trust repair in human–machine interaction, Ergonomics 61 (10) (2018) 1409–1427. Available from: https://doi.org/10.1080/00140139.2018.1457725.
[37] E.K. Chiou, J.D. Lee, Trusting automation: designing for responsivity and resilience, Hum. Factors 65 (1) (2023) 137–165. Available from: https://doi.org/10.1177/00187208211009995.
[38] J. Haspiel, N. Du, J. Meyerson, L.P. Robert Jr., D. Tilbury, X.J. Yang, et al., Explanations and expectations: trust building in automated vehicles (2018).
[39] Q. Zhang, L.J. Robert, N. Du, X.J. Yang, Trust in AVs: the impact of expectations and individual differences (2018).
[40] M.F. Jung, N. Martelaro, P.J. Hinds, Using robots to moderate team conflict: the case of repairing violations (2015).
[41] M. Walch, K. Mühl, J. Kraus, T. Stoll, M. Baumann, M. Weber, From car-driver-handovers to cooperative interfaces: visions for driver–vehicle interaction in automated driving, Automot. User Interfaces (2017) 273–294.

[42] K. Christoffersen, D.D. Woods, 1. How to make automated systems team players, Adv. Hum. Perform. Cognit. Eng. Res. 2 (2002) 1−12. Available from: https://doi.org/10.1016/S1479-3601(02)02003-9.

[43] G. Hoffman, C. Breazeal, Collaboration in human-robot teams. Proceedings of AIAA 1st Intelligent Systems Technical Conference, Chicago, IL, 2004. Available from: https://doi.org/10.2514/6.2004-6434.

[44] M. Saffarian, J.C. De Winter, R. Happee, Automated driving: human-factors issues and design solutions (2012).

[45] O. Benderius, C. Berger, V.M. Lundgren, The best rated human−machine interface design for autonomous vehicles in the 2016 grand cooperative driving challenge, IEEE Trans. Intell. Transport. Syst. 19 (4) (2017) 1302−1307.

[46] A. Kunze, S.J. Summerskill, R. Marshall, A.J. Filtness, Augmented reality displays for communicating uncertainty information in automated driving, Automot. User Interfaces Interact. Veh. Appl. (2018).

[47] M. Tönnis, G. Klinker, Survey and classification of head-up display presentation principles (2023).

[48] G. Wiegand, C. Mai, K. Holländer, H. Hussmann, InCarAR: a design space towards 3d augmented reality applications in vehicles, Automot. User Interfaces Interact. Veh. Appl. (2019).

[49] S. Debernard, C. Chauvin, R. Pokam, S. Langlois, Designing human-machine interface for autonomous vehicles, Ifac Papersonline (19)(2016) 609−614.

[50] J. Rasmussen, Skills, rules, and knowledge; signals, signs, and symbols, and other distinctions in human performance models, IEEE Trans. Syst., Man, Cybern. (3)(1983) 257−266.

[51] B. Lee, M. Cordeil, A. Prouzeau, B. Jenny, T. Dwyer, A design space for data visualisation transformations between 2D And 3D in mixed-reality environments, in: Paper presented at the CHI '22: CHI Conference on Human Factors in Computing Systems, 2022.

[52] T. Munzner, Visualization Analysis and Design, A K Peters/CRC Press, 2014.

[53] T. Müller, M. Colley, G. Dogru, E. Rukzio, AR4CAD: creation and exploration of a taxonomy of augmented reality visualization for connected automated driving, Proc. Acm Human-Comput. Interact. 6 (MHCI) (2022) 1−27. Available from: https://doi.org/10.1145/3546712.

[54] J.A. Wagner Filho, C.M.D.S. Freitas, L. Nedel, VirtualDesk: a comfortable and efficient immersive information visualization approach, Comput. Graph. Forum 37 (3) (2018) 415−426. Available from: https://doi.org/10.1111/cgf.13430.

[55] F. Chevalier, P. Dragicevic, S. Franconeri, The not-so-staggering effect of staggered animated transitions on visual tracking, IEEE Trans. Vis. Comput. Graph. 20 (12) (2014) 2241−2250. Available from: https://doi.org/10.1109/TVCG.2014.2346424.

[56] B. Kondo, C. Collins, DimpVis: exploring time-varying information visualizations by direct manipulation, Ieee Trans. Vis. Comput. Graph. 20 (12) (2014) 2003−2012. Available from: https://doi.org/10.1109/TVCG.2014.2346250.

[57] R. Suzuki, A. Karim, T. Xia, H. Hedayati, N. Marquardt, Augmented reality and robotics: a survey and taxonomy for AR-enhanced human-robot interaction and robotic interfaces, in: Paper presented at the CHI '22: CHI Conference on Human Factors in Computing Systems, 2022.

[58] Z. Wang, X. Bai, S. Zhang, Y. Wang, S. Han, X. Zhang, et al., User-oriented AR assembly guideline: a new classification method of assembly instruction for user cognition, Int. J. Adv. Manuf. Technol. 112 (1) (2021) 41−59. Available from: http://doi.org/10.1007/s00170-020-06291-w.

[59] A. Fuste, B. Reynolds, J. Hobin, V. Heun, Kinetic AR: a framework for robotic motion systems in spatial computing (2020).

[60] S. Winkler, J. Kazazi, M. Vollrath, Distractive or supportive - how warnings in the head-up display affect drivers' gaze and driving behavior, in: 2015 IEEE 18Th International Conference on Intelligent Transportation Systems, 2015, pp. 1035−1040.

[61] K.J. Vicente, Cognitive work analysis toward safe, productive, and healthy computer-based work, IEEE Trans. Profess. Commun. (No. 1)(2003) 63.

[62] J. Keil, A. Korte, A. Ratmer, D. Edler, F. Dickmann, Augmented reality (AR) and spatial cognition: effects of holographic grids on distance estimation and location memory in a 3D indoor scenario, PFG — J. Photogramm., Remote. Sens. Geoinf. Sci. 88 (2) (2020) 165−172. Available from: https://doi.org/10.1007/s41064-020-00104-1.

[63] M.C. Schall, M.L. Rusch, J.D. Lee, J.D. Dawson, G. Thomas, N. Aksan, et al., Augmented reality cues and elderly driver hazard perception, Hum. Factors 55 (3) (2013) 643−658. Available from: https://doi.org/10.1177/0018720812462029.

[64] M. Tönnis, Towards automotive augmented reality (2008).

[65] M. Colley, S. Krauss, M. Lanzer, E. Rukzio, How should automated vehicles communicate critical situations? A comparative analysis of visualization concepts, Proc. ACM Interactive, Mobile, Wearable Ubiquitous Technol. 5 (3) (2021) 91−94. Available from: https://doi.org/10.1145/3478111.

[66] H. Kim, J.D. Isleib, J.L. Gabbard, Virtual shadow: making cross traffic dynamics visible through augmented reality head up display, Proc. Hum. Factors Ergonom. Soc. Annu. Meet. 60 (1) (2016) 2093−2097. Available from: https://doi.org/10.1177/1541931213601474.

[67] M. Colley, M. Rädler, J. Glimmann, E. Rukzio, Effects of scene detection, scene prediction, and maneuver planning visualizations on trust, situation awareness, and cognitive load in highly automated vehicles, ACM 6 (2) (2022) 41−49. Available from: https://doi.org/10.1145/3534609.

[68] J.D. Lee, K.A. See, Trust in automation: designing for appropriate reliance, Hum. Factors 46 (1) (2004) 50−80. Available from: https://doi.org/10.1518/hfes.46.1.50_30392.

[69] T. Von Sawitzky, P. Wintersberger, A. Riener, J.L. Gabbard, Increasing trust in fully automated driving: route indication on an augmented reality head-up display (2019).

[70] A. Pauzie, A method to assess the driver mental workload: the driving activity load index (DALI), IET Intell. Transp. Syst. 2 (4) (2008). Available from: https://doi.org/10.1049/iet-its:20080023.

[71] Y. Forster, F. Naujoks, A. Neukum, Increasing anthropomorphism and trust in automated driving functions by adding speech output, in: Paper presented at the 2017 IEEE Intelligent Vehicles Symposium (IV), 2017.

[72] J.S.U.N. Jian, A.M.O. Bisantz, C.G. Drury, Foundations for an empirically determined scale of trust in automated systems, Int. J. Cognit. Ergonom. (No. 1)(2000) 53−71.

[73] A. Bangor, P. Kortum, J. Miller, Determining what individual SUS scores mean: adding an adjective rating scale, J. Usability Stud. 4 (3) (2009) 114−123.

[74] W. Dong, Q. Ying, Y. Yang, S. Tang, Z. Zhan, B. Liu, et al., Using eye tracking to explore the impacts of geography courses on map-based spatial ability, Sustainability 11 (1) (2018) 76. Available from: https://doi.org/10.3390/su11010076.

[75] S. Hergeth, L. Lorenz, R. Vilimek, J.F. Krems, Keep your scanners peeled: gaze behavior as a measure of automation trust during highly automated driving, Hum. Factors: J. Hum. Factors Ergonom. Soc. 58 (3) (2016) 509−519. Available from: https://doi.org/10.1177/0018720815625744.

[76] A.A. Nofi, Defining and Measuring Shared Situational Awareness, Center For Naval Analyses, Alexandria, VA, 2000.

[77] R. Currano, S.Y. Park, D.J. Moore, K. Lyons, D. Sirkin, Little road driving HUD: heads-up display complexity influences drivers' perceptions of automated vehicles, in: Paper presented at the Proceedings of the 2021 CHI Conference on Human Factors in Computing Systems, Yokohama, Japan (2021).

[78] K. Pammer, J. Bairnsfather, J. Burns, A. Hellsing, Not all hazards are created equal: the significance of hazards in inattentional blindness for static driving scenes, Appl. Cognit. Psychol. 29 (5) (2015) 782−788.

[79] Y. Wang, Y. Wu, C. Chen, B. Wu, S. Ma, D. Wang, et al., Inattentional blindness in augmented reality head-up display-assisted driving, Int. J. Human−Comput. Interact. 38 (9) (2022) 837−850. Available from: https://doi.org/10.1080/10447318.2021.1970434.

[80] C.L. Baldwin, C. Spence, J.P. Bliss, J.C. Brill, M.S. Wogalter, C.B. Mayhorn, et al., Multimodal cueing: the relative benefits of the auditory, visual, and tactile channels in complex environments (2012).

[81] J.M. Henderson, J.R. Brockmole, M.S. Castelhano, M. Mack, Visual Saliency Does Not Account for Eye Movements During Visual Search in Real-World Scenes, Elsevier Ebooks, 2007.

[82] T.J. Wright, S. Samuel, A. Borowsky, S. Zilberstein, D.L. Fisher, Experienced drivers are quicker to achieve situation awareness than inexperienced drivers in situations of transfer of control within a level 3 autonomous environment (2016).

CHAPTER 25

Behavioral indicators affecting driving performance in human-machine interface assessments with simulation

Yukun Xie[1], Tianyang Yue[1], Preben Hansen[2] and Fang You[1]

[1]Car Interaction Design Lab, College of Arts and Media, Tongji University, Shanghai, P.R. China [2]Department of Computer and Systems Sciences, Stockholm University, Stockholm, Sweden

Chapter Outline

25.1 Introduction 391
25.2 Method 392
    25.2.1 Participants 392
    25.2.2 Experimental design 392
    25.2.3 Experimental procedure 394
    25.2.4 Data collection 394
25.3 Result 394
    25.3.1 Vehicle vertical indicators 394
    25.3.2 Vehicle horizontal indicators 394
25.4 Discussion 395
25.5 Conclusion 396
Further reading 396

25.1 Introduction

The human-machine interface (HMI) is an important medium for the digital cockpit to interact with the driver for information. Moreover, with the increase of intelligent networked functions in cars, drivers use the touch screen to accomplish some secondary tasks during the driving process. To ensure the safety and usability of automotive HMIs, they should be continuously evaluated and validated throughout the entire automotive HMI design process, including early prototyping and vehicle mass production.

    Simulated driving evaluation is an important method for evaluating in-vehicle HMI. Driving simulators are used to analyze and validate driver behavior. Simulator evaluations are safe and speed up the evaluation process compared with costly and time-consuming morning real-vehicle evaluations. In many cases, evaluators collect a large number of objective indicators of driving and behavior during simulator experiments. However, the efficiency of HMI evaluation and its contribution to design iterations still deserve attention. Simultaneous operation of primary and secondary tasks is used in the HMI simulator review, and the workload is moved to the appropriate region based on a direct relationship curve between workload and operation level. In this region, the operating characteristics of the primary driving task decline as the workload of the secondary task increases and exceeds the driver's ability to compensate. The driver's driving performance reflects the workload of the secondary task. Therefore focus should be placed on objective indicators that have a significant impact on driving performance.

The purpose of this article is to explore which behavioral objective indicators reviewers need to focus on when evaluating touch-screen HMIs in a driving scenario with constant straight-ahead speed. Finding the important behavioral indicators can yield data useful for HMI design optimization at the lowest possible evaluation cost (including time, labor, etc.), suitable for situations when HMIs require rapid design iterations.

25.2 Method

25.2.1 Participants

The demographic information of the participants is shown in Table 25.1. In this experiment, 26 drivers participated, with 53.84% of the participants being male and 46.15% of the participants being female. All participants held a driver's license and had a mean driving age of 5.67 (SD = 4.11). The mean age was 31.08 (SD = 4.26). Of the participants, 30.77% drove daily, 23.08% drove one to three times a week, 15.28% drove four to six times a week, and 30.77% drove one to three times a month. Regarding the use of in-vehicle HMI, 34.62% of the participants used it often, 34.62% used it frequently, and the remaining participants used it rarely.

25.2.2 Experimental design

As shown in Fig. 25.1, this experiment was conducted on a driving simulator involving an in-vehicle HMI, which includes a center control screen and an air conditioning screen. The study used a within-subjects experimental design in which each participant was required to complete six

Table 25.1 Demographics of the participants.

Characteristic	Descriptive statistics
Gender	Male (53.84%), Female (46.15%)
Age	M = 31.08, SD = 4.26
Driver's license	M = 5.67, SD = 4.11
Frequency of driving	Every day (30.77%), One to three days a week (23.08%), Four to six days a week (15.28%), One to three days a month (30.77%)
Frequency of use of in-car HMI	Hardly used (30.77%), Sometimes used (34.62%), Regularly used (34.62%)

FIGURE 25.1

Experimental equipment and environment.

Table 25.2 Secondary tasks and corresponding screens.

Screen	Secondary tasks
Air conditioning screen	Turn up the air speed and turn down the temperature
Central control screen	FM tuned to 102.0 HZ
Central control screen	Random music playback
Air conditioning screen	Adjusting the wind direction
Air conditioning screen	Turn on the air conditioning internal circulation
Central control screen	Turn up the volume of the music

in-vehicle subtasks while driving. Traffic conditions and other traffic-related factors can complicate the driver's driving situation to a greater or lesser extent. Therefore the difficulty of the driving scenario is controlled by the flow of traffic, roads, and other factors. The driving scenario selected for the study is a two-way straight four-lane city with a certain traffic density, and the experiment requires the driver to drive straight ahead at a speed of about 40 km/hour. The scenario meets the requirements of the main task workload and can give the driver some psychological pressure, while not being too difficult. Therefore, it is realistic to require participants to complete the in-vehicle secondary task in such a scenario.

The six secondary tasks about air conditioning function, vehicle setting function, and multimedia function were obtained from the interviews of the 13 drivers. As shown in Table 25.2, the secondary tasks included three tasks for the center control screen and three tasks for the air conditioning screen, which were air conditioning temperature and air speed adjustment, FM tuning station, music playback mode switching, air conditioning air direction switching, and volume adjustment.

25.2.3 Experimental procedure

The experimenter distributed a demographic questionnaire and an experiment information sheet and explained the background and purpose of the experiment to the participants. Prior to the experiment, participants were allowed to perform simulator driving training for about 10 minutes. Before the start of each task, participants were asked to read aloud the task requirements, including the driving task, secondary tasks, and conditions for the task to start. Once the formal experiment started, the secondary task was completed when the participant drove steadily for 40 km/hour and heard a beep. This task was integral to maintaining driving performance. The task could be terminated under three conditions: when the participant successfully completes the task, when the participant proposes to abandon the task, or when there is a driving accident such as a vehicle collision.

25.2.4 Data collection

To investigate the relationship between behavioral data and driving vehicle data, the experiment collected vehicle- and behavior-related data. Vehicle indicators include vertical control and horizontal control. The vehicle vertical control-related indicators are vehicle speed average, vehicle speed standard deviation, acceleration average, and vehicle speed hold percentage. The vehicle horizontal control-related indicators are horizontal offset mean, horizontal offset standard deviation, and lane-keeping percentage.

The behavioral indicators include task success, reaction time, task completion time, total operation path, operation effective path, number of swipes, number of clicks, number of operations, number of no feedback, invalid operation rate, task type, and effective path rate.

25.3 Result

Data collected for all tasks were analyzed for 26 subjects. The study performed a stepwise regression analysis of the collected behavioral indicators with each vehicle data indicator separately.

25.3.1 Vehicle vertical indicators

The results of stepwise regression analysis on vehicle vertical indicators and behavioral indicators are shown in Table 25.3. The mean of speed was predicted independently by the number of swipes ($\beta = -0.26$, $p < 0.01$). The task success contributed significantly to the prediction of the standard deviation of speed ($\beta = -0.36$, $p < 0.001$). The mean of acceleration was predicted independently by the number of swipes ($\beta = -0.25$, $p < 0.05$) and the total operation path ($\beta = 0.19$, $p < 0.001$). The effective path rate contributed independently to the prediction of the percentage of speed maintenance ($\beta = 0.22$, $p < 0.01$), which means the speeds were between 35 and 45 km/hour.

25.3.2 Vehicle horizontal indicators

Table 25.4 shows the results of stepwise regression analyses on vehicle horizontal indicators and behavioral indicators. The mean of horizontal deviation was predicted independently by the number

Table 25.3 Summary of stepwise regression analysis on vehicle vertical indicators and behavioral indicators.

Dependent variables	Independent variables	Adjusted R^2	β
Speed (M)	Number of swipes	0.04	-0.26^b
Speed (SD)	Task Success	0.12	-0.36^c
Acceleration (M)	Number of swipesTotal operation path	0.05	$-0.25^a 0.19^c$
Percentage of speed maintenance	Effective path rate	0.04	0.22^b

$^a p < 0.05.$
$^b p < 0.01.$
$^c p < 0.001.$

Table 25.4 Summary of stepwise regression analysis on vehicle horizontal indicators and behavioral indicators.

Dependent variables	Independent variables	Adjusted R^2	β
horizontal deviation (M)	Number of swipesTask completion timeEffective path rate	0.14	$0.25^c 0.30^c 0.26^c$
horizontal deviation (SD)	Task completion timeTask successEffective path rate	0.40	$0.56^c - 0.26^c 0.18^a$
Percentage of lane maintenance	Number of swipesReaction time	0.09	$-0.27^c - 0.19^a$

$^b p < 0.01.$
$^a p < 0.05.$
$^c p < 0.001.$

of swipes ($\beta = 0.25$, $p < 0.001$), the task completion time ($\beta = 0.30$, $p < 0.001$), and the effective path rate ($\beta = 0.26$, $p < 0.001$). The standard deviation of horizontal deviation was predicted independently by the task completion time ($\beta = 0.56$, $p < 0.001$), the task success ($\beta = -0.26$, $p < 0.001$), and the effective path rate ($\beta = 0.18$, $p < 0.05$). The percentage of lane maintenance was predicted independently by the number of swipes ($\beta = -0.27$, $p < 0.001$) and the reaction time ($\beta = -0.19$, $p < 0.05$).

25.4 Discussion

Overall, the stepwise regression analysis showed that only some of the behavioral indicators were significant independent predictors of the vehicle indicators. For the vehicle vertical indicators, the number of swipes is an influential factor in predicting the mean of speed, and also one of the influential factors in predicting the mean of acceleration. The total number of operations also affects the mean of acceleration. Task success is a factor that affects the standard deviation of vehicle speed. The speed retention percentage may be influenced by the effective path rate. For the vehicle horizontal indicators, the number of swipes still affects the mean value of horizontal offset and the percentage of lane keeping. In addition, the β of the regression analysis indicates that task completion time is an important factor influencing the mean and standard deviation of horizontal offsets. Also, the effective path rate affects the horizontal offset mean and horizontal offset standard deviation. Although Adjusted R^2 is low, task response time may affect lane-keeping percentage.

This experiment has some practical implications for HMI evaluation and design optimization. It can be found that some behavioral indicators do not have significant effects on the common indicators of vehicle vertical control and horizontal control, such as the number of clicks, number of operations, number of no feedback, invalid click rate, task type, etc. Then the collection or analysis of such indicators can be appropriately reduced according to the actual evaluation situation. Choosing appropriate behavioral objective evaluation indicators can not only reduce the workload of evaluation experiments and data analysis but also effectively reflect the performance of drivers in completing secondary tasks. Therefore, for the evaluation, the above-mentioned indicators related to vehicle control should be collected in the in-vehicle HMI touch screen evaluation experiments with emphasis. Some indicators only have an impact on the vertical control of the vehicle, such as the total task path. Some have only a certain impact on horizontal control, such as task completion time and task response time. Evaluators can focus on collecting relevant behavioral indicators based on the vehicle data of interest.

For optimization of in-vehicle HMI design, it is crucial to consider the indicators mentioned above. For example, the number of swipes may have a negative impact on vehicle control in both directions. Therefore designers should aim to minimize swipe interactions in the in-vehicle HMI design. The increase in effective path rate helps the vehicle control in both directions. Therefore it is recommended to improve the effective path rate of operation by strengthening visual guidance and other means during design.

This experiment only evaluated the touch screen interaction method and did not consider the evaluation of other interaction methods such as voice interaction and gesture interaction, thus involving some limitations. The experimental results are not necessarily applicable to the evaluation of other interaction methods for secondary tasks.

25.5 Conclusion

In this study, six secondary tasks using touch screens were selected for simulator driving evaluation experiments to investigate the relationship between behavioral indicators and driving indicators. The experimental results show that some behavioral indicators may affect driving vertical control and Vehicle horizontal indicators. Among them, task success, number of swipes and effective path rate have significant effects on vehicle control in both directions. The results of the study have positive implications for the selection and collection of evaluation indicators, as it can reduce the workload of evaluators and help designers to iteratively optimize the HMI design.

Further reading

T. Lansdown, N. Brook-Carter, T. Kersloot, Distraction from multiple in-vehicle secondary tasks: vehicle performance and mental workload implications, Ergonomics 47 (1) (2004) 91−104.

R. Häuslschmid, B. Pfleging, A. Butz, The influence of non-driving-related activities on the driver's resources and performance, in: G. Meixner, C. Müller (Eds.), Automotive User Interfaces. HIS, Springer, Cham, 2017, pp. 215–247. Available from: https://doi.org/10.1007/978-3-319-49448-7_8.

I. Alvarez, A. Jordan, J. Knopf, D. LeBlanc, L. Rumbel, A. Zafiroglu, The insight prototype–product cycle best practices and processes to iteratively advance in-vehicle interactive experiences development, in: G. Meixner, C. Müller (Eds.), Automotive User Interfaces. HIS, Springer, Cham, 2017, pp. 377–400. Available from: https://doi.org/10.1007/978-3-319-49448-7_14.

P.D. Spyridakos, N. Merat, E.R. Boer, G.M. Markkula, Behavioural validity of driving simulators for prototype HMI evaluation, IET Intel. Transp. Syst. 14 (6) (2020) 601–610.

L. Banjanovic-Mehmedovic, R. Stojak, S. Kasapovic, Driving behavior simulator of lane changing using user-designed interface, in: R. Stojanovic, L. Jozwiak, B. Lutovac (Eds.), 2016 5th Mediterranean Conference on Embedded Computing, 2016, pp. 493–497.

Y.N. Zhang, Z.Y. Guo, Z. Sun, Driving simulator validity of driving behavior in work zones, J. Adv. Transport, 2020, 2020, pp. 1–10.

R.D. O'Donnell, F.T. Eggemeier, Workload assessment methodology, Handbook of Perception and Human Performance: Cognitive Processes and Performance, 2, John Wiley and Sons, Oxford, 1986, pp. 1–49.

J. Radlmayr, C. Gold, L. Lorenz, M. Farid, K. Bengler, How traffic situations and nondriving related tasks affect the take-over quality in highly automated driving, Proc. Hum. Factors Ergonom. Soc. Ann. Meet 58 (1) (2014) 2063–2067.

Electrodermal activity measurement in a driving simulator

CHAPTER 26

Fang You, Yukun Xie, Yaru Li and Yijun Xu

Car Interaction Design Lab, College of Arts and Media, Tongji University, Shanghai, P.R. China

Chapter Outline

26.1 Introduction 399
26.2 Applications and advantages of electrodermal activity measurement 400
    26.2.1 Applications of electrodermal activity measurement in evaluation 400
    26.2.2 Advantages of experiments combined with electrodermal activity measurement 401
26.3 Teaching objectives and method exploration 401
    26.3.1 Teaching objectives 401
    26.3.2 Method exploration 402
26.4 Course experiment methods 403
    26.4.1 Experimental design 403
    26.4.2 Experimenter 403
    26.4.3 Design scheme 403
    26.4.4 Experimental environment and equipment 404
    26.4.5 Electrodermal activity measurement equipment 405
26.5 Analysis and discussion of experimental results 406
    26.5.1 Electrodermal activity and system availability scale 406
    26.5.2 Electrodermal activity and task behavior time 406
26.6 Conclusion 409
Further reading 409

26.1 Introduction

In the field of automotive human-machine interaction (HMI) design, the role of virtual simulation driving cockpit evaluation experiments is crucial. Effective evaluations can assist designers in identifying design issues, facilitating design optimization. Experimental teaching directly influences the research proficiency and cultivation of innovative capabilities for graduate students. Integrating teaching with virtual simulation and mastering various effective evaluation methods is beneficial for students in the automotive HMI design discipline to engage in learning, research, and scientific work. There are primarily three types of design evaluation metrics [1]: subjective measurement, behavioral performance measurement, and physiological measurement. In traditional research, subjective assessments through survey questionnaires and behavioral performance indicators are commonly used for evaluating design solutions. The introduction of galvanic skin response (GSR) technology can provide new methods and perspectives for teaching virtual simulation driving cockpit evaluation [2]. It helps students explore the cognitive load and experiences of users interacting with in-car devices. The continuous measurement feature of electrodermal activity (EDA) measurement technology enables real-time and objective reflection of dynamic changes. When combined with subjective and behavioral data, it enhances credibility and accuracy, assisting students in the comprehensive analysis of various data.

This chapter discusses an experiment on in-car music evaluation as part of a graduate-level interaction design course. The students' selected project is: Multiscreen interactive music system. In the early stages of the course, students are guided to independently conduct design research and express their design concepts. In the later stages, the course aims to cultivate students' skills in experimental design and design evaluation. The students are tasked with designing an experiment that involves multiple scenarios for evaluating the usability of a HMI interface based on GSR measurement. The experiment is conducted by students in a virtual simulation driving cockpit, allowing them to assess the usability of their designed HMI interface in practical scenarios.

26.2 Applications and advantages of electrodermal activity measurement

26.2.1 Applications of electrodermal activity measurement in evaluation

EDA measurement, as an objective physiological measurement method, can gauge psychological stress levels through physiological activation. It is one of the most commonly used physiological signals in physiological measurement. In the field of HMI design, GSR measurement is primarily applied in the following areas:

1. Skin measurements can reflect emotional arousal. Skin conductance is linearly correlated with arousal levels, and changes in the amplitude of emotional arousal can evoke noticeable variations in skin conductance responses. It is often used as a physiological measurement index in arousal experiments. When studying changes in users' emotions during interaction, it is common to combine measurements with the self-assessment manikin scale (SAM) for self-reported emotional assessment.
2. EDA measurement serves as a usability indicator for interface design. Relevant studies have found that when interface design usability is low, user task performance decreases, leading to increased sympathetic nervous system activity, and skin conductance levels are twice as high as those in a resting state. This phenomenon may be associated with the fight-or-flight response of the participants. EDA measurement can effectively reflect interface usability and is often combined with subjective scales such as the system usability scale (SUS), the questionnaire for user interface satisfaction, the post-study system usability questionnaire, the after-scenario questionnaire, as well as behavioral performance for evaluation.

3. EDA measurement is applied to situational awareness and workload. In a study by Mehler et al. using the classic psychological experiment paradigm of the n-back test, it was found that as task difficulty increased, users' skin conductance responses also increased. GSR responses effectively reflect the dynamic changes in mental workload with task difficulty. GSR measurement is often combined with subjective scales such as the NASA task load index (NASA-TLX), situation awareness rating technique, situation awareness global assessment technique, and others for evaluation [3].

EDA measurement can be applied to traditional HMI scenarios. Some researchers in China have applied GSR measurement to fatigue detection and workload assessment of participants in road driving scenarios. This paper further integrates EDA measurement with subjective assessment and behavioral performance, applying it to the virtual simulation driving cockpit evaluation teaching in automotive interaction design.

26.2.2 Advantages of experiments combined with electrodermal activity measurement

In the instructional experiment for virtual simulation cockpit evaluation in automotive interaction design, combining EDA measurement with subjective and behavioral performance assessment has the following advantages:

1. EDA measurement features real-time detection, enhancing result reliability when analyzed in conjunction with subjective measurement. Subjective measurement, often requiring participants to report product experiences postusage, introduces delays and potential biases. Combining it with electrodermal responses helps validate the accuracy of subjective measurements.
2. EDA measurement is sensitive, and when combined with behavioral performance analysis, it facilitates in-depth exploration of cognitive psychological states. Behavioral performance measurement, reliant on experimenter observation, may lack sensitivity to changes in participants' cognitive states. Integrating electrodermal measurement directly captures effective cognitive changes, revealing associations between behavior and psychology.
3. EDA measurement is easy to operate and integrates seamlessly with subjective and behavioral performance assessments. Noninvasive and with a simple installation process, it does not disrupt participants' driving. Students find it easy to learn, reducing the likelihood of errors during solo experiments and facilitating straightforward integration into teaching.

26.3 Teaching objectives and method exploration

26.3.1 Teaching objectives

To enhance students' in-depth understanding of automotive interaction design, it is essential to emphasize the cultivation of their practical skills based on the foundation of testing and evaluation theory. In line with this approach, a virtual simulation cockpit evaluation experiment course is introduced, targeting graduate students. Leveraging the research background in automotive interaction design accumulated over several years within the institution and the available equipment and environmental conditions, the course encourages students to independently propose topics for evaluation. The objectives of the course include:

1. To allow graduate students to experience the complete virtual simulation cockpit evaluation process, cultivating their interest in learning automotive HMI design and broadening their knowledge, through teaching.
2. To acquire skills in virtual simulation cockpit evaluation experiment design and the preparation of experimental materials, enhancing practical experimentation abilities.
3. To learn various design evaluation methods, including electrodermal, subjective, and behavioral performance assessments. Master the primary operational methods of EDA measurement devices and understand how to analyze electrodermal data in conjunction with subjective and behavioral performance data, obtaining the direction for optimizing design solutions.

26.3.2 Method exploration

The course has transformed the traditional teaching model by establishing a learning-centered relationship with graduate students as the main participants and teachers as guides throughout the interactive design course. This teaching model emphasizes student initiative and is integrated into the entire design process, primarily manifested in the following three aspects.

26.3.2.1 Independent proposition and design

Based on the research background, the teacher establishes the overarching theme and guides students to independently propose topics within this framework. The teacher provides assistance in formulating topic suggestions. Each student group conducts research using suitable design methods, with the teacher guiding the design direction throughout the course. The teacher supports students in completing the interactive design proposals and development work.

26.3.2.2 Independent design and evaluation experiment

The teacher introduces the evaluation environment, methods, and equipment usage. Students independently design experiments based on their design proposals, selecting appropriate measurement metrics. The teacher offers suggestions for improving the experimental design. Students select participants based on their experimental design, with course participants serving as test subjects and others assisting the main researcher in completing the experiment. Students are responsible for a reasonable division of labor during the evaluation.

26.3.2.3 Independent data analysis and optimization

Students are required to analyze electrodermal data, subjective data, and behavioral performance data to enhance their data processing skills. They link EDA measurement data with subjective and behavioral performance data, identify correlations and patterns in the data, comprehensively analyze the reasons behind the results, and finally propose directions for optimizing the design proposal.

26.4 Course experiment methods

26.4.1 Experimental design

The experiment design is proposed by students, and the purpose of the experiment is to conduct a usability evaluation of the car HMI interface design using physiological measurement methods when switching music scenes. The experiment is divided into static and dynamic parts. Participants, in both stationary and driving states, complete the task of switching music and announcing the music number using two interaction methods: physical buttons on the steering wheel or the central control touchscreen. During the experiment, the EDA of the participants is continuously collected in real time. After completing the tasks, participants provide SUS ratings. To minimize the impact of the experimental sequence on participants (static using steering wheel button group, static using central control touchscreen group, driving using steering wheel button group, driving using central control touchscreen group), it is necessary to achieve complete balance across the four intervention conditions. The experiment adopts a within-group design and utilizes a Latin square balance to set the task sequence, as shown in Fig. 26.1.

26.4.2 Experimenter

To ensure the participation of all students in the course, the main experimenters for this experiment are students who are part of the design team or other course students without a driver's license. There are 12 participants who hold a driver's license, consisting of course students and other individuals on campus. The average age is 24 years, with a standard deviation (SD) of 2.42. The average driving experience is 1.85 years (SD = 1.14).

26.4.3 Design scheme

The design proposal is based on a multiscreen interactive entertainment information system developed by course students using the Kanzi software platform. The system takes the in-car music application as an example. In the preliminary phase, research and analysis were conducted to understand the driver's behavior and music entertainment needs in different scenarios. The analysis included the positioning of various screens in the in-car information system, interface layout, and exploration of multiscreen interactive collaboration methods within the vehicle. The entertainment

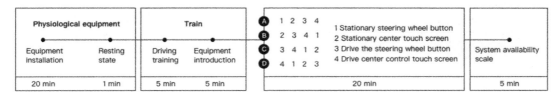

FIGURE 26.1

Experimental task sequence.

features are tailored to the in-car driving scenario, presenting music and entertainment content on multiple screens in the vehicle. Figs. 26.2 and 26.3 depict the design of the instrument panel and central control double-screen music application interface. During the initial user research, students identified two frequently used interaction methods: physical buttons on the steering wheel and touchscreen interaction on the central control panel. Students aim to investigate which interaction method is more suitable for users to perform in-car entertainment functions in both stationary and driving scenarios. Under driving conditions, participants need to ensure the safe completion of primary driving tasks while quickly obtaining and operating information systems. This aims to meet people's social needs and the requirements of safe driving, providing participants with a better user experience.

26.4.4 Experimental environment and equipment

As shown in Fig. 26.4, leveraging the research background of the Automotive Interaction Lab over the years, the design evaluation experiment will be conducted on a self-developed driving simulation platform in the laboratory. This platform aims to create a realistic driving space for

FIGURE 26.2

Instrument interface design.

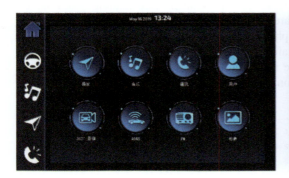

FIGURE 26.3

Central control interface design.

26.4 Course experiment methods 405

FIGURE 26.4

Driving simulation bench.

participants. The driving simulator utilizes the Fanatec Porsche 918 RSR steering wheel and related accessories, along with a computer equipped with a three-channel fused panoramic display. The simulation of the vehicle's surroundings is developed using Unity software, and the virtual scenario for evaluation involves a self-developed bidirectional two-lane program. The in-car interactive display screens can be freely replaced according to the experimental requirements. The design includes instrument and central control double screens, simulated using two Surface Pro tablets. The corresponding interface programs are developed using Kanzi, and the driving data from the Unity program is transmitted through the message queuing telemetry transport telemetry transmission protocol.

26.4.5 Electrodermal activity measurement equipment

The EDA measurement equipment includes the BIOPAC Smart Center physiological recorder and related materials from Biopac company. The original sampling rate is 2 kHz. As shown in Fig. 26.5, to ensure accurate data collection, the experiment involves first removing the calluses from the participant's foot and applying conductive gel, such as GEL101. Subsequently, electrode patches are attached to the participant's left foot.

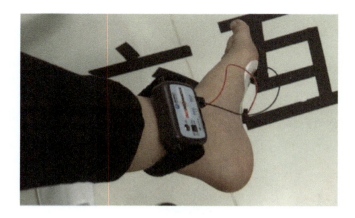

FIGURE 26.5

Electrodermal activity electrode installation process and position.

26.5 Analysis and discussion of experimental results

26.5.1 Electrodermal activity and system availability scale

The EDA measurement data is combined with subjective data, represented by the SUS, to provide a detailed analysis and measurement of the system's usability and effectiveness. The SUS consists of 10 items rated on a 5-point scale, which need to be converted to scores on a percentage scale. Among these, 8 items form the Usability subscale, and 2 items form the Learnability subscale. Fig. 26.6 shows the results of the SUS, where the average learnability score for the Center Console Touchscreen group is 91.67 (SD = 9.73), higher than the Steering Wheel Button group with an average score of 69.79 (SD = 27.42). The average system usability score for the Center Console Touchscreen group is 76.46 (SD = 11.45), higher than the Steering Wheel Button group with an average score of 73.33 (SD = 18.54). These results align with the data on electrodermal responses and resting skin conductance level differences, as depicted in Fig. 26.7. Combining static and driving conditions, the average EDA difference for the Center Console Touchscreen group is 3.25 (SD = 3.23), which is higher than the average EDA difference of 3.48 (SD = 3.23) for the Steering Wheel Button group. This suggests that participants experience a lower mental load when switching songs using the center console touchscreen. The electrodermal response validates the system's usability results, indicating that the HMI interface design of the Center Console Touchscreen group is more usable than the Steering Wheel Button group. In summary, considering both electrodermal and system usability data, participants in the experiment had a better experience when using the center console touchscreen to switch songs compared with using the steering wheel physical buttons.

26.5.2 Electrodermal activity and task behavior time

The EDA data is analyzed in conjunction with task behavior time. All participants successfully completed the tasks, so task behavior time is used as a measure of behavioral performance.

26.5 Analysis and discussion of experimental results

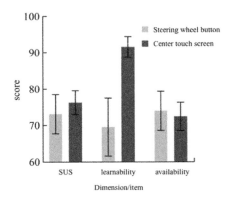

FIGURE 26.6

System availability scale statistics.

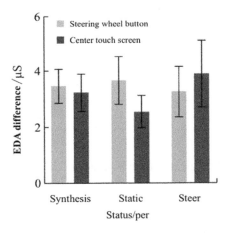

FIGURE 26.7

Statistical chart of electrodermal activity measurement level.

Previous studies have found that users in a relaxed state exhibit a gradual decrease in skin conductance levels when using systems with good usability. In contrast, when using systems with poor usability, skin conductance levels initially increase for a period and then stabilize or decrease, indicating the presence of a certain level of psychological load. As shown in Fig. 26.8, participants received specific task requirements before the task began, with the time of task prompt as the zero point on the x-axis and EDA conductance on the y-axis. Both the Steering Wheel Button group and the Center Console Touchscreen group show a trend of initial increase followed by a slow decrease,

408 Chapter 26 Application of EDA measurement in a driving simulator

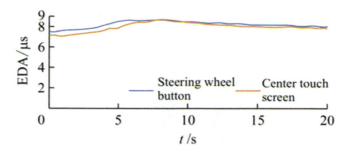

FIGURE 26.8

Electrodermal activity level statistical chart.

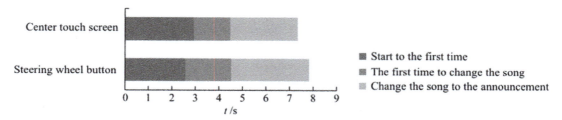

FIGURE 26.9

Task behavior time statistics chart.

indicating the presence of psychological load in both groups. However, the initial skin conductance level of the Steering Wheel Button group is significantly higher than that of the Center Console Touchscreen group, suggesting that participants using the steering wheel buttons maintain higher alertness upon hearing specific task information, indicating a higher psychological load. In Fig. 26.9, considering both static and dynamic testing states, the Steering Wheel Button group takes less time from the task prompt to the first key press than the Center Console Touchscreen group. This aligns with the sequence of the initial increase and peak of skin conductance in the Steering Wheel Button group, and the overall time for 2 key presses and task completion in the Steering Wheel Button group is longer than in the Center Console Touchscreen group. The trend of a slightly smaller decrease in skin conductance after the initial rise in the Steering Wheel Button group is consistent with the trend in the Center Console Touchscreen group. Combining task behavior with skin conductance data, it can be concluded that the Steering Wheel Button group has lower task performance, reflecting better usability in the Center Console Touchscreen group. These results are consistent with subjective measurement results. Students analyze that the main reason is that the two buttons on the steering wheel are too far apart, with poor indicators and a high learning cost. They further suggest optimizing the position and style of the steering wheel buttons to improve learnability and reduce mental load.

26.6 Conclusion

The article focuses on a student interaction design course with the topic: Multiscreen Interactive Music System. The design involves two interaction methods: steering wheel buttons and a central control touchscreen. The experiment utilizes EDA measurement methods to evaluate the human-computer interaction interface in both stationary and driving states on a simulator. A comparison is made with the SUS and task behavior time. The study suggests that human-computer interaction interfaces with lower usability lead to increased excitement in the sympathetic nervous system, causing changes in skin conductance. The experiment demonstrates that EDA measurement is an objective and accurate method for detecting changes in skin conductance, indicating the effectiveness of this measurement approach.

References

[1] N. Meshkatin, P.A. Hancock, M. Rahimi, et al., Techniques in mental workload assessment, Evaluation of Human Work: A Practical Ergonomicsmethodology, 2nd ed., Taylor & Francis, Philadelphia, PA, US, 1995, pp. 749–782.
[2] A.C. Dirican, M. Goktuürk, Psychophysiological measures of human cognitive states applied in human computer interaction, Procedia Computer Sci. (3) (2011) 1361–1367.
[3] S. Rubio, E. Diaz, J. Martin, et al., Evaluation of subjective mental workload: a comparison of SWAT, NASA-TLX, and workload profile methods, Appl. Psychol. 53 (1) (2004) 61–86.

Further reading

M.M. Bradley, P.J. Lang, Measuring emotion: behavior, feeling, and physiology [M]Edn Cognitive Neuroscience of Emotion, Oxford University Press, New York, NY, US, 2000.

Z. Mao, Research on the Recognizing Arithmetic of Fatigue Driving Based on the Physiologic Factor Analysis, Wuhan University of Technology, Wuhan, 2006.

B. Mehler, B. Reimer, Y. Wang, A comparison of heart rate and heart rate variability indices in distinguishing single-task driving and driving under secondary cognitive workload, in: Proceedings of the Sixth International Driving Symposium on Human Factors in Driver Assessment, Training and Vehicle Design. Lake Tahoe, CA, 2011, pp. 590–597.

Q. Min, M. Li, From usability to adoption: new m-commerce adoption study framework, Appl. Res. Computers 26 (5) (2009) 1799–1802.

P. Muter, J.J. Furedy, A. Vincent, et al., User-hostile systems and patterns of psychophysiological activity, Computers Hum. Behav. 9 (1) (1993) 105–111.

B. Reimer, B. Mehler, J.F. Coughlin, et al., Theimpact of a naturalistic hands-free cellular phone task on heart rate and simulated driving performance in two age groups, Transportation Res. Part. F: Traffic Psychol. Behav. 14 (1) (2011) 13–25.

B. Reimer, B. Mehler, The impact of cognitive workload on physiological arousal in young adult drivers: a field study and simulation validation, Ergonomics 54 (10) (2011) 932–942.

D. Shi, The Analysis of User Interaction Behavior Based on EDA Experiment, Zhejiang University, Hangzhou, 2017.

F. Steinberger, R. Schroeter, C.N. Watling, From road distraction to safe driving: evaluating the effects of boredom and gamification on driving behaviour, physiological arousal, and subjective experience, Computers Hum. Behav. 75 (2017) 714–726.

C. Stickel, A. Scerbakov, T. Kaufmann, et al., Usability Metrics of Time and Stress - Biological Enhanced Performance Test of a University Wide Learning Management System., 5298, Springer, Berlin Heidelberg, 2008, pp. 173–184.

B. Sun, J. Yang, Y. Sun, et al., Color design of human vehicle interaction based on eye-tracking experiment, Packaging Eng. 40 (2) (2019) 23–30.

Z. Tan, M. Ma, J. Sun, et al., Study on user experience of web page chart rate based on physiological electrical technology, Packaging Eng. 37 (22) (2016) 97–101.

A.P. O.S. Vermeeren, E.L.-C. Law, V. Roto, et al., User experience evaluation methods: current state and development needs, in Proceedings of the 6th Nordic Conference on Human-Computer Interaction: Extending Boundaries, Reykjavik, Iceland, Association for Computing Machinery, 2010, pp. 521–530.

H. Wang, H. Wen, M. Liu, Experimental evaluation of nighttime driver's physiological characteristics in driver simulator, J. Jilin Univ. (Eng. Technol. Ed.) 47 (2) (2017) 420–428.

Y. Wang, T. Gu, Y. Chen, et al., Research on virtual simulation replacing real car test in university teaching, Res. Exploration Laboratory 39 (2) (2019) 79–82. 130.

R.D. Ward, P.H. Marsden, Physiological responses to different web page designs, Int. J. Human-Computer Stud. 59 (1–2) (2003) 199–212.

Y. Zhang, M. Li, L. Yang, Research and practice of open research-based experiment teaching method for engineering graduate students, J. High. Educ. (4)(2019) 11–13.

CHAPTER 27

Using eye-tracking to help design HUD-based safety indicators for lane changes

Fang You[1], Jürgen Friedrich[2], Yang Li[1], Jianmin Wang[1] and Ronald Schroeter[3]

[1]Car Interaction Design Lab, College of Arts and Media, Tongji University, Shanghai, P.R. China [2]Center for Computing Technologies, University of Bremen, Bremen, Germany [3]Centre for Accident Research and Road Safety-Queensland (CARRS-Q), Queensland University of Technology, Brisbane, Australia

Chapter Outline

27.1 Introduction and related works 411
27.2 Methods 412
    27.2.1 Experimental setup 412
    27.2.2 Data collection 412
    27.2.3 Data analysis 413
27.3 Results 413
27.4 Discussion 414
27.5 Lessons learned 416
Further reading 417

27.1 Introduction and related works

Accident statistics confirm the need for systems that assist car drivers during lane changes. Lane changes increase the overall workload. The driver has to deal with multiple tasks, gathering detailed information from the road scenario ahead, from mirrors, the blind spot, and the dashboard within the vehicle, to obtain information about current road signs, distances to surrounding cars, current speed, etc.

    Head-up displays (HUDs) are extremely well-researched in the automotive domain. According to Lamble et al., it is recommended that information should be positioned as closely as possible to the driver's normal line of sight, that is, typically on top of the dashboard. Tretten et al. noted that HUDs allow "for the display of safety systems information in an easily noticeable location" and suggest positioning the display just below the line of sight at approximately 5 degrees as the most favorable placement. As HUDs become more advanced, offering full windscreen capabilities, the question of where to best position driving task-specific HUD-based warning indicators or decision aids needs to be addressed. To address this, we leverage eye-tracking methodologies.

    Eye-tracking studies illuminate the visual and attentional processes involved in driving, enabling researchers to identify visual demands and detect drivers' distraction processes. Drivers use fixation for scene interpretation and decision-making. In this context, fixations, saccades, and smooth pursuits are extensively studied. However, in our study, we only consider fixation and saccades. We refer to saccades between fixations as moving paths when investigating the driver's normal lines of sight during lane changes.

As such, the overall aim of this project is to enable a good view management concept for HUDs that provides rules for the location of information specific to different driving scenarios, avoiding clutter and overlap, which could distract the driver and impact driving performance. Our pilot study presented here aims to identify where to display what information to the driver during lane changes. It tests our methods and obtains preliminary verification of our hypothesis that eye-tracking can become a general method for informing HUD design.

27.2 Methods

Our pilot study took place in standard cars on regular streets. Rather than assuming existing typical HUD placements, our research takes a step back, employing a user-centered "in the wild" approach by investigating the drivers' natural eye movements during lane changes on semiopen roads.

27.2.1 Experimental setup

The experiment took place within our comparatively large (and road-rich) Tongji University campus. The roads are standard two-lane China motorway roads with a stable (low) level of traffic. The lane-changing experiment was part of a longer drive, totally lasting 30 minutes on average and including several typical driving tasks (reversing, parking, crossing intersections, etc.) with 17 test drivers (9 male, 8 female; aged 25–50 years with driving experience of 1–13 years) (mean: 6.8 years, standard deviation: 17.29). The test vehicle was an SAIC Tiguan SUV.

For the lane changes under study, an accompanying test supervisor asked participants to move into an adjacent lane. The number of lane changes was limited to two per driver. All drivers performed the lane changes along the same stretch of road. Being a quiet university campus, the traffic conditions were similar during all drives, but being an open road, they could not be controlled. It represented a compromise between a fully controlled but unnatural test track and an uncontrolled public road.

27.2.2 Data collection

We used the wearable eye tracker Tobii Glasses (Analyzer_ 1.41.2285) in binocular tracking mode. An accompanying test supervisor used the Tobii Pro Glasses Controller Software on a Pro tablet to manage participants, control the eye tracker, and view both real-time and recorded eye-tracking data. Parallax compensation was set to automatic. We used tinted lenses in case of strong sunshine to decrease noise in the data. The algorithm was based on angular velocity to make eye recognition more reliable. Calibration was done once at the beginning and live viewing was used during the experiment to ensure the system performed well.

27.2.3 Data analysis

After data collection, we used Tobii Pro Glasses Analyzer Software to export and interpret the data with a sampling rate of 50 Hz; the duration of fixation was more than 200 msecond in our experiment. The Tobii software automatically creates an eye-movement video, which contains the fixation points (Fig. 27.1). To reduce their number and complexity, we manually grouped fixation points located close to each other into five fixation areas. Through further analysis of the eye movement video, we obtained the fixation moving paths, i.e., the eye's way from one fixation area to another. We then analyzed the different characteristics of the fixation moving paths quantitatively during (a) left-to-right and (b) right-to-left lane changing. We also analyzed the eye-movement video qualitatively through detailed observations to obtain what information the driver appears to search for and needs during lane changes.

27.3 Results

Our participants relied on fixation to obtain traffic information. In this report, the visual characteristics of the driver are mainly identified by two aspects, fixation moving path and fixation area.

The fixation moving path is the direction of the driver's gaze moving from one fixation area to another. In Fig. 27.2, we use arrow thickness to illustrate the number of runs through fixation moving paths and circle thickness to show the number of visits to a fixation area. Fig. 27.2 shows fixation moving paths and fixation areas superimposed on all participants in right-to-left lane changing.

Fig. 27.3 shows the mean values of the eight most frequent eye movements between fixation areas (fixation moving paths) per participant for right-to-left-lane changing. We obtained 14 moving paths. The highest number of movements stems from the current lane to the left mirror (CL→LM: 1.82), the second highest is from the left mirror to the current lane (LM→CL: 1.47), and the third highest is from the left mirror to the target lane (LM→TL: 1.24). The visual weight, that is, the sum of these three values ($\Sigma = 4.52$), is much higher than the visual weight of all other (11) values together ($\Sigma = 3.02$). The analysis means that movements

FIGURE 27.1

Fixation points (examples). *Top*: target lane; *Bottom*: left mirror, red dots mark fixation points.

Chapter 27 Using eye-tracking to help design hud-based safety indicators

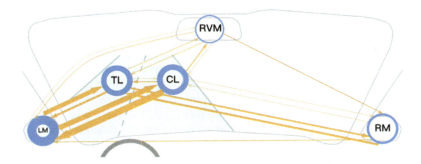

FIGURE 27.2

Fixation moving paths and fixation areas on right-to-left lane change.

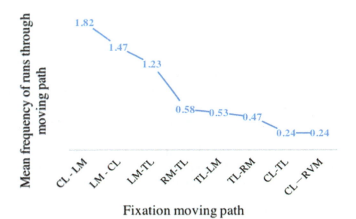

FIGURE 27.3

Fixation moving paths on right-to-left lane change.

between the left mirror and lanes occur most frequently and that this region attracts the most attention. This result is confirmed by the number of visits to the respective fixation areas (Fig. 27.4). For left-to-right lane changing, the results are similar.

Observing the eye-movement video in detail, we found that the driver checked mirrors several times or with long glances until an object's intention seemed to be resolved.

27.4 Discussion

In our study, information organization is the task of visually structuring and defining interface information, workflow, and interaction for the driver so that they can quickly and accurately find

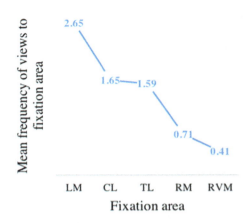

FIGURE 27.4

Visits of fixation areas on right-to-left lane change.

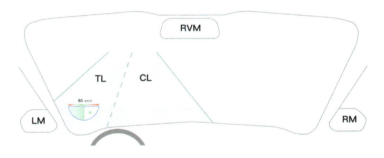

FIGURE 27.5

Indicator location on right-to-left lane change (green here means that in this situation the driver can change lanes).

the information they need. Subsequently, we discuss how our findings from the experiment influenced design choices for supporting drivers in lane changes.

As shown in Fig. 27.2, attention attraction in lane changing from right to left is maximized in the LM/CL/TL region. Considering the relevant fixation areas, LM is more often visited than CL and TL. We therefore recommend putting the HUD indicator in this region, leaning slightly in the direction toward LM, as illustrated in Fig. 27.5. For left-to-right lane changing, we suggest putting the HUD indicator leaning slightly in the direction toward TL (Fig. 27.6). In either case the indicators are displayed in such a way that they virtually appear on the respective target lanes. This position avoids visual distraction because the driver will look toward the target lane anyway. The preliminary HUD design (Fig. 27.7) shows how to potentially help drivers get more information about the most likely projections/intentions of the rear car.

416 **Chapter 27** Using eye-tracking to help design hud-based safety indicators

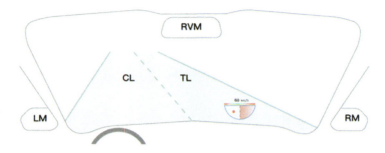

FIGURE 27.6

Indicator location on left-to-right lane change (red here means that in this situation the driver must not change lanes).

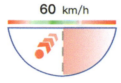

FIGURE 27.7

Potential information of indicator on left-to-right lane change.

27.5 Lessons learned

We found some very interesting results about positioning safety indicators for lane changing on a HUD. The results were achieved in a semiopen traffic environment and therefore may be more valid than those from closed traffic tracks. Nevertheless, our results have a preliminary character only because we did not control our (pilot) experiment, and we did not yet accompany our eye-tracking method with a systematic collection and interpretation of user experience data. Therefore, it is worth thinking about conclusions we can draw from the pilot for a larger main study. What worked well, and what did not? What will we do differently when continuing our research?

We also consider using a second car in addition to the test car to create special situations, such as overtaking the test car. This could create a more complex and realistic situation in a controlled way. In general, we want to find out if eye-tracking is an appropriate method for the overall design of HUD safety support.

Positioning indicators was just the start of a more complex consideration of the HUD-based visual aid to improve safety. Good warnings require advanced sensors and software. Besides speed and acceleration information, indicator systems will need data about distances between cars, image recognition of camera data, driver reaction time, etc. Considering these factors into account, the content of the HUD design should involve dynamic levels of detail. In our design (Fig. 27.5), we use simple color-coded warnings, but if the traffic environment changes suddenly and frequently, the flickering warning may confuse drivers, so we need to consider what other details should be offered to the driver, such as the potential direction and prediction time of other cars. This functionality is already available in highly automated cars. In this context, information-driven indicators used as HUD safety elements could aid situational awareness and trust.

Further reading

T. Friedrichs, M.-C. Ostendorp, A. Lüdtke, Supporting drivers in truck platooning: development and evaluation of two novel human-machine interfaces, in: Proceedings of the 8th International Conference on Automotive User Interfaces and Interactive Vehicular Applications (AutomotiveUI '16), Ann Arbor, MI, USA, October 24–26, 2016, pp. 277–284.

Joseph L. Gabbart, Gregory M. Fitch, Hyungil Kim, Behind the glass: driver challenges and opportunities for ar automotive applications. In, Proc. IEEE 102 2 (2014) 124–136.

Steven Feiner Tobias Höllerer, Blaine Bell Drexel Hallaway, Dennis Brown Marco Lanzagorta, Yohan Baillot Simon Julier, Lawrence Rosenblum, User interface management techniques for collaborative mobile augmented reality, Computers & Graph. 25 (2001) 799–810.

S. Hurtado, S.. Chiasson, An eyetracking evaluation of driver distraction and unfamiliar road signs, in: Proceedings of the 8th International Conference on Automotive User Interfaces and Interactive Vehicular Applications (AutomotiveUI '16), Ann Arbor, MI, USA, October24–26, 2016, pp. 153–160.

Enkelejda Kasneci, Gjergji Kasneci, Thomas C. Kübler, Wolfgang Rosenstiel, Online recognition of fixations, saccades, and smooth pursuits for automated analysis of traffic hazard perception, in: Petia Koprinkova-Hristova, et al. (Eds.), Artificial Neural Networks, 4, Springer Series in Bio-/ Neuroinformatics, 2015, pp. 411–434.

Vassiliki Filippakopoulou Vassilios Krassanakis, Byron Nakos, EyeMMV toolbox: an eye movement post-analysis tool based on a two-step spatial dispersion threshold for fixation identification, J. Eye Mov. Res. 7 (1) (2014) 1–10.

Y.S. H. HyunSuk Kim, C.H. P. DaeSub Yoon, An analysis of driver's workload in the lane change behavior, in: Proceedings of International Conference on ICT Convergence, 2013, pp. 242–247.

Dave Lamble, Matti Laakso, Heikki Summala, Detection thresholds in car following situations and peripheral vision: implications for positioning of visually demanding in-car displays, Ergonomics 42 (6) (1999) 807–815.

L. Lorenz, P. Kerschbaum, J. Schumann, Designing take over scenarios for automated driving: How does augmented reality support the driver to get back into the loop? in: Proceedings of the Human Factors and Ergonomics Society Annual Meeting, Vol. 58, issue 1, 2014, pp. 1681–1685.

R. Schroeter, F. Steinberger, Pokémon DRIVE: towards increased situational awareness in semi-automated driving, in: Proceedings of the 28th Australian Conference on Computer-Human Interaction (OzCHI '16). Launceston, Tas, 2016, pp. 25–29.

Statistisches Bundesamt: Verkehr – Verkehrsunfälle, Fachserie 8, Reihe 7, Wiesbaden 2016, 2014.

P. Tretten, A. Gärling, R. Nilsson, T.C. Larsson, An on-road study of head-up display: preferred location and acceptance levels, in: Proceedings of the Human Factors and Ergonomics Society 55th Annual Meeting, 2011, pp.1914–1918.

Marc Wittmann, Miklós Kiss, Peter Gugg, Alexander Steffen, Martina Fink, Ernst Pöppel, et al., Effects of display position of a visual in-vehicle task on simulated driving, Appl. Ergonom. 37 (2006) 187–199.

CHAPTER 28

The situation awareness and usability research of different HUD HMI designs in driving while using adaptive cruise control

Jianmin Wang[1], Wenjuan Wang[1], Preben Hansen[2], Yang Li[1] and Fang You[3]

[1]Car Interaction Design Lab, College of Arts and Media, Tongji University, Shanghai, P.R. China [2]Department of Computer and Systems Sciences, Stockholm University, Stockholm, Sweden [3]Shenzhen Research Institute, Sun Yat-Sen University, Shenzhen, GD, P.R. China

Chapter Outline

28.1 Introduction 420
    28.1.1 Adaptive cruise control 420
    28.1.2 Head-up display: windshield and augmented reality 420
    28.1.3 Situation awareness 421
    28.1.4 Usability test 421
28.2 Research propositions 422
28.3 Experimental method 422
    28.3.1 Participants 422
    28.3.2 Technical equipment 422
    28.3.3 Experimental design 423
    28.3.4 Experimental tasks 425
    28.3.5 Experimental process 425
28.4 Results and preliminary findings 425
    28.4.1 Situation awareness 426
    28.4.2 Usability 427
28.5 Discussion 428
28.6 Limitations 429
28.7 Conclusion 429
Further reading 429

28.1 Introduction

28.1.1 Adaptive cruise control

During the last few years, an increasing number of advanced driver assistance systems (ADAS) have been developed to improve driving safety and comfort. Adaptive cruise control (ACC) is one of the most important semiautomated functions, which is a longitudinal support system that can adjust the speed and maintain a time-based separation from the vehicle in front.

ACC systems can have positive and negative effects on driving safety [1]. ACC provides a potential safety benefit in helping drivers maintain a constant speed and headway. Therefore the driver has more resources available to attend to other tasks, such as looking at route signs and traffic signals. Also, when ACC is provided, drivers pay more attention to lateral control than when ACC is not provided. However, drivers may glance "off-road" more frequently and longer when using ACC, which decreases safety. As drivers rely on the ACC system, they do not monitor their surroundings as carefully and might thus lose some of their situation awareness (SA). Overall, human drivers and vehicles need to drive together when using ACC, which requires an exchange of information between the person and the car. Therefore, for the designer, it is necessary to consider the general mental model of the driver when using ACC to design better HMIs for the function of ACC.

28.1.2 Head-up display: windshield and augmented reality

HUD technology in aviation has been known since the 1940s. It keeps the pilot's attention focused outside of the cockpit and supports aircraft control by information visualization within the pilot's main sight line on a transparent combiner. In recent years, with the development of augmented reality (AR) technology, more and more car manufacturers have begun to work on the development of augmented reality head-up display (AR-HUD). AR-HUD images can be shown not only in the drivers' line of sight, but also matching the real traffic environment and providing a true augmented experience.

It is found that HUD allows improving "eyes on the road" time by reducing the number of glances to the in-vehicle. HUD allows more time to scan the traffic scene and enhances understanding of the vehicle's surrounding space, particularly under low visibility conditions. This benefit in terms of increasing SA can impact the probability a driver will succeed in detecting a time-critical event. However, Stanton and Young find that the provisions of a HUD reduce reported SA. Perhaps one of the reasons for this finding is that with the instrument cluster display, drivers can have discretion over when they want to sample the information, whereas with the HUD the data is displayed all the time. Therefore the HUD might have made the driving task more visually complex and reduced overall SA.

In some studies, researchers also give some design suggestions for automotive HUD. According to Wang et al. [2], the information elements of HUD-HMI are mainly presented in two forms: text and graphic symbols. It is also found that prompt and warning information should be in the state of low brightness or no brightness in general, and the driver's attention should be aroused by increasing brightness, flashing, or making prompt sounds when necessary. One of the main differences is that if the vehicle is equipped with a driving assistance system, the design elements of driver assistance features will be supplemented in HUD, such as lane departure warning. Another main difference is that the HUD design of different cars may have different color combinations, for example, BMW uses white, green, and orange, while Benz uses white, blue, and red.

28.1.3 Situation awareness

When driving, SA can be defined as the driver's mental model of the current state of the vehicle and mission environment. This indicates that the driver's SA of the ACC system and the road environment is very important for driving performance during the use of ACC. SA can be formally defined as the "perception of the elements in the environment within a volume of time and space, the comprehension of their meaning, and the projection of their status in the near future." The SA model encompasses three levels of SA. Level 1 includes the perception of elements in the current situation (e.g., reading the set speed and time gap), level 2 is the comprehension of the current situation (e.g., knowing whether the vehicle is following a leading vehicle), and level 3 is the projection of future status (e.g., anticipating status changes of vehicle, and avoiding conflict). Driving can be thought of as a dynamic control system in which system input variables change over task time. In theory, the construct of SA in dynamic systems fits very well in this domain. Though advanced automation technologies have been expected to improve system and operator safety, efficiency, and comfort, such technologies may also generate negative effects on driver behavior. As such, some researchers put forward the view that increasing the driver's SA is a key to successful automation.

There are four classes of approaches to evaluating SA: process, direct, behavioral, and performance measures. Process measures include eye tracking, information acquisition, and analysis of communications. Direct measures of SA attempt to directly assess a person's SA through subjective measures and objective measures. Subjective measures ask the users to rate their SA on a scale. Objective measures collect data from users on their perceptions of the situation and compare them to what is actually happening to score the accuracy of their SA. Behavior measures infer the operators' level of SA based on their behavior. Performance measures infer SA based on how well an operator performs a given task as compared with a predetermined standard, or how good the outcomes of a scenario are (e.g., the number of successful missions).

In this experiment, performance measures and objective measures are adopted. We measure the driving performance by the driver's lateral control of the vehicle; we measure objective SA by the situation awareness global assessment technique (SAGAT). The SAGAT assesses level 1 (perception), level 2 (comprehension), and level 3 (projection) SA by asking the driver questions related to the relevant features of the car and the external environment necessary for safe driving. In the experiment of this paper, because the program pause will affect the execution of the task, participants complete the SAGAT immediately after each task. Endsley's study has shown that SA data are readily obtainable through SAGAT for a considerable period after a stop in the simulation (up to 5–6 minutes). In this experiment, the execution time of each task is about 1 minute, so SAGAT can be used after each task without affecting the SA inquiry effect as much as possible.

28.1.4 Usability test

For the development and design of ADAS, system usability should be improved while ensuring safety. The ISO DIS 9241-11 defines usability as the "effectiveness, efficiency and satisfaction with which specified users can achieve specific goals in particular environments." A system with high usability can perform functions effectively, by which users can complete tasks efficiently and have a high degree of satisfaction with the interaction process. In the usability study, the system usability scale (SUS) is a classic scale and consists of 10 topics, including a positive statement of odd items and a negative statement of even items, requiring participants to score 5 points for each topic after using the system or product. Several empirical studies have shown that SUS works well. There are also large sample studies indicating that the reliability coefficient of SUS is 0.91. Our usability test uses the SUS framework.

Endsley argues that interface design should ideally provide an overview of the situation and support projection of future events, as well as provide cues of current mode awareness. Thus, in our case, it is important to know the exact influences of the HUD design on SA when using ACC.

28.2 Research propositions

In this study, we evaluated the HUD design under the ACC function mainly based on the following two hypotheses:

1. HUD designs can increase drivers' SA and system usability when using ACC.
2. Compared with static HUD design, HUD design with dynamic guidance can increase SA and system usability better.

Based on the outcome of the research propositions above, we will make suggestions for HUD design under the ACC function.

28.3 Experimental method

28.3.1 Participants

Eight participants, including students, teachers, and other school staff, from Tongji University, China were recruited. Among them one was male and seven were female. Their ages range between 20 and 50 (M = 26.13, SD = 7.27). All participants had valid driving licenses, with five (62%) having licenses for 1–5 years and three (38%) for 6–10 years. One person (12%) had no knowledge about ACC, five (63%) knew something about ACC but never used it, and two (25%) had never used ACC.

In this study, we explained everything about ACC and the simulator operations to all participants before the experiment. The traffic conditions in the experimental tasks were relatively simple, so the participants could handle them with common real-life driving experiences. To minimize the impact of driving experience on the experimental results, the subjects underwent a complete training session to be fully familiar with the driving simulator before starting the experimental tasks.

28.3.2 Technical equipment

The experiments with the driving simulator were conducted in a laboratory at Tongji University, Shanghai, China. There are two driving simulators: one serves as the main test vehicle and the other as the auxiliary test vehicle. The main test vehicle consisted of a longitudinally adjustable seat, Logitech G27 force-feedback steering wheel, and pedals. During the experiment, three LED screens were designed and used to display the driving scene. The auxiliary test vehicle

28.3 Experimental method

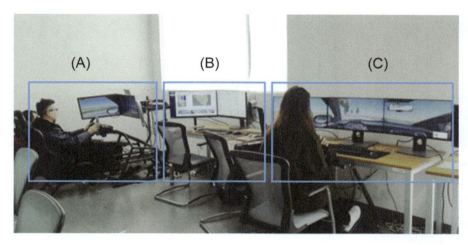

FIGURE 28.1

Experimental equipment. (A) The main test vehicle. (B) Supervisory control and data acquisition. (C) The auxiliary test vehicle.

consisted of a Logitech G27 force-feedback steering wheel, pedals, and three LED screens. The programs in both driving simulators are developed based on Unity software. They all have basic driving functions, and the two test vehicles run in the same traffic scene and can see each other. In addition to the basic driving functions, the main test vehicle has the function of ACC (including adjusting the set speed, adjusting the time gap, normal follow-up, etc.) with a dashboard display and HUD. When the auxiliary test vehicle is running, the experimenter can see the value of relative distance and time headway (THW), which are used for the experimental conditions.

In the experiment setting (Fig. 28.1), the participants carried out the test task by driving the main test vehicle, and the experimenter completed the cut in working condition with the main test vehicle by driving the auxiliary test vehicle. The supervisory control and data acquisition (SCADA) system was also used in the experiment to record and save the driving data, such as speed and steering wheel angle, to the CSV file in real time, so that the data can be statistically analyzed after the experiment.

28.3.3 Experimental design

A dual-task within-subject approach was applied. The experimental dependent variables are drivers' SA and system usability. The experimental independent variables are three different HMI design schemes, dashboard-HMI design, dashboard and HUD1 HMI design, and dashboard and HUD2 HMI design (referred to as dashboard-HMI, HUD-HMI1, and HUD-HMI2).

1. Dashboard-HMI (Fig. 28.2): The HMI of ACC is only displayed on the dashboard. Design elements of ACC have been adapted and displayed with reference to vehicles equipped with ACC on the market. These design elements include speed, ACC logo, setting speed, identifier of leading vehicle, and time gap.

424 Chapter 28 The situation awareness and usability research

FIGURE 28.2

Dashboard-human-machine interface (HMI) (baseline HMI that design elements are displayed only on the dashboard).

FIGURE 28.3

Head-up display-human-machine interfaces 1 (the augmented reality time-gap bar is relatively stationary with the main test vehicle).

FIGURE 28.4

Head-up display-HMI2 (the augmented reality time-gap bar dynamically points to the leading vehicle).

2. HUD-HMI1 (Fig. 28.3): Compared with dashboard-HMI, the display mode of HUD (including W-HUD and AR-HUD) is added. And we use the same design elements for the dashboard and the HUD.
3. HUD-HMI2 (Fig. 28.4): Compared with HUD-HMI2, the design form of AR time-gap bar is different. The time-gap bars in the two design schemes display the time-gap information of the

ACC. The difference is that the AR time-gap bar in the HUD-HMI1 is relatively stationary with the main test vehicle, and the AR time-bar in the HUD-HMI2 dynamically points to the leading vehicle.

28.3.4 Experimental tasks

The cut-in event is a typical scenario of ACC and it is fairly common in day-to-day driving. In this experiment, each participant performs three tasks corresponding to three different HMI design schemes, all of which are in the cut-in scenario involving the ACC function enabled. When the auxiliary test vehicle is cutting in, the speed of both vehicles is 30 km/hour and the distance is about 20 meters (THW $\approx$ 2.4 m/second). To avoid a learning effect, these three HMI designs are presented to each participant at random.

28.3.5 Experimental process

The experimental procedure is as follows:

1. The demographic information was collected from the participants, including name, age, time of possession of driver's license, frequency of driving, and understanding of ACC.
2. The researcher introduced the purpose of the experiment and the general experimental procedure to the participants.
3. After the introduction, the participants were allowed to sit in the driver seat of the simulator for 10–15 minutes to familiarize themselves with the basic driving controls and the operation of the ACC (turning on/off, adjusting the set speed, adjusting the time gap). The researcher then introduced the HMI display of the ACC on the instrument panel. To avoid possible learning effects, the traffic scene during the practice session was different from the traffic scene of the formal test session.
4. Each participant was asked to keep the simulated vehicle straight in the lane under ACC while braking if any unexpected events should emerge. Each participant completed the experimental tasks after they familiarize themselves with the ACC function.
5. Between each task, the participant was asked to rate the SA and system usability via commonly recommended SAGAT and SUS.
6. The process of performing the task on the screen was computer captured and the whole user-simulator interaction process was videotaped to facilitate the search and verification after the experiment.

28.4 Results and preliminary findings

To investigate the effects of different HMI designs on SA and system usability, we collected user feedback scores of SAGAT and SUS for three different HMI conditions. We also recorded the lateral control data—the steering angle of the steering wheel through SCADA. Only descriptive statistics were reported, given that there was not enough data to justify confirmatory analyses. The statistical analysis results are discussed in the following section.

28.4.1 Situation awareness

SAGAT. Average SAGAT scores were broken down by design schemes (Fig. 28.5). As Fig. 28.5 shows, the SA scores of HUD (HUD-HMI1 and HUD-HMI2) are both higher than dashboard-HMI, indicating that the subject is aware of fewer of the environment and car under HUD conditions, which is consistent with Hypothesis 1. Compared to HUD-HMI1, the SA score of HUD-HMI2 is higher, indicating that dynamic HUD design can increase SA, which is also consistent with Hypothesis 2. Moreover, the SAGAT is designed to evaluate SA from three dimensions: level 1 (perception), level 2 (comprehension), and level 3 (projection). The driver's scores in different dimensions are shown in Fig. 28.6. In contrast to the best perception (level 1 SA), participants have the worst projection (level 3 SA) under the dashboard-HMI condition. Participants have the same understanding (level 2 SA) of the three different design schemes. Generally speaking, HUD-HMI2 reports higher overall SA.

28.4.1.1 Driving performance

During the execution of the experimental task, ACC longitudinally controls the vehicle, liberating the driver's feet, and the driver only needs to operate the steering wheel to laterally control the vehicle. Therefore we analyzed driving performance by the driver's lateral control of the vehicle. In this experiment, we calculated the overall standard deviation (SD) of the steering wheel angle from the time the auxiliary vehicle starts to cut into that the main test vehicle follows the leading vehicle steadily, which is used to characterize the driver's lateral control. The greater the value of the SD, the more unstable the driver's lateral control of the vehicle. As shown in Fig. 28.7, the standard deviation of the steering wheel angle of dashboard-HMI is the highest, and that of HUD-HMI2 and HUD-HMI1 are the lowest. We found that HUD conditions report more stable lateral control than dashboard conditions and HUD-HMI1 condition reports the best driving performance.

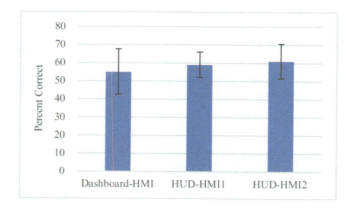

FIGURE 28.5

Average overall situation awareness global assessment technique percent correct score. Error bars represent 95% confidence intervals.

28.4 Results and preliminary findings

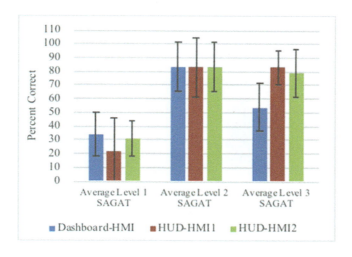

FIGURE 28.6

Situation awareness global assessment technique score of three levels of situation awareness. Error bars represent 95% confidence intervals.

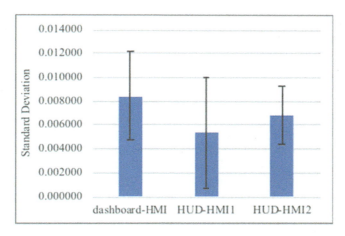

FIGURE 28.7

Standard deviation of steering wheel angle. Error bars represent 95% confidence intervals.

28.4.2 Usability

As shown in Fig. 28.8, the HUD-HMI2 score is higher than both dashboard-HMI and HUD-HMI1, and the scores of dashboard-HMI and HUD-HMI1 are the same. This result is consistent with Hypothesis 2. As far as the level of usability is concerned, both dashboard-HMI and HUD-HMI1 are level D. However, HUD-HMI2 is level C, indicating that different HMI designs of HUD may result in different results, improving or reducing usability. Taken together, the usability of the ACC system is better under HUD-HMI2 in this study.

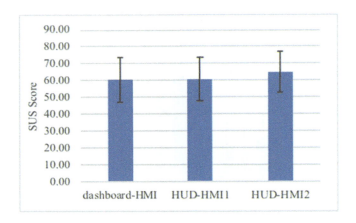

FIGURE 28.8

System usability scale score of three human-machine interfaces. Error bars represent 95% confidence intervals.

28.5 Discussion

The result indicates that the participants demonstrated lower overall SA with the dashboard-HMI than HUD-HMI1 and HUD-HMI2. Previous studies found that since the HUD data were displayed all the time, the HUD may have made the driving task more visually complex and reduced overall SA. Our experimental results show another conclusion. Similar to what Rutley found, the HUD speedometer could increase the awareness of speed. From the analysis of the three different levels of SA, participants had a lower perception of relevant elements in the driving environment with HUD, but at the same time, they had higher projection. It is possible that HUD may capture too much attention from the drivers, a phenomenon known as cognitive capture. This may lead drivers to focus more on the elements of the HUD, than on the elements in the real world. However, since HUD highlights more driving information related to the ACC status, drivers begin to process the information contained in the ACC status earlier and more often, allowing them to be better prepared for upcoming traffic scenarios. This may lead to a higher level of SA. In our study, we also found that the SA of HUD-HMI2 was higher than HUD-HMI1. This suggests that dynamically changing elements that followed the real traffic scenario could further increase the drivers' SA.

Accidents are caused by drivers who fail to perceive important elements in the driving environment, do not understand how these elements interact with each other, and/or are unable to predict what will happen in the near future. In other words, a lack of SA leads to unsafe driving. In the same way, low SA leads to unstable lateral control when using ACC. In our study, participants have more stable lateral control with HUD than only the dashboard, which is consistent with the self-reported SA result. However, the driving performance and SAGAT score are inconsistent under HUD-HMI1 and HUD-HMI2 conditions. One reason for this could be that the small sample size of the data may cause inaccurate statistical results. Moreover, slight numerical changes may not represent the difference in actual driving performance.

In the usability study, dashboard-HMI and HUD-HMI1 were found to have the same usability score. However, the HUD-HMI2 had the highest usability score. This indicates that, based on the display mode of the instruments, the addition of HUD would not necessarily improve system usability, but the design elements presented on the HUD would determine system usability. Finally, we found that the HUD design with dynamic directivity could increase usability.

28.6 Limitations

The most obvious limitation of this study was the size of the participant pool. A larger sample size could potentially yield more definitive results. In addition, this experiment relied on the cooperation of the experimenter. Although the experimenter had rich experience, manual control could have introduced invariability into the experimental task, potentially impacting their stability.

28.7 Conclusion

For the ACC functions that are increasingly used in automobiles, we have designed three HMI display solutions. To come up with better design improvements, we conducted an experiment in a driving simulator to assess the SA level and usability of different HMI designs in cut-in scenarios of ACC. Our results show that HUD affects the SA of the participants and system usability. Our study also shows that different HUD designs have different effects on the SA of drivers and system usability. Based on the research results, HUD design suggestions are proposed.

In this experiment, we used two HUD designs that were compared with a baseline dashboard design. Compared with the dashboard-HMI, the HUD design that is static relative to the main vehicle (HUD-HMI1) increases drivers' SA and reports the same system usability, while the HUD design with dynamic guidance effect (HUD-HMI2) increases both SA and system usability.

In summary, the design of the HUD has an important impact on driving, and an important contribution of this study is to propose that the HUD design with dynamic directivity of augmented reality should be recommended as a way of increasing drivers' SA and usability.

References

[1] L.C. Davis, Effect of adaptive cruise control on traffic flow, Phys. Rev. E 69 (6) (2004) 066110.
[2] J. Wang, W. Luo, B. Cao, et al., Design of the human-machine interface for automobile HUD, Process. Autom. Instrum. 36 (7) (2015) 85–87.

Further reading

A. Bangor, P. Kortum, J. Miller, Determining what individual SUS scores mean: adding an adjective rating scale, J. Usab. Stud. 4 (3) (2009) 114–123.

J. Brooke, SUS-A quick and dirty usability scale, in: P.W. Jordan, B. Thomas, B.A. Weerdmeester, A.L. McClelland (Eds.), Usability Evaluation in Industry, Taylor and Francis, London, 1996, pp. 189–194.

V. Charissis, S. Papanastasiou, Human–machine collaboration through vehicle head up display interface, Cogn. Technol. Work. 12 (2010) 41–50. The Situation Awareness and UsabilityResearch 247.

M.R. Endsley, D.G. Jones, Design for Situation Awareness-An Approach to User-Centered Design, 2nd edn., CRC Press of Taylor and Francis Group, London, 2004, pp. 259–266.

M.R. Endsley, D.G. Jones, Design for Situation Awareness-An Approach to User-Centered Design, 2nd edn., CRC Press of Taylor and Francis Group, London, 2004, p. 58.

M.R. Endsley, Design and evaluation for situation awareness enhancement, Proc. Hunan Factors Soc. 32nd Annu. Meet. 32 (2) (1988) 97–101.

M.R. Endsley, Measurement of situation awareness in dynamic systems, Hum. Factors J. Hum. Factors Ergon. Soc. 37 (1) (1995) 65–84.

M.R. Endsley, Toward a theory of situation awareness in dynamic systems, Hum. Factors J. Hum. Factors Ergon. Soc. 37 (1) (1995) 32–64.

Feng, Z., Ma, X., et al.: Analysis of driver initial brake time under risk cut-in scenarios, in: The 14th International Forum of Automotive Traffic Safety, 2018.

J.L. Gabbard, G.M. Fitch, H. Kim, Behind the glass: driver challenges and opportunities for ar automotive applications, Proc. IEEE 102 (2) (2014) 124–136.

K.W. Gish, L. Staplin, Human factors aspects of using head up displays in automobiles: a review of the literature, DOT. HS 808 (1995) 320.

R. Gu, X. Zhu, Summarization of typical scenarios of adaptive cruise control based on natural drive condition, in: Proceedings of the 11th International Forum of Automotive Traffic Safety, 2014, pp. 387–393.

J.S. Hicks, D.B. Durbin, A.W. Morris, Davis, B.M. Davis, A summary of crew workload and situational awareness ratings for U.S. Army Aviation Aircraft. ARL-TR-6955 (2014).

W.J. Horrey, C.D. Wickens, A.L. Alexander, The effects of head-up display clutter and invehicle display separation on concurrent driving performance. In: Proceedings of the Human Factors and Ergonomics Society Annual Meeting, p. 1880 (2003).

ISO, Ergonomics of human-system interaction - Part 11: usability: definitions and concepts (ISO9241-11), 2008. <https://www.iso.org/obp/ui/#iso:std:iso:9241:-11:ed-2:v1:en>.

P.W. Jordan, An Introduction to Usability, CRC Press, London, 1998, pp. 5–7.

R.J. Kiefer, Effects of a head-up versus head-down digital speedometer on visual sampling behavior and speed control performance during daytime automobile driving (SAE Tech. Paper 910111), Society of Automotive Engineers, Warrendale, 1991.

Y.C. Liu, Effect of using head-up display in automobile context on attention demand and driving performance, Displays 24 (2003) 157–165.

R. Ma, D.B. Kaber, Situation awareness and workload in driving while using adaptive cruise control and a cell phone, Int. J. Ind. Ergon. 35 (10) (2005) 939–953.

S.S. MarkVollrath, C. Gelau, The influence of cruise control and adaptive cruise control on driving behaviour – a driving simulator study, Accid. Anal. Prev. 43 (3) (2011) 1134–1139.

H. Ohno, Analysis and modeling of human driving behaviors using adaptive cruise control, Appl. Soft Comput. 1 (3) (2001) 237–243.

K. Rutley, Control of drivers' speed by means other than enforcement, Ergonomics 18 (89) (1975) 100.

N.A. Stanton, M.S. Young, Driver behaviour with ACC, Ergonomics 48 (10) (2005) 1294–1313.

Y. Sun, Symbol design of interface information based on user cognitive psychology, Art. Technol. 4 (2014) 303.

L.K. Thompson, M. Tönnis, C. Lange, H. Bubb, G. Klinker, Effect of active cruise control design on glance behaviour and driving performance, in: Paper presented at the 16th World Congress on Ergonomics, Maastricht the Netherlands, 10–14 July 2006.

D.R. Tufano, Automotive HUDs: the overlooked safety issues, Hum. Factors J. Hum. Factors Ergon. Soc. 39 (2) (1997) 303–311.

T.S. Tullis, J.N. Stetson, A comparison of questionnaires for assessing website usability, in: Proceedings of Usability Professionals Association (UPA) 2004 Conference, 2004.

G.-D. Tuzar, A. Van Laack, White paper. Augmented reality head-up display HMI impacts of different field-of-views on user experience, 2016. http://docplayer.net/33909881 Whitepaper-augmented-reality-head-up-displays-hmi-impacts-of-different-field-of-views-on-userexperience.html.

N.J. Ward, Automation of task processed: an example of intelligent transportation systems, Hum. Factors Ergon. Manuf. Serv. Ind. 10 (4) (2000) 395–408.

Index

Note: Page numbers followed by "*f*" and "*t*" refer to figures and tables, respectively.

A

AAM. *See* Automation acceptance model (AAM)
Abstract design, 247
Abstraction hierarchy (AH), 158–159
　analysis, 158–159, 159*t*
　　of shared situation awareness human-machine interaction interfaces, 159–160
ACC system. *See* Adaptive cruise control system (ACC system)
Acceptance, 345–346
Accidents, 428
　statistics, 411
Action anthropomorphism, 331
Adaptive cruise control system (ACC system), 115, 184, 188–190, 420
　engagement flowchart, 191*f*
　limitations, 120
　method, 116–119
　　experimental design, 116, 116*t*
　　experimental procedure, 117–119, 117*f*, 118*f*
　　participants, 116
　　research question, 116
　results, 119–120, 119*f*
Adaptive dynamic system, 138–139
ADASs. *See* Advanced driver assistance systems (ADASs)
Adjustment phase vision scanning analysis, 128
Advanced automated systems, 156
Advanced driver assistance systems (ADASs), 99–100, 115, 135, 152, 420–422
AFD. *See* Average fixation duration (AFD)
After-scenario questionnaire (ASQ), 90, 93, 163–165, 266–267
Age on response time, influence of, 40–41
AH. *See* Abstraction hierarchy (AH)
AI. *See* Artificial intelligence (AI)
ALMA. *See* Layered model of affect (ALMA)
American Society of Automotive Engineers (SAE), 156
Analysis of variance (ANOVA), 268, 292, 343
Analytical hierarchy process, 185
Anger, 227
ANOVA. *See* Analysis of variance (ANOVA)
Anthropomorphic design, 247
Anthropomorphism, 261, 276, 282
Anxiety, 177–178, 227
AOI. *See* Area of interest (AOI)
APD. *See* Average pupil dilation rate (APD)
Application scenario design, 188–189
AR. *See* Augmented reality (AR)

Area of interest (AOI), 126, 376
AR-HMI. *See* Augmented reality-human machine interface (AR-HMI)
AR-HUD. *See* Augmented reality head-up display (AR-HUD)
Artificial intelligence (AI), 175, 184
ASQ. *See* After-scenario questionnaire (ASQ)
Attentional processes, 411–412
Auditory approaches, 338
Augmented reality (AR), 176, 420
　human-machine interface
　　cognitive information visualization, 368–370, 369*t*
　　collaborative contextual analysis, 365–368
　　component adjustment and overlay, 370–371
　　design method for, 364–371, 365*f*
　information elements, 362–364, 363*f*
　　knowledge-based information, 364
　　rule-based information, 363–364
　　skill-based information, 363
　technology, 420
Augmented reality head-up display (AR-HUD), 46, 152, 190, 420
　case 1, 66–67
　case 2, 67–69
　　details and demonstration graphs of three augmented conditions, 68*t*
　case study 1, 55–59
　　experimental equipment and materials, 55–56
　　experimental variables, 56–58, 57*t*
　　objectives and hypotheses, 55
　　participants, 55
　　procedure, 58–59
　case study 2, 59–61
　　experimental equipment and materials, 59
　　experimental variables, 60–61
　　objectives and hypotheses, 59
　　participants, 59
　　procedure, 61
　context and theoretical framework, 47–52
　　automation surprises and trust, 51–52
　　cognitive information design, 50–51
　　human-vehicle teaming and cooperative interface, 47
　　situational awareness in cooperative driving, 48–49, 58*t*
　designing HMI, 52–54
　method and materials, 55–61
　results and analysis, 61–66
　results of case 1, 61–64
　　reaction time, 64
　　situational awareness, 62

433

434 Index

Augmented reality head-up display (AR-HUD) (*Continued*)
    usability, 62−63
    workload, 63
    results of case 2, 64−66
        reaction time, 65
        situational awareness, 66
        trust, 66
        usability, 66
Augmented reality-human machine interface (AR-HMI), 358
Automated driving, 156
    function, 163
    handovers in, 84
    systems, 95, 124
        human-computer collaborative interaction in, 185
        taxonomy of, 184−185
Automated system, 84
Automated vehicles (AVs), 358, 372
Automatic driving system, 66
Automatic systems, 51−52
Automation, 82
    surprises, 52
    systems, 135
    technology, 125
Automation acceptance model (AAM), 326, 326*f*
Automobile intelligence, 135
Automotive augmented reality head-up display design (automotive AR-HUD), 152−154
    project description, 152, 152*f*
    project implementation process, 152−154
        business model and concept design, 153
        design evaluation and user testing, 154
        information architecture and design implementation, 153
        market research and design research, 153
        system development and operation tracking, 154
        user research, 153
Automotive design process, 12−13, 22
Automotive head-up display interaction design, 178−181
    head-up display information filtration and organization, 179
    head-up display interface design, 179−181
Automotive interaction design, 401−402
Automotive Interaction Lab, 404−405
Automotive security, 123
Autonomous driving
    function, 100−101
    system, 67
    technology, 99−100
Autonomous vehicles (AVs), 75, 106, 112−114, 337−338
    method, 76
    results and analysis, 76−78
    socially incapable autonomous vehicles, 75−76
Auxiliary test vehicle, 422−423, 425
Average fixation duration (AFD), 376
Average pupil dilation rate (APD), 376
AVs. *See* Automated vehicles (AVs); Autonomous vehicles (AVs)

B

Behavioral indicators, 392, 394
BibExcel, 205
Bibliometrics, 204−205
    results and analysis, 205−218
        distribution map of annual publication volume, 206*f*
        distribution of social robot interaction design literature sources, 215−216, 216*t*
        evolution and trend of social robot interaction design research hotspots, 210−211
        high-impact author analysis, 217−218
        high-impact countries and research institutions, 216−217
        knowledge base of SR-human-robot interaction research, 211−215
        research framework, 206*f*
        research hotspots of social robot interaction design, 207−210
        trend analysis of annual outputs of social robot interaction design literature, 205−207
Bidirectional information flow, 364−365
Big data, 153, 184
Bipolar pleasure-arousal scale, 227
Blind-vision time, 178−179
Blindfolded testers, 180
Blinking dynamic effect, 317
Blocks, 112
Bonferroni corrections, 343
Box plots, 378
Brand-new design project, 152
Business model, 150

C

CA driving. *See* Conditional automated driving (CA driving)
Car intelligence degree, 123
Car Interaction Design Lab of Tongji University, 289−291
Car manufacturers, 176
Car technology, 175
Car-machine interface, 7
CASA paradigm. *See* Computer is a social actor paradigm (CASA paradigm)
Case analysis, 306
CAVs. *See* Connected and automated vehicles (CAVs)
Central consoles, 12
Cerebrovascular system, 176
Chinese characters, 21
    guidelines, 29−30
    limitations, 31
    method, 22−25
        experimental environment, 22, 23*f*

Index

participants, 22
procedure, 24
questionnaires and data analysis, 24–25
task and stimuli, 23–24
related works, 20–22
classification of visual signals and human-computer interaction model, 20–21
driving cognitive model, 21–22
standards and guidelines about size and length of text, 21
results, 25–29
reaction time, 25, 26f
usability scores and workload, 27–29
visual subjective rating, 26–27, 28f
CINEMA 4D software, 341
CiteSpace, 205
City Rig 1.7 plug-in, 317–318
CL-LM. *See* Current lane to left mirror (CL-LM)
Classic research model, 13
Classification model, 227
Clear organization system, 153
Cloud computing, 184
Cognitive ability, 136
Cognitive analysis methods, 362
Cognitive capture, 428
Cognitive dimension cooperation, 138
Cognitive information, 140–142
decision, 141–142
design, 50–51
comparison of information processing between humans and intelligent systems, 50t
proposed information design for cognitive cooperation, 51f
intention, 140
situation awareness, 140–141
visualization, 368–370, 369t
Cognitive interface, 362–364, 363f
knowledge-based information, 364
rule-based information, 363–364
skill-based information, 363
Cognitive law, 38–40
Cognitive process, 21–22, 24, 35
Cognitive resources, 35, 38–40
Cognitive systems, 137
Cognitive theory, 282
Collaboration, 330
contextual analysis, 365–368
driving, 188
Color contrasts, 4
Color scheme, 317
Communication strategies, 339
Communication transparency, 261
Companion robots for elderly rehabilitation, 209–210
Compatibility, 326
Compliance, 343–344
Complicated traffic conditions, 128
Comprehensive design method model, 150
Computational cognitive models, 21
Computer is a social actor paradigm (CASA paradigm), 208
Computer technology, 175
Conditional automated driving (CA driving), 128, 184–185
Conductive gel, 405
Connected and automated vehicles (CAVs), 152
Consciousness perception, 283
Content technology, 152
Context-based AR, 382
Cooperative driving, 135–136
Cooperative interface, 139–140
for collaborative driving
dynamic external factors, 144
environment, 144
framework of cooperative interface, 137–144, 138f
human autonomy team, 138–143
level of interaction, 143–144
structure comparison between manual driving and human-machine co-driving, 136f
traffic conditions, 144
vehicle, 144
human-machine cognitive information exchange, 139f
Coordinate distance of emotional space, 232
Correlation analysis, 111
interview data, 112, 113t
trust and situational awareness, 111, 112t
trust and workload, 111, 112t
Correlation calculation, 111
Course experiment methods, 403–405
design scheme, 403–404
central control interface design, 404f
instrument interface design, 404f
electrodermal activity measurement equipment, 405
experimental design, 403, 403f
experimental environment and equipment, 404–405
driving simulation bench, 405f
experimenter, 403
Crisis management, 157
Cross-dissolve effect, 317
Crosswalk scenario, 340–341
Cultural context, 79
Current lane to left mirror (CL-LM), 413–414
Cut-in conditions, 116
Cut-in event, 425
Cut-in situation, 120

D

DA. *See* Driving assistance (DA)
DALI. *See* Driving activity load index (DALI)
Dark environment, 5
Dashboard-HMI, 423, 426

Data sources, 204−205
Data technology, 152
Decision-making, 50, 101, 136−137, 142
   phase vision scanning analysis, 127
   process, 52, 141−142
   visual scanning mode, 127f
Deep learning, 184
Degree of automation, 82
Demographic information, 24
Depression, 227
Design elements, 111
Design evaluation
   methods, 402
   and user testing, 154
Design implementation, 153
Design methods, 67, 153, 158−160
   abstraction hierarchy analysis, 158−159, 159t
     of shared situation awareness human-machine interaction interfaces, 159−160, 159f
   four factors of shared situation awareness interface, 160
Design practice of environment information surrounding vehicles in parking scenarios, 313−318
   color scheme, 317
   design output of scenario, 317−318
   display scheme design, 315−318
   physical prototype, 318
   scenario analysis, 314
     conceptual storyboards of future parking scenario, 315f
     design demands of parking scenario, 315t
     user journey map of parking process, 314f
   user activity analysis, 313−314
Design research, 153
Design space of environment information surrounding vehicles, 307−310
   case analysis, 308−310
   future trends of environment information surrounding vehicles, 307−308
     attractive display, 307−308
     expanded display of environment information, 307
     personalized information content, 307
Design strategy, 329
Designers, 29−30
Designing HMI, 52−54
   for Case 1, 52−54, 58f
   for Case 2, 54, 58f
Dimensional emotion model, 227
Discrete emotion model, 227
Disgust, 227
Display scheme design, 315−318
   graphs and dynamic effects, 316−317
   text design for ground display, 316
Display system, 176
Distraction events, 130
Distress, 227

Distribution of social robot interaction design literature sources, 215−216, 216t
Driver assistance, 187
Driver scenario, 125
Driver-vehicle interface, 21
Driver's decision-making process, 79
Driver's mental comfort, 76−77
Driver's trust, 110
Driver's visual scanning mode, 126
Driving activity load index (DALI), 56−57, 63, 266−267, 289, 375−376
Driving assistance (DA), 184−185
Driving cognitive model, 21−22
Driving process, 66, 118, 137
Driving simulator, 108−109, 116, 154, 171−172, 374, 422−423
Driving-assistance technology, 175
Dual-task within-subject approach, 423−425
Dynamic Street, 310, 310f
Dynamic testing, 42

E

Early warning system, 124
EDA. See Electrodermal activity (EDA)
Educational social robotics research, 210
Ego vehicle, 86
eHMI. See External human-machine interfaces (eHMI)
Electrodermal activity (EDA), 399−401
   advantages of experiments combined with electrodermal activity measurement, 401
   analysis and discussion of experimental results, 406−408
     electrodermal activity and system availability scale, 406
     electrodermal activity and task behavior time, 406−408
   applications of electrodermal activity measurement in evaluation, 400−401
   course experiment methods, 403−405
   measurement, 401
     equipment, 405
   method exploration, 402
     independent data analysis and optimization, 402
     independent design and evaluation experiment, 402
     independent proposition and design, 402
   teaching objectives, 401−402
Electronic mobile, 130
Electronic reading state, 130
Emojis, 229
Emoticons, 229, 247−249
Emotional design of social robots, 209
Emotional experience, 243
Emotional recognition in social robots, 209
Emotional space, 244−245
Emotionalization, 330
Emotions, 243−245

expressions, 225
    design in virtual images, 226, 228−229
    model, 225−226
    perception, 283
    set, 246−247
Endsley's situational awareness model, 101
Environment, 144, 177
    information, 306
Ergonomic design, 115
Euclidean distance algorithm, 250−251
Execution phase vision scanning analysis, 126−127
Experimental design method, 116
Explicit communications, 337−338
Expression design, 246
External human-machine interfaces (eHMI), 337−338
Eye movement, 166−167
    subject's eye movement heat zone, 166*f*
    video, 413−414
Eye tracking
    devices, 163−164
    method, 416
    physiological data, 377
    studies, 411−412

F

FA driving. *See* Fully automated driving (FA driving)
Facial emotion evaluation model, 227−228
Facial expressions, 225, 249
    design evaluation of virtual image, 229−233, 230*t*
Fear, 227
FFD. *See* First fixation duration (FFD); Fixation to first displacement (FFD)
Field research, 306, 312
First fixation duration (FFD), 376
First-edition expression design, 234, 234*f*
5-point Likert scale, 311, 406
5G technology, 3
Fixation areas, 413
Fixation moving paths, 413
Fixation to first displacement (FFD), 379
Flemisch's hierarchical framework, 137
Friedman analysis of variance, 379
Friedman ANOVAs, 343
Frustration, 91
Full automation, 187
Full human, 186−187
Fully automated driving (FA driving), 184−185

G

Galvanic skin response (GSR), 399−400
Gender distribution, 126
Gesture interaction, 396

Ghost in the Shell (2017), 307−308, 308*f*
Global perception, 370
Global workload rating, 14
Graphic method, 332
Gray car model, 104−105
Greenhouse-Geisser corrections, 343
GSR. *See* Galvanic skin response (GSR)

H

HA driving. *See* Highly automated driving (HA driving)
Handovers
    in automated driving, 84
    process, 84
Happiness, 227
HCI. *See* Human-computer interaction (HCI)
HEAD framework. *See* Human-engaged automated driving framework (HEAD framework)
Head-up display (HUD), 36, 179, 411, 420. *See also* Augmented reality head-up display (AR-HUD)
    designs, 422
    HUD-based warning indicators, 411
    HUD-HMI design
        adaptive cruise control, 420
        dashboard-human-machine interface, 424*f*
        experimental design, 423−425
        experimental method, 422−425
        experimental process, 425
        experimental tasks, 425
        head-up display, 420
        head-up display-HMI2, 424*f*
        head-up display-human-machine interfaces 1, 424*f*
        limitations, 429
        participants, 422
        research propositions, 422
        results and preliminary findings, 425−427
        situation awareness, 421, 426
        technical equipment, 422−423
        usability, 427
        usability test, 421−422
    HUD-HMI1, 424
    HUD-HMI2, 424−425
    information filtration and organization, 179
    interface design, 179−181
        visual color, 180−181
        visual motion, 181
        visual symbols, 179−180
    optical machine, 36
    speedometer, 428
    technology, 175
    visual design of
        experimental environment, 36−37, 37*f*
        influence of age on response time, 40−41
        influence of weather conditions on reaction time, 38−40

Head-up display (HUD) (*Continued*)
   method, 36−38
   results, 38−42
   subjects, 38
   test contents and steps, 38
   trends of workload, 41−42, 42*f*
High-impact author analysis, 217−218
High-impact countries and research institutions, 216−217
Highly automated driving (HA driving), 184−185
Highway hazard scenario analysis, 125−128
   adjustment phase vision scanning analysis, 128, 129*f*
   decision-making phase vision scanning analysis, 125
   execution phase vision scanning analysis, 126−127, 128*f*
   lane changing visual scanning analysis, 126−127, 126*t*
   scenario analysis, 125
HistCite, 205
HMCD. *See* Human-machine collaborative driving (HMCD)
HMI. *See* Human-machine interaction/interface (HMI)
Holographic displays, 308
Holographic information, 308
Horizontal deviation, 394−395
HRIs. *See* Human-robot interactions (HRIs)
HUD. *See* Head-up display (HUD)
Human and machine cognitive dimensions, 136
Human autonomy team, 138−143
   cognitive information, 140−142
   cooperative interface, 139−140
   information transparency, 142
   interactive mode, 143
   team, 138−139
Human information processing, 185
Human supervision, 187−188
Human visual perception, 69
Human visual system, 22
Human-artificial systems design, 152, 154
   automotive augmented reality head-up display design, 152−154
   interaction design method framework, 150−152
   projects, 150
   related works, 150
Human−AV collaboration, 78−79
Human-computer collaborative interaction in automated driving, 185
Human-computer interaction (HCI), 21, 84, 184, 226
Human-engaged automated driving framework (HEAD framework), 186−188
   collaboration driving, 188
   driver assistance, 187
   full automation, 187
   full human, 186−187
   human supervision, 187−188
Human-machine co-driving, 156
   cooperation, 137
   interface designs of, 158

Human-machine cognitive information communication, 139
Human-machine collaboration, 102, 358−359
Human-machine collaborative driving (HMCD), 188
Human-machine collaborative interfaces, 156, 160−161
Human-machine cooperation, 66−67, 84, 156
   human-machine cooperative driving, 137
   human-machine cooperative interface, 139
   strategy model, 185
Human-machine integration, 135−136
Human-machine interaction/interface (HMI), 3, 11−12, 19−20, 47, 52−54, 82, 100, 115, 124, 137, 157, 184, 358, 362, 380, 391
   case, 188−194
      application scenario design, 188−189
      information architecture design, 190−193
      interface design, 193−194
   dangerous scenarios with missing information, 53*f*
   design case study, 102−104
   designs, 101−102, 104−107, 175, 399−400
      interface element design, 104−107
      scenario analysis, 104
   HEAD framework, 186−188
   interface experiment, 194−199
   method, 392−394
      data collection, 394
      experimental design, 392−393, 393*f*
      experimental procedure, 394
      participants, 392, 392*t*
   related work, 184−185
      human-computer collaborative interaction in automated driving, 185
      taxonomy of automated driving in China, 185*t*
      taxonomy of automated driving systems, 184−185
   requirements of system transparency, 83−84
   research for autonomous vehicles, 103*t*
   result, 394−395
      vehicle horizontal indicators, 394−395, 395*t*
      vehicle vertical indicators, 394, 395*t*
Human-machine situation awareness, 141
Human-machine team, 139
Human-robot emotional communication
   emotion expression design of virtual image, 228−229
   facial expression design evaluation of virtual image, 229−233, 230*t*
   literature review, 226−228
      emotion expression design in virtual images, 226
      facial emotion evaluation model, 227−228
   method, 233−239
      design of experiment, 233−234
      participants, 233
      procedure, 235
      results, 236−239
Human-robot interactions (HRIs), 203−204, 226, 260, 282
   anthropomorphism, 276

future directions, 277
implications for, 276–277
limitations of current study, 277
methods, 265–267
   apparatus and materials, 266–267
   chatting, 273–275
   design, 265–266
   incoming call, 268–272
   participants, 265
   procedure, 267
   results, 268–275
   welcome, 268
outcomes, 262–264
   affect, 263–264
   current study, 264
   information transparency and anthropomorphic designs, 264f
   safety, 262
   trust, 263
   usability, 262
   workload, 262–263
robot-to-human communication, 260–261
transparency, 275–276
variable relationships, 276
Human-robot interface levels, 285
Human-robot relationship, 208
Human-system cooperation, 158
Human-vehicle collaboration, 358
Human-vehicle collaborative driving safety
attention and cognitive demands, 382–383
cognitive interface and augmented reality information elements, 362–364, 363f
context and theoretical framework, 359–364
design method for augmented reality human-machine interface, 364–371, 365f
empirical evaluation, 371–376
   hypotheses, 372
experimental design, 372–374
   apparatus and materials, 374–375
   measures and analysis, 375–376, 376t
   participants, 374
   procedure, 375
   scenarios and tasks, 372–373
fixation metric, 378–379
   fixation heatmap, 380, 380f
   fixation to first displacement, 379
   interviews, 380–381
   time to first fixation, 379
human-vehicle team and cooperation mechanism, 359–362
   collaborative ladder, 360–361
   content clues, 361
   perspective components, 361–362
limitations, 384
results, 377–381

eye-tracking data used for analysis, 377t
subjective scales data used for analysis, 377t
subjective scales, 377–378
   subjective workload, 377–378
   trust, 378
   usability, 378
training and mental models, 383
visualization components and quantities, 381–382
Human-vehicle cooperation, 47
Human-vehicle team collaboration, 359–360
Human-vehicle teaming
and cooperative interface, 47
developed cognitive information support for humans and machines through HMI, 48f
framework, 47
Human-vehicle-environment interaction, 35

I

IAs. *See* Intelligent agents (IAs)
Icon design recommendations for central consoles of intelligent vehicles
method, 12–14
   experimental environment, 12–13, 13f
   participants, 12
   procedure, 13–14
   questionnaires and data analysis, 14
   task, 13, 14f
results, 15–16
   impact of icon thickness on usability and workload, 12–13, 16
   influence of icon size on usability and workload, 15, 15f
IDMF. *See* Interaction design method framework (IDMF)
Immersive driving simulator, 124
Implicit communications, 337–338
In-car interactive display screens, 404–405
In-depth interview on users' demand for information in parking scenarios, 311–312
   content of interview, 311
   result analysis, 312
In-vehicle HMI design, 396
In-vehicle infotainment, 11–12
In-vehicle interaction system, 125
In-vehicle interface design, 11–12
In-vehicle robots, 281
experiment, 287–292
   apparatus and materials, 289–291
   design of experiment, 287–288
   experiment environment, 290f
   expression design set used on robot's face screen, 291f
   measurement, 289
   participants, 287
   procedure, 291–292
   robot for experiment, 290f

In-vehicle robots (*Continued*)
    speeding task experimental group setup, 288*f*
    telephone task experimental group setup, 288*f*
  related works, 282−284
    proactive interaction, 283
    transparency, 283−284
  results, 292−297
    speeding task results, 295−297
    telephone task results, 292−294
  transparency design, 282
    of proactive interaction, 284−287
In-vehicle screen, 20
In-vehicle visual presentation, 20
Influence legibility, 20
Influential related theory, 141
Information architecture, 153
  cruise original setup, 192−193
  cut-in A, 193
  cut-in B, 193
  design, 190−193
  following, 193
Information design method, 47
Information filtration, 179
Information organization, 179, 414−415
Information recognizability, 329−330
Information technology, 152
Information theory, 21
Information transformation process, 365−366
Information transparency, 142
Intelligence, 184−185
Intelligent agents (IAs), 359
Intelligent car, 159−160
Intelligent driving, 156
  system, 82
  technologies, 156
Intelligent machines, 85, 142
Intelligent planning agent, 83
Intelligent products, 149
Intelligent robot, 142
Intelligent systems, 46, 95, 135−136, 156
Intelligent vehicles, 11−12, 21, 184
  central consoles
    data collection, 6
    driving scene settings, 5
    experiment execution, 6, 6*f*
    experimental design, 4−5
    experimental results, 6−9
    objective data, 8−9
    preliminary preparations, 4−5, 4*f*
    simulator settings, 5, 5*f*
    subjective data, 6−7
  intelligent vehicle-mounted system, 160
Intensity, 227
Intention, 140

Interaction design, 204−205, 282, 306. *See also* Social robot interaction design (SRID)
  design practice of environment information surrounding vehicles in parking scenarios, 313−318
  design space of environment information surrounding vehicles, 307−310
  methods, 149
  research methodology, 306−307
    case analysis, 306
    field research, 306
    physical prototype, 306
    scenario analysis, 306
    usability test, 307
    user research, 306
  results and recommendations, 323
  usability test of design, 319−323
  user research of environment information demand in parking scenarios, 310−312
Interaction design method framework (IDMF), 149−152
  six-phase model, 150−151, 151*f*
  three perspectives, 151−152
Interaction methods, 396, 403−404
Interaction-prone encounters, 76
Interactive mode, 143
Interactive prototype design based on vehicle diplomacy mutual strategy, 332−333
Intercultural study
  interaction messages of autonomous delivery vehicle, 341*t*
  limitations, 348−349
  method, 339−342
    materials, 341
    procedure, 340−341
    questionnaires, 342
    results, 343−346
    sample, 339
  practical implications, 348
  related works, 338−339
Interface design, 61, 193−194
  of human-machine co-driving, 158
Interface element design, 104−107
  human-machine interface, 107*f*, 108*f*
  information requirements and interface design elements based on transparency, 106*f*
  interactive scenes based on transparency during overtaking, 105*f*
Interface experiment, 194−199
  apparatus, 195
  experimental design, 194
  measures, 196
  mode identification, 199
  participants, 195
  results, 196−198
    detection information, 197
    mode identification, 198

set distance, 198
set speed, 197
set distance, 198
task and procedure, 195–196
Interface information, 137
Interface transparency
design, 94
information, 113–114
Internet of Things (IoT), 11–12
Inverse morphisms, 21
IoT. *See* Internet of Things (IoT)
Iterative expression design, 234, 235f

K

KI. *See* Knowledge-based information (KI)
Knowledge base of SR-human-robot interaction research, 211–215
Knowledge-based information (KI), 364, 370
Kolmogorov–Smirnov test, 196
Kruskal–Wallis test, 14

L

Laboratory test, 117–118
Lane changing, 126–127
automotive head-up display interaction design, 178–181
behaviour, 178
decision-making stages, 126–127
lessons learned, 416–417
methods, 412–413
data analysis, 413, 413f
data collection, 412
experimental setup, 412
processes, 129
related works, 411–412
research on lane-changing scenario, 176–178
observation, 178
preliminary investigations, 176–177
study on environment, behavior, and psychology, 177–178
results, 413–414
fixation moving paths and fixation areas on right-to-left lane change, 414f
fixation moving paths on right-to-left lane change, 414f
visits of fixation areas on right-to-left lane change, 415f
scenarios, 176, 181, 412
visual scanning analysis, 126–127
Language anthropomorphism, 331
Layered model of affect (ALMA), 245
Left mirror to current lane (LM-CL), 413–414
Left mirror to target lane (LM-TL), 413–414
Left-to-right lane changing, 415
Levene test, 378
Light projection, 308
Likert scale, 289
LM-CL. *See* Left mirror to current lane (LM-CL)
LM-TL. *See* Left mirror to target lane (LM-TL)
Locally weighted scatterplot smoothing method (LOWESS method), 64
LOWESS method. *See* Locally weighted scatterplot smoothing method (LOWESS method)
Lyons's research, 82
Lyons's task model of transparency, 83

M

Machine cognitive information, 100
Machine learning, 49, 184
Machine transparency module, 142
Manipulation check, 343
Market research, 153
Mass-produced vehicles, 103–104
Massachusetts Institute of Technology (MIT), 217
Mauchly test, 343
Mental workload measurement method, 35–36
Military teamwork, 157
MIT. *See* Massachusetts Institute of Technology (MIT)
Modern reading behaviour, 20
Modern vehicles, 100–101
Motor lane, 79
Multidimensional scoring process, 36
Multidirectional method, 6
Multimodal design of external vehicle interaction, 332, 333t
Multimodal interaction methods, 143, 283
Multimode cross-sensory systems, 143
Multiple dynamic hazard sources, 368
Music cards, 102–103

N

NASA task load index scale (NASA-TLX), 91, 93, 401
NASA-TLX. *See* NASA task load index scale (NASA-TLX)
National Highway Traffic Safety Administration (NHTSA), 262
NCSSs. *See* Noncritical spontaneous situations (NCSSs)
Negative emotions, 228–229
Negotiation, 75–76
Network(ing), 184–185
resources, 4
technology, 175
NHTSA. *See* National Highway Traffic Safety Administration (NHTSA)
Noncritical spontaneous situations (NCSSs), 137
Nonparametric pairwise tests, 61–62
Nonverbal communication, 75–76
Nonverbal cues, 209
Nonverbal emotion measurement tool, 245–246

Nonverbal text report questionnaire, 245
Nuclear power plant management, 157

O

Objective data, 8–9
OCC model. *See* Ortony, Clore, and Collins model (OCC model)
On-road vehicles, 124
Onboard intelligent robots, 260
Onboard robots, 260
One-way analysis of variance (ANOVA), 40, 165
Optical machine, 36
Ortony, Clore, and Collins model (OCC model), 233, 243–244, 246

P

PA driving. *See* Partially automated driving (PA driving)
Pairwise comparison analysis, 379
Paper, 82, 104, 149
   prototyping, 153
Parking scenarios
   design practice of environment information surrounding vehicles in, 313–318
   environment information in, 313*t*
   in-depth interview on users' demand for information in, 311–312
   questionnaire survey on users' demand for information in, 311
   summary of users' demands for information in, 313*f*
   user research of environment information demand in, 310–312
Partially automated driving (PA driving), 184–185
Pedestrians, 325, 337–338
   hazards, 368
   pedestrian-unmanned vehicles
      behavioral characteristics and demand analysis of, 328
      in-depth interview and user research on, 327–328
PEOU. *See* Perceived ease of use (PEOU)
Perceived ease of use (PEOU), 326
Perceived usefulness (PU), 326
Perception layer, 363
Perceptual cycle model, 48
Perceptual cycle theory, 48–49
Performance metrics, 267
Physical prototype, 306
Physiological measurement method, 400–401
Pic-a-mood scale, 245–246
Pilot study, 412
Pleasure–arousal–dominance (PAD)
   dimension space, 243–245
   emotion, 225–227
   scores, 271

Polite strategy, 345
Politeness, 339
Positive emotions, 228–229
Posthoc tests, 346
Pre-recorded video segments, 374–375
Predictive information, 59, 112
PrEmo. *See* Product emotion measure (PrEmo)
Proactive interaction, 283
Proactivity, 282
Product development process, 150
Product emotion measure (PrEmo), 245–246
   PrEmo2, 245–246
Prototypes, 89–90, 150
   experimental prototype of designs, 89*f*
   interface, 163
Psychological anthropomorphism, 331
Psychological confrontation, 177–178
PU. *See* Perceived usefulness (PU)
Purpose, process, and performance (3Ps), 283–284

Q

Questionnaires, 90
   survey on users' demand for information in parking scenarios, 311
   questionnaire setting, 311
   result analysis, 311

R

Railway operations, 157
Reaction time (RT), 38, 64
   indicator comparisons related to reaction time, 64*f*
   influence of weather conditions on, 38–40
      average reaction time for elements on sunny and snowy days, 39*f*
      one-way ANOVA analysis of weather conditions, 40*t*
Real-world HMI tests, 115
Red, Green and Blue value (RGB value), 7
Research hotspots of social robot interaction design, 207–210
Research methods, 205
Research questions (RQ), 46–47
Respiratory system, 176
Response time
   influence of age on, 40–41
   one-way ANOVA analysis of age groups on
      snowy day, 41*t*
      sunny day, 41*t*
   recognition response time of elements for age groups on
      snowy day, 41*f*
      sunny day, 40*f*
RGB value. *See* Red, Green and Blue value (RGB value)
Right of way, 330
Road elements, 59

Road traffic safety, 177
Road Traffic Signs and Markings, 315–316
Robot anthropomorphism, 261
Robot-to-human communication, 260–261
    anthropomorphism, 261
    communication transparency, 261
RQ. *See* Research questions (RQ)
RT. *See* Reaction time (RT)
Rule-based information, 363–364

S

SA. *See* Situation (al) awareness (SA)
Sadness, 227
SAE. *See* American Society of Automotive Engineers (SAE)
SAET. *See* Self-assessment emotion tool (SAET)
Safety, 262
    and driving-assistance system, 179
    information, 87
SAGAT. *See* Situation awareness global assessment technique (SAGAT)
SAM. *See* Self-assessment manikin (SAM)
SAT. *See* Situation awareness theory (SAT)
Satisfaction, 345–346
SCA. *See* Shared cooperative activity (SCA)
SCADA system. *See* Supervisory control and data acquisition system (SCADA system)
Scenario analysis, 125, 306, 314
Scenario-based pedestrian behavior research, 327–328
    behavioral characteristics and demand analysis of pedestrian-unmanned vehicles, 328
    design background, 327
    in-depth interview and user research on pedestrian-unmanned vehicle interaction, 327–328
    interview and questionnaire quantitative research results, 328f
    pedestrian behavior characteristics and needs analysis, 329t
Screen displays HMI, 110
SD. *See* Standard deviation (SD)
Self-assessment emotion tool (SAET), 243–244
    design, 246–249
        emotion design, 247–249
        emotion set, 246–247
    methods of application, 257
    pilot study, 250–252
        apparatus, 250
        participants, 250
        procedure, 250
        results, 250–251
    related works, 244–246
        emotion and emotional space, 244–245
        nonverbal emotion measurement tool, 245–246
    validation, 250–256
    validation study, 253–254

Self-assessment manikin (SAM), 229, 245–246, 265
    scale, 232f, 289, 400
Self-developed driving simulation system, 163–164
Self-driving cars, 123
Self-driving vehicles, 123
Self-report metrics, 266–267
Self-reported measurement instruments, 243–244
Semiautomatic system, 115
Semicircular arc symbol, 370
Sensors, 54, 85
Sensory dimensions, 20
7-point Likert scale, 342
Shanghai Municipal Public Security Bureau Traffic Police Corps, 176
Shapiro-Wilk tests, 62, 343, 378
Shared cooperative activity (SCA), 359–360
Shared mental model, 282
Shared situation awareness (SSA), 156–157
    abstraction hierarchy analysis of, 159–160
    design methods, 158–160
    factors of, 160
    interface designs of human-machine co-driving, 158
    related works, 156–158
    results, 165–170
        eye movement, 166–167
        interview, 169–170, 170t
        situation awareness, 167, 167f
        task response time and task accuracy, 168–169
        usability, 167–168, 168f
    selection of experimental scenarios and requirement analysis of, 161–163
        descriptions of levels of abstraction, 162t
        elements of human-machine cooperation interface in rear car scene, 162f
    in semiautomatic driving
        after-scenario questionnaire, 90, 93f
        apparatus and simulated environment, 90
        experiment, 87–91
        framework, 85–86, 86f
        handovers in automated driving, 84
        human-machine interface requirements of system transparency, 83–84
        Lyons's task model of transparency, 83
        materials, 90–91
        NASA-TLX, 91
        participants, 88
        procedure, 90
        prototype, 89–90, 89f
        related works, 82–84
        relationship between transparency, performance, and trust, 83
        results, 91–94, 92t
        scenario, 84–85
        shared situation representation, 82

Shared situation awareness (SSA) (*Continued*)
  situation awareness, 83, 92
  situation awareness global assessment technique, 91
  study, 84–87
  transparency, 91–92
  transparency design, 86–87, 87f
  usability, 92
  workload, 93–94
 simulation experiment, 160–165
 situation awareness, 156–157
Shared situation representation, 85
SI. *See* Skill-based information (SI)
Simulated driving evaluation, 391–392
Simulation experiment, 160–165
  experimental design, 161
  experimental environment, 163–164
  experimental evaluation method, 164–165
   after-scenario questionnaire, 165
   quick interview, 165
   situation awareness global assessment technique, 165
   task response time and task accuracy, 165
  experimental process, 164
  experimental subjects, 160
  selection of experimental scenarios and requirement analysis of shared situation awareness information, 161–163
Simulators, 12
  evaluations, 391–392
  testing, 116
Situation (al) awareness (SA), 46, 62, 66, 83, 92, 94–95, 100–102, 111, 140–141, 156–157, 167, 261, 282–283, 420–421, 426
 average situation awareness global assessment technique percent correct score, 426f
 in cooperative driving, 48–49
  conceptual human-vehicle teaming perceptual cycle framework, 49f
 driving performance, 426, 427f
 global assessment technique score of three levels of situation awareness, 427f
 team situation awareness model, 157f
Situation awareness global assessment technique (SAGAT), 58, 61, 91, 163–165, 421, 425–426
Situation awareness theory (SAT), 82, 261, 282, 284
Six-phase model of interaction design method framework, 150–151
Skill-based information (SI), 363
Small text test, 40
Smart cars, 138
Smartphone, 20
Social intelligence, 76
Social robot interaction design (SRID), 203–204
 bibliometric results and analysis, 205–218
 effective measurement of research indicators, 218
 longitudinal trial, 219
 psychological cognitive mechanisms, 219
 realistic research scenarios, 218
 research design, 204–205
  data sources, 204–205
  research methods, 205
 specific design strategies, 219
Social robots, 203–204
 for children's psychotherapy, 209
Software engineering discipline, 150
Spearman's rank correlation coefficient, 25
Speeding task results, 295–297
 emotion score, 297f
 saccade time and fixation time in, 296f
 standard deviation of driveway offset in, 295f
 workload in, 296f
SRID. *See* Social robot interaction design (SRID)
SSA. *See* Shared situation awareness (SSA)
Standard deviation (SD), 108, 403, 426
Starling Crossing, 308–310
 graph design, 309f
 smart road solutions, 309t
Statistical significance, 384
Steering Wheel Button group, 406–408
Storyboard techniques, 153
Street scenario, 340–341
Subjective data, 6–7
 color contrast evaluation, 7f
 recommended color contrast and corresponding RGB value
  day, 7t
  night, 8t
 Spearman correlation coefficient: color contrast and, 6t
Subjective workload, 63
Sunny scenarios, 41–42
Supervisory control and data acquisition system (SCADA system), 423
Surprise, 227
SUS. *See* System usability scale (SUS)
Symbolic design, 247
System transparency, human-machine interface requirements of, 83–84
System usability scale (SUS), 195–196, 375–376, 378, 400, 406, 421–422, 425
Systematic design method model, 150

T

T-test, 196, 292
Tablet computer, 12–13
Take-over requests
 application and evaluation, 129–131
  evaluation procedure, 129–130, 130t
  results, 130–131
 elements, 125

highway hazard scenario analysis, 125–128, 125f
TAM. *See* Technology acceptance model (TAM)
Task accuracy, 165, 168–169
Task Load Index (TLX), 35–36
Task model, 94–95
Task response time and task accuracy, 165, 168–169
   average reaction time and correct rate results, 169f
Team situation awareness, 141
Technology acceptance model (TAM), 326
Telephone task results, 292–294
   emotion score of, 294f
   saccade time and fixation time in, 293f
   standard deviation of driveway offset in, 292f
   usability in, 293f
   workload in, 294f
Tesla's interface, 89
Test contents, 13–14
Text-based eHMIs, 338–339
TFD. *See* Total fixation duration (TFD)
3D
   PAD emotion model, 225–227
   Software1, 55–56
   theory of emotion, 227
THW. *See* Time headway (THW)
Time constraint-driven transparency framework, 142
Time headway (THW), 116, 422–423
Time to first fixation (TTFF), 376, 379
TLX. *See* Task Load Index (TLX)
Tobii I-VT filter, 374
Tobii Pro wearable eye tracker, 374
Total fixation duration (TFD), 376
Touch screen interaction method, 396
Traditional implicit communication methods, 325
Traffic accidents, 176
Traffic conditions, 144, 392–393, 422
Traffic encounters, 75
Traffic environment, 144, 157
Traffic system, 144
Transparency, 82, 91–92, 100–101, 261, 275–276, 282–284
   analysis of experimental results, 110–114
   correlation analysis, 111
   design, 85–87, 93–94, 103–104
      design assumptions, 285–287
      human-robot interface levels, 285
      method, 82
      of proactive interaction, 284–287
      in proactive interaction condition of in-vehicle robot, 286t
   elements, 83
   experimental evaluation, 108–109
      experimental environment, 108–109, 109f
      experimental process, 109
      participants, 108
   human-machine interface design, 104–107

   Lyons's task model of, 83
   scale data, 110
      trust, 110
      workload, 110
   theoretical basis, 100–104
      human-machine interface design case study, 102–104
      situational awareness and transparency, 100–102
      trust, 102
Trend analysis of annual outputs of social robot interaction design literature, 205–207
Trust, 102, 110–111, 115, 263, 329, 360
   in automation, 346
   model, 102
   scale, 102–104
   scale score, 110f
   situational awareness, 111
   workload, 111
TTFF. *See* Time to first fixation (TTFF)

U

"Uncanny valley" effect, 331
Unmanned vehicle external interaction
   AAM, 326, 326f
   design test evaluation, 334
   information architecture, 332f
   scenario-based pedestrian behavior research, 327–328
   vehicle diplomacy mutual strategy, 329–333
Usability, 262, 289, 427
   evaluation, 403
   study, 429
   test, 307, 421–422
      basic information of test, 319–320
      comparison of display positions of environment information, 322–323
      comparison of text scheme and graphic scheme, 320–322
      of design, 319–323
      participants, 320
      test methods, 319–320
Usefulness, 345–346
User activity analysis, 313–314
User experience (UX), 225–226
User research, 306
   of environment information demand in parking scenarios, 310–312
   field research, 312
   in-depth interview on users' demand for information in parking scenarios, 311–312
   questionnaire survey on users' demand for information in parking scenarios, 311
UX. *See* User experience (UX)

V

v-box technology, 178
VAs. *See* Virtual assistants (VAs)
Vehicle, 144
  approaching, 104
  automation, 102
  environment, 157
  hazards, 368
  horizontal control, 396
    vehicle horizontal control-related indicators, 394
  horizontal indicators, 394–395, 395*t*
  intelligence, 47
  module, 163–164
  safety technology, 175
  vehicle-mounted robots, 143
  vehicle-machine interface, 3
  vehicle-to-vehicle distance, 123
  vertical control, 396
    vehicle vertical control-related indicators, 394
  vertical indicators, 394, 395*t*
Vehicle diplomacy mutual strategy based on automation acceptance model, 329–333
  anthropomorphic design promotes trust and emotion, 330–331
    action anthropomorphism, 331
    language anthropomorphism, 331
    psychological anthropomorphism, 331
    visual anthropomorphism, 331
  classification and visualization of information outside vehicle, 331–332
  construction, 329–330
    collaboration, 330
    information recognizability, 329–330
    right of way and emotionalization, 330
    trust, 329
  high-fidelity prototype of car exterior interaction, 333*f*
  interactive prototype design based on vehicle diplomacy mutual strategy, 332–333
  multimodal design of external vehicle interaction, 332, 333*f*
Velocity, 116
Verbal cues, 209
Verbal report questionnaire, 245
Video analysis approach, 76
Video materials, 59
Virtual assistants (VAs), 226
Virtual environments, 47, 49
Virtual robots, 143
Vision cameras, 176
Vision scanning, 127
Vision test, 38
Visual anthropomorphism, 331
Visual approaches, 338
Visual color, 180–181
Visual evaluation, 24–25
Visual interaction, 36
Visual interfaces, 35, 46
Visual motion, 181
Visual processes, 411–412
Visual resource theory, 382–383
Visual signals, 20
  classification of visual signals and human-computer interaction model, 20–21
Visual subjective rating, 26–27
  average usability and workload for sizes of Chinese characters, 28*f*
Visual symbols, 179–180
  acceleration, 180*f*
  BMW-head-up display interface design, 180*f*
Visual weight, 413–414
Visualization
  approach, 369
  components and quantities, 381–382
Voice chat, 130
Voice communication, 260
Voice interaction, 396
VOSviewer, 205
VRUs. *See* Vulnerable road users (VRUs)
Vulnerable road users (VRUs), 337–338

W

W-HUD. *See* Windshield head-up display (W-HUD)
Waterfall model, 150
Weather conditions, 177–178
  influence of weather conditions on reaction time, 38–40
Web of Science (WOS), 204–205
WestboroPhotonics, 5
Wilcoxon posthoc tests, 343
Windscreen, 129
Windshield, 420
Windshield head-up display (W-HUD), 190
"Wizard of Oz" approach, 327, 327*f*
Workload, 16, 27–29, 63, 93–94, 110, 262–263
  impact of icon thickness on usability and, 12–13, 16
    Kruskal–Wallis test results of icons line thickness, 16*t*
  influence of icon size on usability and, 15
    Kruskal–Wallis test for icon size, 15*t*
  Pearson correlation analysis, 93–94
  rating of DALI scale, 63*f*
  remarkable otherness analysis, 94*t*
  trends of, 41–42
WOS. *See* Web of Science (WOS)

X

XiaoV robot, 266, 266*f*

Printed and bound by CPI Group (UK) Ltd, Croydon, CR0 4YY
02/12/2024
01798480-0016